Springer Series in

OPTICAL SCIENCES 131

founded by H.K.V. Lotsch

Springer Series in

OPTICAL SCIENCES

The Springer Series in Optical Sciences, under the leadership of Editor-in-Chief *William T. Rhodes*, Georgia Institute of Technology, USA, provides an expanding selection of research monographs in all major areas of optics: lasers and quantum optics, ultrafast phenomena, optical spectroscopy techniques, optoelectronics, quantum information, information optics, applied laser technology, industrial applications, and other topics of contemporary interest.
With this broad coverage of topics, the series is of use to all research scientists and engineers who need up-to-date reference books.

The editors encourage prospective authors to correspond with them in advance of submitting a manuscript. Submission of manuscripts should be made to the Editor-in-Chief or one of the Editors. See also www.springeronline.com/series/624

Mark L. Brongersma   Pieter G. Kik
(Editors)

Surface Plasmon Nanophotonics

With 147 Figures

Springer

A C.I.P. Catalogue record for this book is available from the Library of Congress.

ISBN: 978-1-4020-4349-9 (HB)
ISBN: 978-1-4020-4333-8 (e-book)

Published by Springer,
P.O. Box 17, 3300 AA Dordrecht, The Netherlands.

www.springer.com

Printed on acid-free paper

CONTENTS

PREFACE

At the moment the first issue of this book appears, hundreds of groups around the world are pushing the boundaries of the field of surface plasmon nanophotonics. The newfound ability to use metallic nanostructures to manipulate light at a length scale far below the diffraction limit has opened a myriad of exciting opportunities. Based on the exponential increase in the number of published papers every year (Chapter 1), it is clear that we are at the eve of a new revolution that will impact many fields of science and technology, including photonics, computation, the Internet, biology, medicine, materials science, physics, chemistry, and photovoltaics.

It has been a great pleasure and honor to work with some of the leading scientists in the field during the preparation of this work. The book truly reflects the present status of this rapidly developing area of science and technology and highlights some of the important historic developments. Most of the chapters discuss ongoing scientific research, and promising future directions are identified. Plasmon excitations in single and periodic arrays of metallic nanostructures are discussed in Chapters 2 and 3. The unique properties of metallic waveguides and metallo-dielectric photonic crystal structures that can route information on a chip are treated in Chapters 3 through 7. Surface Plasmon Mediated Field Concentration and Imaging methods, including superlenses and nanoscale optical antennas, are described in Chapters 8 through 10. The rapid developments in nanoscale optical probes that can visualize the flow of light and new, powerful electromagnetic simulation tools are treated in Chapters 11 through 13. The final chapters (Chapters 14–17) analyze a set of exciting applications of surface plasmon nanophotonics with tremendous commercialization potential, ranging from biology to data storage, and integrated optics.

We would like to thank all of the contributing authors for providing us with such excellent snapshots of the current state of the field of plasmonics. We would also like to acknowledge Kathleen Di Zio and Beatriz Roldán Cuenya for boundless mental and moral support. Kathy also contributed in a significant way by carefully proofreading many of the chapters and providing useful editorial comments. It has been a great pleasure to work with each and every one of you.

The Editors
Mark L. Brongersma
Pieter G. Kik

CHAPTER ONE

SURFACE PLASMON NANOPHOTONICS

PIETER G. KIK[1] AND MARK L. BRONGERSMA[2]
[1]CREOL, College of Optics and Photonics, University of Central Florida, Orlando, FL 32816, USA
[2] Geballe Laboratory for Advanced Materials, Stanford University, Stanford, CA 94305, USA

1.1. INTRODUCTION

In recent years, we have witnessed a flurry of activity in the fundamental research and development of surface plasmon based structures and devices. Surface plasmons are collective charge oscillations that occur at the interface between conductors and dielectrics. They can take various forms, ranging from freely propagating electron density waves along metal surfaces to localized electron oscillations on metal nanoparticles. Their unique properties enable a wide range of practical applications, including light guiding and manipulation at the nanoscale, biodetection at the single molecule level, enhanced optical transmission through subwavelength apertures, and high resolution optical imaging below the diffraction limit. This book is intended for people entering this diverse and rapidly growing field, recently termed "Plasmonics". It covers the fundamentals of surface plasmon science as well as some of the exciting new applications. The contributing Authors include world leaders in the field. Together they provide an overview of the current state-of-the art and their personal views on where the field is heading. The Editors hope that by reading this book you will get caught up in the excitement and join us to define and shape the future of Plasmonics.

1.2. SURFACE PLASMONS—A BRIEF HISTORY

Well before scientists set out to study the unique optical properties of metal nanostructures, they were employed by artists to generate vibrant colors in glass artifacts and in the staining of church windows. One of the most famous examples is the Lycurgus cup dating back to the Byzantine Empire (4th century AD). Some of the first scientific studies in which surface plasmons were observed date back to the beginning of the twentieth century. In the year 1902 Prof. Robert W. Wood observes unexplained

M.L. Brongersma and P.G. Kik (eds.), Surface Plasmon Nanophotonics, 1–9.

features in optical reflection measurements on metallic gratings.[1] Around that same time, in 1904, Maxwell Garnett describes the bright colors observed in metal doped glasses[2] using the then newly developed Drude theory of metals, and the electromagnetic properties of small spheres as derived by Lord Rayleigh. In an effort to develop further understanding, in 1908 Gustav Mie develops his now widely used theory of light scattering by spherical particles.[3]

Some fifty years later, in 1956, David Pines theoretically describes the characteristic energy losses experienced by fast electrons traveling through metals,[4] and attributes these losses to collective oscillations of free electrons in the metal. In analogy to earlier work on plasma oscillations in gas discharges, he calls these oscillations "plasmons". Coincidentally, in that same year Robert Fano introduces the term "polariton" for the coupled oscillation of bound electrons and light inside transparent media.[5] In 1957 a study is published by Rufus Ritchie on electron energy losses in thin films,[6] in which it is shown that plasmon modes can exist near the surface of metals. This study represents the first theoretical description of surface plasmons. In 1968, nearly seventy years after Wood's original observations, Ritchie and coworkers describe the anomalous behavior of metal gratings in terms of surface plasmon resonances excited on the gratings.[7] A major advance in the study of surface plasmons is made in 1968 when Andreas Otto as well as Erich Kretschmann and Heinz Raether present methods for the optical excitation of surface plasmons on metal films,[8] making experiments on surface plasmons easily accessible to many researchers.

At this point the properties of surface plasmons are well known, however the connection to the optical properties of metal nanoparticles has not yet been made. In 1970, more than sixty years after Garnett's work on the colors of metal doped glasses, Uwe Kreibig and Peter Zacharias perform a study in which they compare the electronic and optical response of gold and silver nanoparticles.[9] In their work, they for the first time describe the optical properties of metal nanoparticles in terms of surface plasmons. As the field continues to develop and the importance of the coupling between the oscillating electrons and the electromagnetic field become more apparent, Stephen Cunningham and his colleagues introduce the term surface plasmon-polariton (SPP) in 1974.[10]

Another major discovery in the area of metal optics occurs in that same year, when Martin Fleischmann and coworkers observe strong Raman scattering from pyridine molecules in the vicinity of roughened silver surfaces.[11] Although it was not realized at the time, the Raman scattering—an exchange of energy between photons and molecular vibrations—was enhanced by electromagnetic fields near the rough silver surface due to the presence of surface plasmons. This observation led to the now well established field of Surface Enhanced Raman Scattering (SERS). All these discoveries have set the stage for the current surge in surface plasmon nanophotonics.

1.3. SURFACE PLASMONS—PRESENT AND FUTURE

Since the early days of surface plasmon optics there has been a gradual transition from fundamental studies to more application driven research. The present surge in plasmon based research is happening at a time where crucial technological areas such as optical lithography, optical data storage, and high density electronics manufacturing are approaching fundamental physical limits. Several current technological challenges may be overcome by utilizing the unique properties of surface plasmons. Thanks to many recent studies, a wide range of plasmon-based optical elements and techniques have now been developed, including a variety of passive waveguides, active switches, biosensors, lithography masks, and more. These developments have led to the notion of *plasmonics*, the science and technology of metal-based optics and nanophotonics.[12]

The growth of the field of plasmonics is clearly reflected in the scientific literature. Figure 1.1 shows the annual number of publications containing the words "surface plasmon" in the title or the abstract. Since 1990 the annual number of papers on surface plasmons has doubled every five years. This rapid growth is stimulated by the development and commercialization of powerful electromagnetics simulation codes, nanofabrication techniques, and physical analysis techniques, providing researchers and engineers with the necessary tools for designing, fabricating, and analyzing the optical properties of metallic nanostructures. A major boost to the field was given by the development of a commercial surface plasmon resonance (SPR) based sensor in 1991. At present an estimated fifty percent of all publications on surface plasmons involve the use of plasmons for biodetection.

Most recently, metal nanostructures have received considerable attention for their ability to guide and manipulate "light" (SPPs) at the nanoscale, and the pace of new

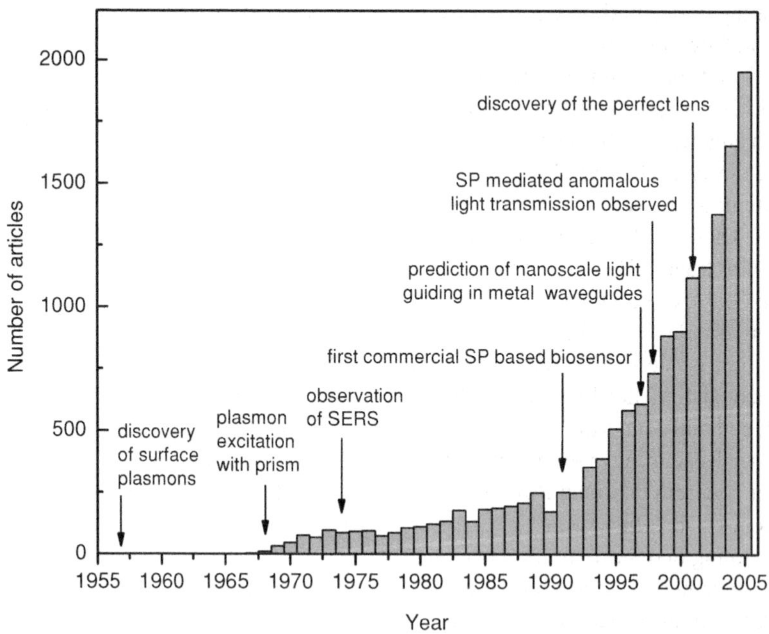

Figure 1.1. The growth of the field of metal nanophotonics is illustrated by the number of scientific articles published annually containing the phrase "surface plasmon" in either the title or abstract (based on data provided on www.sciencedirect.com).

inventions in the area has accelerated even further. In 1997 Junichi Takahara and co-workers suggest that metallic nanowires enable the guiding of optical beams with a nanometer scale diameter,[13] in 1998 Thomas Ebbesen and coworkers report on the extraordinary optical transmission through subwavelength metal apertures, and in 2001 John Pendry suggests that a thin metallic film may act as a "perfect lens".[14,15] All these findings have motivated a tremendous amount of new research, captured in a number of exciting review articles.[16,17] It should be noted that the brief overview provided here cannot do justice to all the different research directions the field of Plasmonics. Rather than making an attempt to be exhaustive, we have tried to present a short historical perspective including the fundamentals and some of the current hot topics. In the following section we highlight the various topics covered in this book. Together, these represent the true state-of-the art in the field of plasmonics.

1.4. THIS BOOK

This book contains seventeen chapters broadly divided into five major topical areas, which are indicated in boldface below. Although, all the chapters are self-contained, the topical areas help to provide connections between the different research directions.

Surface plasmon excitations in isolated and periodic metal nanostructures are discussed in the first two chapters. This section forms a solid basis for understanding several concepts used throughout this book.

Chapter 2 deals with the optical near-field and far-field properties of isolated metal nanoparticles and periodic metal nanoparticle arrays (Fig. 1.2). First, a qualitative introduction to surface plasmon excitations in metal nanoparticles is given and their resonant behavior is discussed in detail. In the latter part of the chapter, the properties

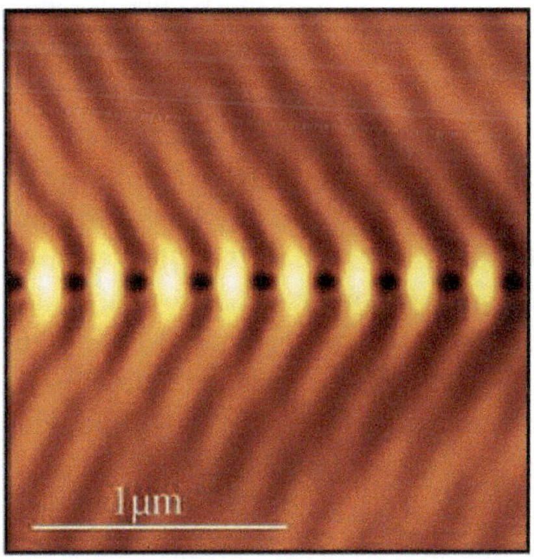

Figure 1.2. Optical near fields of metal nanoparticle chains (Chapt. 2) with a grating constant of 400 nm. The chains are excited under total internal reflection from the right at 800 nm. The image was taken using a photon scanning tunneling microscope. The circles indicate the nanoparticles.

of metallic nanoparticle arrays (extinction spectra, optical near fields, non-linear optical properties and particle interactions) are treated. Metallic nanoparticle arrays played an important role in the early days of this field and hold tremendous promise for the development of new applications, from nanoparticle-based plasmonic waveguides discussed in Chapt. 7 to advanced substrates for surface enhanced Raman spectroscopy, as shown in Chapt. 14. A deep understanding of the optical properties of metallic particles has also provided an intuitive vantage point from which the resonant behavior exploited in optical nanoscale antennas (Chapt. 9) and tip-enhanced spectroscopy (Chapt. 10) can be understood.

Chapter 3 provides the theoretical foundation of the extraordinary optical transmission (EOT) phenomenon observed in thin metallic films perforated with a two-dimensional array of subwavelength holes. It also explains the appearance of EOT in single apertures surrounded by periodic corrugations in the input side and the observation of beaming effects for the case in which the periodic corrugation is placed in the output surface. This phenomenon has given the field of plasmonics a tremendous boost in attention in recent years and the role of SPPs in this phenomenon is discussed.

Surface plasmon waveguides are analyzed in detail in Chapts 4 through 7 and 15. Such waveguides can manipulate and route "light" at the nanoscale. A number of different geometries are discussed, each with their individual strengths and weaknesses.

Chapter 4 shows how near-field optical techniques can be employed to characterize the properties of SPP waveguides. It starts with an explanation of the operation of a photon scanning tunneling microscope with which the propagation of SPP patterned metal films can be imaged. The authors apply this technique to analyze straight waveguides, bent waveguides, and structures with Bragg gratings. These measurements provide a deeper understanding of the properties of patterned, stripe waveguides and will be invaluable in designing improved more complex devices, including biosensors, interferometers, and active plasmonic structures.

Chapter 5 discusses the behavior of plasmonic transmission lines consisting of metal strips embedded in an isotropic medium. To this end, a computational method is presented that relies on the use of Eigensolutions available from many commercial solvers. Particular attention is given to the long-range modes supported by the strips and important practical issues such as the effect of surface roughness and smooth, curved edges on the propagation distance are analyzed. Practical device geometries such as periodically corrugated waveguides and waveguide bends are treated. These form important building blocks for plasmonic devices.

Chapter 6 demonstrates the feasibility of employing photonic band gap effects for realizing miniature photonic circuit elements by nano-patterning metal surfaces. In particular, the Authors show that SPP waveguiding and routing structures can be realized by defining line-defects in periodically corrugated metal surfaces. Numerical methods are used to simulate the behavior of these structures by solving the famous Lippmann-Schwinger equation. The reader is provided with intuitive insights into the operation of these devices and with a working knowledge on how to choose the bump spacing, height, and type of lattice for obtaining high transmission at specific operating wavelengths.

Chapter 7 discusses how metallic waveguides can be constructed that enable deep subwavelength spatial confinement of optical modes. Two waveguide types are investigated in detail, namely nanoparticle-based waveguides and metal-insulator-metal waveguides. Although the propagation distances are short (a few microns), they offer exciting opportunities for the development of truly nanoscale optical and photonic components.

Surface Plasmon Mediated Field Concentration and Imaging methods are discussed in Chapts 8 through 10. Such imaging techniques include the now famous perfect lens and the apertureless near-field optical microscope. Both optical elements rely on the use of metallic nanostructures and the excitation of SPPs to enable deep subwavelength operation.

Chapter 8 reviews the intriguing properties of superlenses. A superlens is an optical element capable of imaging objects of subwavelength dimensions. First the theory behind the superlens is discussed and it is shown how a thin slab of metal is capable of enhancing a broad band of evanescent waves in the wave vector spectrum. Such enhanced evanescent waves can be used to reconstruct the image of a subwavelength object. The theory is followed by a series of experiments that provide information on the transfer function of a silver superlens and show true subwavelength imaging with imaging resolution of 60 nm or $\lambda/6$. Super lenses exhibit enormous potential for commercial applications including ultra-high resolution imaging, high-density memory storage devices, and nanolithography.

Chapter 9 analyzes the intriguing properties of resonant bowtie antennas that can cause enormous field concentration (Fig. 1.3 (a)). Experimental and theoretical results demonstrate that local optical field enhancements greater than 1000 times can be obtained compared to the incident light wave. Physical explanations are given for the resonant behavior and its dependence on the dimensions of the antennas.

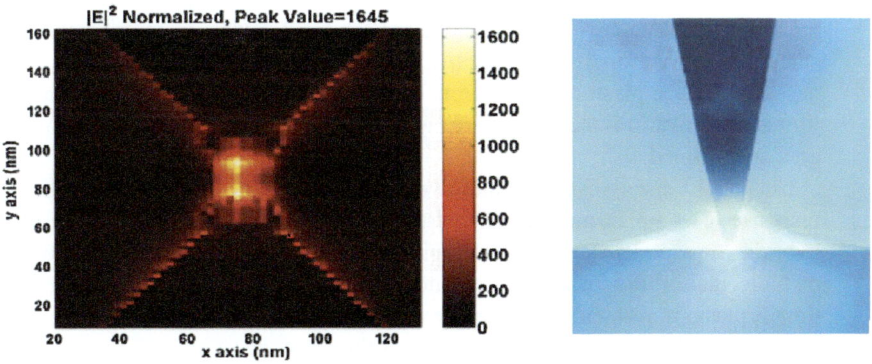

Figure 1.3. (a) Simulation of the intensity, $|E|^2$, enhancement in the 16 nm wide gap of a bowtie antenna, as measured 4 nm above the antenna (Chapt. 9). (b) Artist impression of the optical field enhancement underneath the metallic tip used in apertureless, near-field optical microscopy (Chapt. 10).

Chapter 10 reviews the application of apertureless, near-field optical microscopy for the local excitation and detection of SPPs. In such a microscope a sharp metallic tip is used for the imaging. Similar to the bow-tie antennas in Chapt. 9, such tips possess the ability to convert incident free-space light waves to strongly confined fields at the end of the tip (Fig. 1.3(b)). It is explained in detail how a combination of lightning-rod and surface-plasmon resonance effects gives rise to the localization. It is this light confinement that allows for the use of these tips as tiny (secondary) light sources with which structures can be imaged with nanoscale spatial resolution. To illustrate this, it is shown how SPP interference, decay, and scattering on patterned metal surfaces can be visualized. The strong field enhancement near the end of metal tips can give rise to a range of exciting non-linear effects, including second-harmonic and continuum generation. This is demonstrated experimentally and analyzed in theory. These microscopic techniques are expected to play a key role in the analysis of future nanoscale plasmonic devices.

Experimental characterization and simulation techniques for plasmonic structures are treated in Chapts 11 and 12. Our current understanding of plasmonic devices is largely based on the development of powerful far-field and near-field optical analysis tools and numerical simulation techniques. The unique ability of SPP-based structures to manipulate light at the nanoscale has generated substantial and fascinating challenges in the analysis and prediction of their behavior. Some of these challenges and their solutions are described in the following three chapters.

In Chapt. 11 Alain Dereux discusses ways to map the optical near-field in the vicinity of nanoscale optical structures. It is shown that a true near-field measurement does not allow the simultaneous detection of the electric field and the magnetic field. Calculational methods to predict near-field distributions on scattering structures are described, and practical examples are given that prove the validity of the developed approach under specific experimental conditions.

Chapter 12 provides a valuable overview of numerical simulation techniques that are suitable for predicting the behavior of plasmonic structures and devices. The authors first describe the challenges we are faced with in the numerical modeling of metallic nanostructures. These include the arbitrary device geometries, the complicated dispersive properties of metallic materials at optical frequencies, and the rapid decay of surface plasmons away from metal-dielectric interfaces. The strengths and limitations of various techniques are highlighted and future research directions in this area are outlined.

Chapter 13 discusses a very intuitive model that can be used to predict the optical response of complex metallic nanostructures and systems. It is called the plasmon hybridization model and shows intriguing similarities to the well-established theories on molecular orbital formation from atomic orbitals. This model can predict the location of plasmon resonances in complex systems based on knowledge of the resonant behavior of elementary building blocks. It thus provides a powerful tool for optical engineers in the design of future functional plasmonic nanostructures. The basic theory is explained and applied to a number of important examples, including multilayer concentric gold nanoshells and multi-particle geometries.

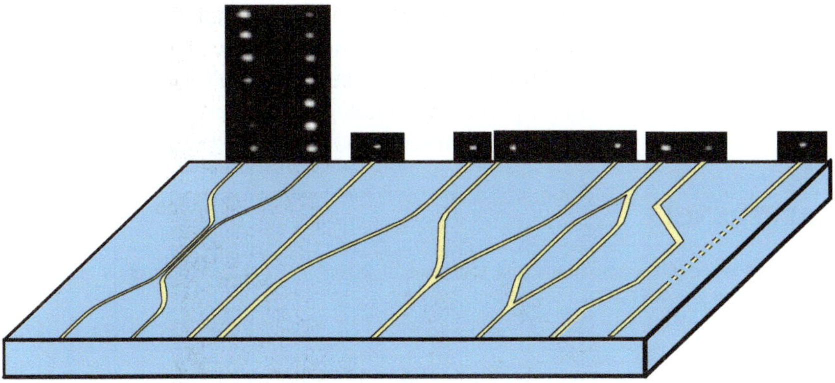

Figure 1.4. A schematic representation of several integrated optical devices based on Long Range Surface Plasmon Polaritons that are described in Chapt. 15. The insets represent optical images recorded at the exit facet of a real device.

Applications of surface plasmon nanophotonics are analyzed in Chapts 14 through 17. In these chapters, several exciting applications of SPPs that have already been commercialized or have tremendous potential to be commercialized are discussed.

Chapter 14 is devoted to the applications of Raman spectroscopy and way to improve the power of this technique with the use of metallic nanostructures. Raman scattering spectra enable molecular "fingerprinting", which is of great importance in the fields of molecular sensing and biology. Carefully engineered metallic nanostructures can enable surface enhanced Raman scattering (SERS) to provide a far greater detection sensitivity than conventional Raman spectroscopy. It is further shown that under special conditions, nanoparticulate silver films allow for fine rearrangement of their local structure under protein deposition, causing substantial SERS enhancements. Such newly discovered adaptive, metal nanostructures allow for new way to perform protein sensing and can be applied in protein micro-arrays.

Chapter 15 introduces a variety of integrated optics components based on long-ranging surface plasmon polaritons (Fig. 1.4). The current state-of-the art is reviewed, with an emphasis on passive elements including straight and curved waveguides, s-bends, y-junctions, four-port couplers, Mach-Zehnder interferometers and Bragg gratings.

Chapter 16 discusses a unique near-field based technology that can push the current limits in optical data storage. It is based on a so-called "super-resolution near-field structure" that enables the generation of strong, localized optical fields and localized surface plasmons in a thin film that exhibits a giant optical nonlinearity. The use of strongly localized fields occurring in nonlinear resonant systems could lead to optical data storage and readout with a capacity exceeding 1 terabyte.

Chapter 17 treats the use of metallic structures to modify and control the fluorescence properties of dyes. Particular attention is paid to a recently reported phenomenon that allows surface plasmon-coupled emission of excited fluorophores into a cone-like directional beam in a glass substrate. This allows for simple and efficient collection of the emitted photons. The authors expect that such near-field manipulation of light in combination with other nanophotonic technology will open a new era for biophysical and biomedical applications of fluorescence.

REFERENCES

1. R.W. Wood: On a remarkable case of uneven distribution of light in a diffraction grating spectrum, Phil. Mag. **4**, 396 (1902).
2. J.C. Maxwell Garnett: Colours in metal glasses and in metallic films, Philos. Trans. R. Soc. London **203**, 385 (1904).
3. G. Mie: Beiträge zur Optik trüber Medien, speziell kolloidaler Metallösungen, Ann. Phys. (Leipzig) **25**, 377 (1908).
4. D. Pines: Collective energy losses in solids, Rev. Mod. Phys. **28**, 184–198 (1956).
5. U. Fano: Atomic Theory of electromagnetic interactions in dense materials, Phys. Rev. **103**, 1202 (1956).
6. R.H. Ritchie: Plasma losses by fast electrons in thin films, Phys. Rev. **106**, 874 (1957).
7. R.H. Ritchie, E.T. Arakawa, J.J. Cowan, R.N. Hamm: Surface-plasmon resonance effect in grating diffraction, Phys. Rev. Lett. **21**, 1530–1532 (1968).
8. A. Otto: Excitation of nonradiative surface plasma waves in silver by the method of frustrated total reflection, Z. Phys. **216**, 398 (1968), Kretschmann, E. and Raether, H., Radiative decay of non-radiative surface plasmons excited by light, Z. Naturf. **23A**, 2135 (1968).
9. U. Kreibig, P. Zacharias: Surface plasma resonances in small spherical silver and gold particles, Z. Physik **231**, 128 (1970).
10. S.L. Cunningham, A.A. Maradudin, R.F. Wallis: Effect of a charge layer on the surface-plasmon-polariton dispersion curve, Phys. Rev. B **10**, 3342 (1974).
11. Fleischmann, M., P.J. Hendra, A.J. McQuillan: Raman spectra of pyridine adsorbed at a silver electrode, Chem. Phys. Lett. **26**, 163 (1974).
12. Brongersma, M.L., J.W. Hartman, H.H. Atwater: Plasmonics: electromagnetic energy transfer and switching in nanoparticle chain-arrays below the diffraction limit. in: *Molecular Electronics. Symposium*, 29 Nov.–2 Dec. 1999, Boston, MA, USA. 1999: Warrendale, PA, USA: Mater. Res. Soc., 2001, (This reference contains the first occurrence of the word "Plasmonics" in the title, subject, or abstract in the Inspec ® database).
13. J. Takahara, S. Yamagishi, H. Taki, A. Morimoto, T. Kobayashi: Guiding of a one-dimensional optical beam with nanometer diameter, Opt. Lett. **22**, 475–478 (1997).
14. T.W. Ebbesen, H.J. Lezec, H.F. Ghaemi, T. Thio, P.A. Wolff: Extraordinary optical transmission through subwavelength hole arrays, Nature (London) **391**, 667–669 (1998).
15. J. Pendry: Negative refraction makes a perfect lens, Phys. Rev. Lett. **85**, 396–3969 (2000).
16. W.L. Barnes, A. Dereux, T.W. Ebbesen: Surface plasmon sub-wavelength optics, Nature **424**, 824–830 (2003).
17. R. Zia, J.A. Schuller, M.L. Brongersma: Plasmonics: the next chip-scale technology, Materials Today **9**, 20–27 (2006).

CHAPTER TWO

NEAR-FIELD AND FAR-FIELD PROPERTIES OF NANOPARTICLE ARRAYS

ANDREAS HOHENAU, ALFRED LEITNER AND FRANZ R. AUSSENEGG
Institute of Physics and Erwin Schrödinger Institute for Nanoscale Research at the Karl-Franzens University, Graz, Austria; http://nanooptics.uni-graz.at

2.1. INTRODUCTION/MOTIVATION

The reason for the strong interest in the electrodynamic properties of metal nanoparticles (i.e., particles with sub μm dimensions) is based on their surface plasmon modes (see Sect. 2.2.2). A metal nanoparticle can in some sense be seen as a resonator for surface plasmons and, like any (moderately damped) resonator, if excited resonantly the oscillation amplitude can overcome the excitation amplitude by orders of magnitude. For surface plasmons on metal nanoparticles this means a strong enhancement of the local electromagnetic field compared to the exciting electromagnetic field.[1–4]

Applications of metal nanoparticles are based on this resonance behavior in different ways. Biological labels[5,6] (Chapt. 14), optical filters[7] or product labels[5,8] use their extinction and absorption properties and waveguide applications[9,10] are based on the enhanced near fields at frequencies close to the resonance (Chapt. 7). Surface enhanced effects, like surface enhanced Raman scattering (Chapt. 14) or surface enhanced fluorescence,[4,11,12] take advantage of both, the near field and the far field of resonantly excited metal nanoparticles: on one hand, the enhanced near fields lead to a higher excitation efficiency and on the other hand, the nanoparticles act as a transmitting antenna and enhance the coupling of atomic or molecular resonances to the optical far field.

The application of regular arrays of metal nanoparticles this chapter deals with is mainly to be seen for scientific purposes. This is due to the rather complicated and expensive fabrication scheme for such arrays, which requires the use of electron beam lithography. However, the huge advantage above other common fabrication schemes for metal nanoparticles (chemical production or thermally vacuum deposited metal island films) is the unsurpassed design flexibility and the particles monodispersity and homogeneity achieved by this method.

M.L. Brongersma and P.G. Kik (eds.), Surface Plasmon Nanophotonics, 11–25.

2.2. SURFACE PLASMONS ON ISOLATED METAL NANOPARTICLES

2.2.1. Optical Properties of Metals

Metals are defined by their quasi-free electrons in the ground state, which are not bound to single atoms but to the metal bulk. These free electrons are responsible for the well-known properties of high electric conductivity and high optical reflectivity.

Qualitatively, the free electrons of the metal behave like a gas of free charge carriers (a plasma) and can be excited to sustain propagating plasma waves.[13] Plasma waves are longitudinal electromagnetic charge density waves and their quanta are referred to as plasmons. They exist in two forms: bulk plasmons in the volume of a plasma and surface plasmons, which are bound to the interface of a plasma and a dielectric. Both modes cannot directly couple to propagating electromagnetic modes (light) in adjacent dielectrics. Surface plasmons exist for metals like Au, Ag, Al, and Cu from DC up to optical and near UV frequencies, depending on the dielectric function of both, the metal and the neighboring dielectric.

2.2.2. Qualitative Description of Surface Plasmons on Metal Nanoparticles

For small, isolated metal particles, with sizes in the range of the penetration depth of an electromagnetic field into the metal (e.g., ~20 nm for silver in the optical spectral range), the clear distinction between surface and bulk plasmons vanishes. In contrast to bulk metals, an external electromagnetic field can penetrate into the volume of the metal nanoparticle and shift the free conduction electrons with respect to the particles' metal ion lattice. The resulting surface charges of opposite sign on the opposite surface elements of the particles (see Fig. 2.1) produce a restoring local field within the particle, which rises with increasing shift of the electron gas relative to the ion background. The coherently shifted electrons of the metal particle together with the restoring field consequently represent an oscillator, whose behavior is defined by the electron's effective mass and charge, the electron density and the geometry of the particle. Throughout this text its resonances are called *surface plasmons on metal nanoparticles*.

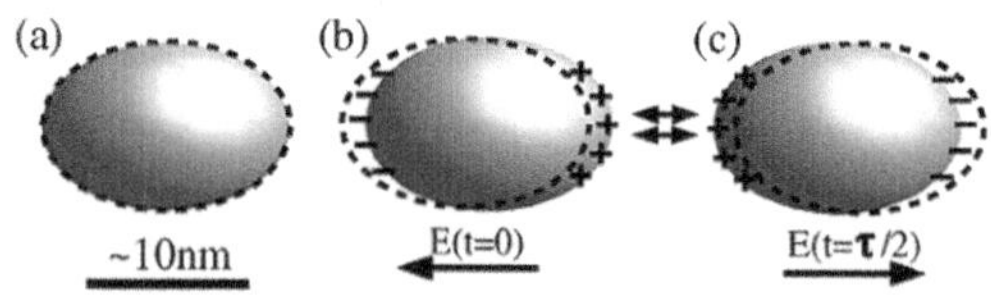

Figure 2.1. If a metal nanoparticle (a) is exposed to an electric field, the electrons are shifted with respect to the ion-lattice (b–c). The polarization charges on opposite surface elements apply a restoring force on the electrons. This, together with the mass of the electron gas leads to coherent, resonant oscillations of the electron gas.

Most of the physical effects associated with surface plasmons on metal nanoparticles (SPN) can qualitatively be understood in this simple oscillator model. Spectrally, the SPN resonances are found somewhere in the visible to near-infrared spectral region, depending on the particle shape, the environment and the metal. If excited at resonance, the amplitude of the induced electromagnetic SPN-field can exceed the exciting fields by a factor in the order of 10 (i.e., optical near field enhancement) and in analogy to a classical oscillator, the damping of the SPN limits the maximum amplitude of the resonance and determines its spectral width.

It is important to consider this model just as a qualitative description of the SPN. Especially for the damping mechanisms described below, the quantum mechanical view of the SPN is important, since at usual excitation conditions and usual values for the damping at optical frequencies, hardly more than a single SPN quantum is present on a nanoparticle at a given time.* Further, metal nanoparticles usually have more than a single oscillation mode. The different modes differ in their charge- and field-distribution.[14] For the lowest (or dipolar) SPN-mode, the distributions are dominated by a dipolar character; the higher energetic modes can be associated with multipolar charge distributions of higher order.

2.2.3. Theoretical Descriptions of SPN

Mie Theory

An exact analytical theoretical description of surface plasmons of spherical metal nanoparticles is part of Mie's theory for the scattering and absorption of light by spheres.[4,14,15] According to Mie-theory, the different eigenmodes of the spherical particles are dipolar or multipolar in character and their excitation strength is determined by the expansion of the exciting electromagnetic field into vector spherical harmonic functions.

Quasistatic Approximation

For particles small compared to local variations of the involved electromagnetic fields, the quasistatic approximation[1,4] leads to results in good agreement with the experiment. It assumes the exciting field to be homogeneous and not retarded over the particle's volume. Under these assumptions, the results of electrostatics can be applied by using the corresponding frequency depended dielectric function instead of the electrostatic values. The polarizability α_i (in SI-units) of an ellipsoidal metal nanoparticle parallel to the ellipsoid axis i follows from this theory as:

$$\alpha_i = \frac{4\pi}{3} abc \frac{\varepsilon_m - \varepsilon_e}{\varepsilon_e + A_i\left[\varepsilon_m - \varepsilon_e\right]} \tag{2.1}$$

*This can be estimated from the SPN-lifetime ($\sim$10 fs), the absorption cross-section (e.g., in the order of 10^{-10}cm^2) and the light intensity (e.g., 100 kW/cm^2 at a wavelength of 800 nm, photon flux $\sim$4 10^{23}s^{-1}cm^{-2}).

ε_m and ε_e are the frequency dependent dielectric functions of the metal and the environment, respectively, and A_i is the shape or depolarization constant. a, b and c are the ellipsoid's half axes. A_i depends on the axis ratio of the ellipsoidal particle and the polarization of the exciting field with respect to the ellipsoid axis i. The value of A_i is determined by the ratio of the ellipsoid axes lengths[14] and ranges from 0 to 1. For spherical particles with $a = b = c$, $A_i = 1/3$ in all directions.

Despite its simplicity the quasistatic approximation already reveals the major parameters influencing the surface plasmon resonance:

- The particle shape
- The dielectric function of the metal
- The dielectric function of the environment

It is important to note that the quasistatic theory does not account for re-radiation (scattering) of light since the interaction of the induced polarization with the retarded electromagnetic field caused by it is ignored in the quasistatic formulation. In an improved form, scattering is approximately included in a complex *effective* depolarization constant A_i.[15] This correction is only introduced phenomenologically and not strictly justified. Additionally, the quasistatic theory allows only the description of dipolar surface plasmon modes.

Dipole Approximation

To a good approximation, the electromagnetic field of the dipolar surface plasmon modes on ellipsoidal metal nanoparticles outside the particles can be analytically described by the electromagnetic field of an oscillating point dipole located in the center of the particle.[13] This is in accordance with the results of the quasistatic approximation for spherical particles, but also holds for dipolar surface plasmon resonances on arbitrarily shaped metal nanoparticles, if the distance from the particle is much larger than the particle itself. The major conclusion from this approximation is the existence of two limiting spatial domains of the electromagnetic field of SPN's:

- The near field: It dominates in the vicinity of the particle for distances much smaller than the wavelength of the electromagnetic fields. For spherical particles it falls off with distance r from the particle center proportional to r^{-3}. Retardation effects play a minor role in this regime.
- The far field: For distances from the particle much larger than the wavelength of the electromagnetic fields, spherical waves dominate. In this range of distance r, the field falls off proportional to r^{-1}.

2.2.4. Damping of SPN-Resonances

According to the theoretical descriptions, the spectral width and the amplitude of the SPN-resonances as well as the SPN temporal decay are influenced by the damping of the surface plasmons. The mechanisms for the damping or decay, respectively, can be categorized into radiation damping, energetic relaxation and pure dephasing.

Radiation-Damping—Scattering

Scattering of an exciting electromagnetic wave by SPN's results from the re-radiation of electromagnetic waves by the oscillating SPN charge distribution. The radiated power is taken from the power stored in or pumped into the SPN and therefore, leads to the so-called radiation damping.[16]

Energetic Relaxation—Absorption

In addition to radiation damping, also internal (i.e., ohmic) loss mechanisms of the surface plasmons exist. The plasmon relaxes energetically and after different intermediate processes it heats the metal nanoparticle.

SPNs typically have their resonances in the visible or near-infrared spectral range. The interaction of an electromagnetic field in this energy range with a metal leads to the creation of electron-hole pairs.[16] Since also surface plasmons are electromagnetic modes, they will decay by the same process, the creation of an electron hole pair.**

After being excited by the decay of a surface plasmon, the electron-hole pair has all usual decay channels known from solid state physics: electron-electron scattering, electron–phonon scattering, surface scattering etc.[17]

Pure Dephasing

Pure dephasing of the plasmon describes the elastic scattering of the surface plasmon quanta themselves, which destroys their phase relationship to the exciting electromagnetic field. Experiments show that this contribution to the overall damping of surface plasmons on metal nanoparticles is negligible.[17]

Size Dependence of Absorption and Scattering Cross-sections

From the point dipole approximation together with the results of quasistatic theory (Eq. (2.1)), relatively simple equations for the absorption and scattering cross-sections C_{abs} and C_{scat} (in SI-units) of a particle can be found.[14]

$$\begin{aligned} C_{\text{abs}} &= k\,\text{Im}(\alpha) \\ C_{\text{scat}} &= \frac{k^4}{6\pi}\,|\alpha|^2 \end{aligned} \tag{2.2}$$

The complex polarizability of the particle α is proportional to the particle volume within the quasistatic approximation (Eq. 2.1). $|\alpha|$ denotes the modulus and $\text{Im}(\alpha)$ the imaginary part of the polarizability. The important conclusion from Eq. (2.2) is that, the radiation damping or scattering, respectively depends on the square of α and therefore on the square of the particle volume whereas the absorption depends only linearly on α and the particle volume. Therefore, the SPN-damping for small particles is dominated by absorption but for large particles it is dominated by scattering. This

**The SPN has only discrete quantum states with energy-quanta of $h\nu$ (h is the Planck-constant and ν the frequency) and there is no possibility for the scattering of *single* electrons out of their coherent oscillation.

is responsible for the broader spectra observed for larger particles compared to very small particles.[4]

2.3. FAR FIELD EXTINCTION SPECTRA OF NANOPARTICLE ARRAYS

2.3.1. Spectral Position of the Surface Plasmon Resonance

The spectral position of the extinction maximum depends on the dielectric function of the metal and the dielectric environment and, most prominently, on the particle's shape. All these dependencies are qualitatively described by the results from the quasistatic model of the particles (see Quasistatic Approximation in Sect.2.2.3).

Influence of the Particle Shape

Considering ellipsoidal particles, the spectral resonance position shifts from the value for a sphere depending on the asymmetry of the particle. Compared to the SPN-resonance of spherical particles, the SPN-resonance of ellipsoidal particles is shifted to the red for a polarization parallel to the long ellipsoid axis and to the blue for a polarization parallel to the short ellipsoid axis. Figure 2.2 shows the extinction spectra

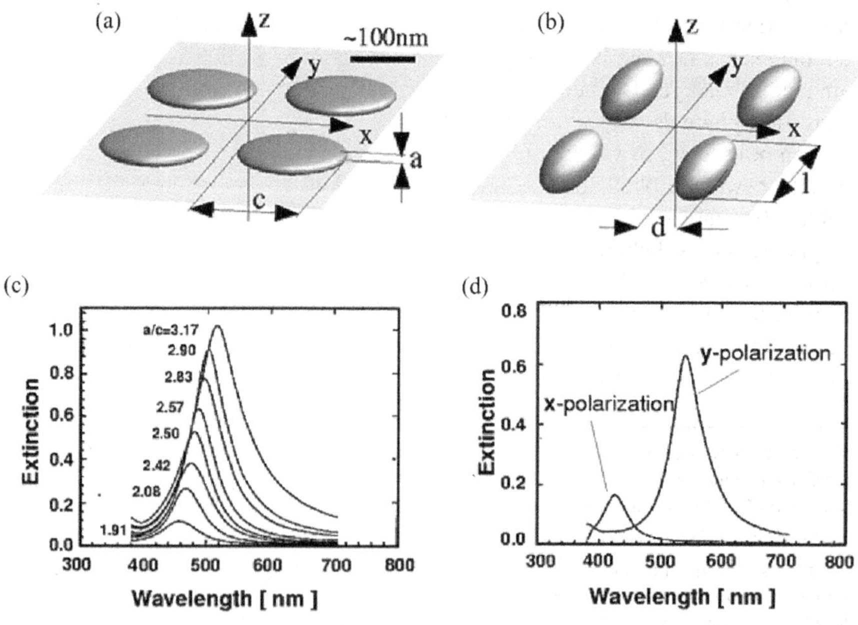

Figure 2.2. Extinction ($\log_{10}$ (transmission)) spectra of arrays of metal nanoparticles with varying shape. The extinction spectra (c) of rotationally symmetric oblate particles on a glass substrate (a) shift to the red with increasing axis ratio a/c and are independent of polarization for light incident parallel to the direction of z. For prolate particles (b) the extinction is red shifted with respect to spherical particles for a polarization parallel to the long particle axis and blue shifted for a polarization parallel to the short axis (d).

of arrays constituted of differently shaped oblate and prolate metal nanoparticles on a glass substrate.[19,20]

Influence of the Dielectric Functions

Ag nanoparticles have their surface plasmon resonance at shorter wavelength than Au particles of the same shape due to the difference in the dielectric function of these metals.[21] Additionally to this material dependence of the dielectric function, also its size dependence plays a role for small particles. If the particle size gets below the dimension of the mean free path of the electrons in the metal ($\approx$10 nm) the electron scattering at the particle surface leads to a modification of the dielectric function, mainly increasing its imaginary part.[4] At even lower particle dimensions ($\approx$1 nm) the spill-out of the electrons from the particle surface cannot be neglected anymore and can only be described by an anisotropic and inhomogeneous dielectric function.[4] The concept of a continuous medium describable by its dielectric function starts loosing its validity in this size regime.

Also the dielectric environment influences the spectral position of the surface plasmon resonance. An increase in the refractive index leads to a red-shift of the SPN resonance.[4]

2.3.2. SPN Spectral Width and Decay Time

For single particles and arrays of identical particles, the spectral width of the surface plasmon resonance is directly related to its decay time, in analogy to a damped oscillator.[22,23] The decay time in turn is determined by the damping of the surface plasmon resonance and consequently, a change in the surface plasmon damping will lead to a change in the spectral width of the surface plasmon resonance.

Influence of Radiation Damping, Grating Orders

A simple possibility for the investigation of the relationship between damping and spectral width is the study of regular arrays of identical particles with varying grating constant.[22] Depending on the grating constant and the wavelength of the exciting light, different grating orders exist, and the light scattered from the particle array is confined in its propagation to certain directions and is partly prohibited. For grating constants d in the range of the light wavelength λ, there are three relevant domains (see Fig. 2.3).

For $d > \lambda$ (Fig. 2.3(c)), due to the coherent excitation and scattering of the individual nanoparticles, the light is scattered only at defined angles, in analogy to light diffraction from a grating. For $d = \lambda$ (Fig. 2.3(b)) the light scattered to the first grating order is emitted parallel to the plane of the particle grating. For (theoretically considered) infinitely large gratings, the partial waves scattered from the single particles sum up and lead to a singularity in the function of the scattered intensity vs. the grating constant or wavelength respectively. Experimentally, these so-called Wood-anomalies show up as sharp dips or peaks in the extinction or reflection spectra, but for gratings of metal nanoparticles they can be seen only for arrays of larger particles with relatively strong scattering cross-sections.[24]

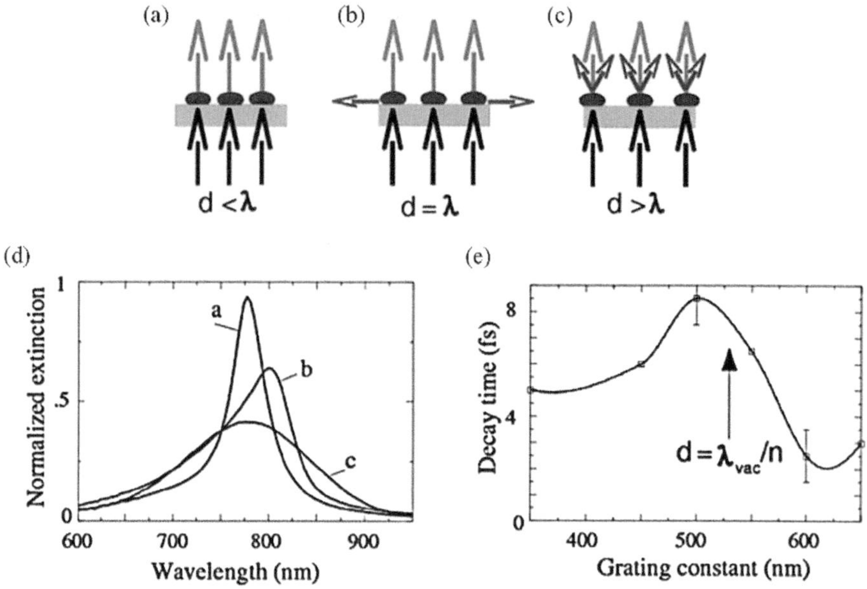

Figure 2.3. Different domains of the grating constant concerning the scattering to the far field ((a), (b) and (c)), SPN far field extinction spectra (d) and SPN decay time if excited at $\lambda = 800$ nm (e) of regular square gratings of oblate gold nanoparticles (diameter 150 nm, height 14 nm) on a glass substrate. The SPN-resonance is at $\lambda_{res} = 780$ nm (vacuum wavelength). At a vacuum wavelength of 800 nm, the first grating order (in glass) is grazing at a grating constant of 533 nm (indicated by the arrow in (e)).

For $d < \lambda$ (Fig. 2.3(a)), scattering is mostly prohibited and only possible in the direction of the zeroth grating order (i.e., the transmission direction).

Since scattering is a relevant contribution to the damping of surface plasmons on larger metal nanoparticles, the suppression of scattering by the suppression of higher grating orders leads to a reduced surface plasmon damping and a narrower spectral width of the surface plasmon resonance.

In Fig. 2.3(d), the extinction spectra measured on gratings of identical metal nanoparticles but different grating constants are depicted. If the center-to-center distance of the particles $d < \lambda_{res}$ (λ_{res} is the wavelength of the SPN-resonance) there is only zero order transmission (curve a). For $d = \lambda_{res}$ the first grating order exists only for wavelength below λ_{res} which leads to the asymmetric shape of the SPN-peak (curve b). For $d > \lambda_{res}$, grating orders exist over the whole spectral range (curve c). The SPN-resonance is again symmetric, but broader and weaker as compared to curve a.[22]

Direct Measurements of SPN Decay Times

Besides the indirect determination of the decay times of surface plasmons on metal nanoparticles by the spectral width of their resonance, there is also the possibility for its direct measurement by femtosecond laser pulses.[25] The measurement is based on a non-linear autocorrelation of the surface plasmon field excited by an fs-laser pulse. The width of the recorded autocorrelation function contains information on the SPN-lifetime, which can be extracted from the measurement by a fitting procedure.

The important results from these investigations are the directly measured SPN decay times in the order of 10 fs and the proof, that the relation of SPN spectral width and decay time can be understood qualitatively and quantitatively in terms of a simple harmonic oscillator model.[25,26]

One application of this method is the study of the influence of radiation damping on the SPN decay time in particle arrays. As shown in Sect. 2.3.2, the radiation damping of SPN can be influenced by the grating constant. The more grating orders exist, the larger is the radiation damping and the shorter the SPN decay time. Figure 2.3(e) depicts the results of the measured decay times vs. grating constant.

2.4. OPTICAL NEAR-FIELDS OF NANOPARTICLE ARRAYS

If excited, SPN not only scatter light to the far field, but they also possess near fields, which can be much stronger than the exciting optical fields (see Sect. 2.2.2). For spherical particles, to a first approximation, these near fields can be described by the near field of oscillating point dipoles located at the center of the particles. If the particles are non-spherical, an external field can lead to strongly inhomogeneous polarization states in the particle and especially in close proximity to tips and corners of particles, the near fields can be strongly enhanced. Qualitatively, this can be viewed as the optical pendant to the electrostatic field enhancement in the strongly inhomogeneous field close to a sharp metal tip (lightning rod effect).[27]

The optical near fields of SPN excited by light can be measured in a photon scanning tunneling microscope (PSTM) setup (see Fig. 2.4(a)). Typically, the particles are excited in total internal reflection geometry and the optical near fields are probed

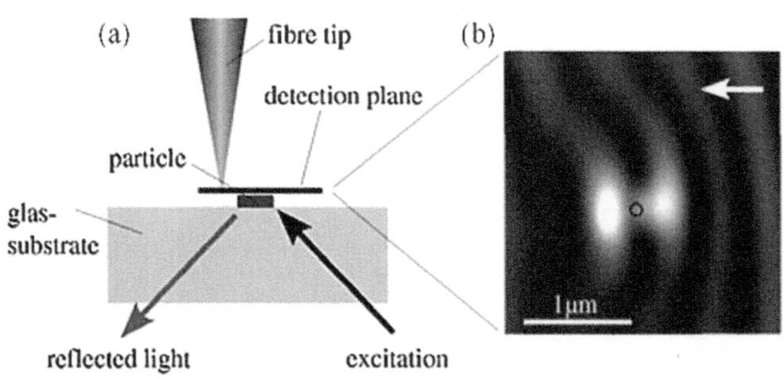

Figure 2.4. Sketch of the PSTM with excitation under total internal reflection (a) and optical near field image of a SPN on a gold nanoparticle (b) outlined by the circle. The particle diameter is ~ 80 nm and its height is ~30 nm. The excitation wavelength is 633 nm. The white arrow in (b) indicates the direction of the exciting light.

by the tip of a sharpened optical fiber. The optical field at the fiber tip is partly scattered into guided modes within the fiber which is attached to a photo-detector. The recorded intensity is proportional to the electric field intensity at the tip to good approximation.[28] Typical tip sizes are in the range of 50–100 nm, which limits the spatial resolution for this type of microscope.

2.4.1. Optical Near fields of an Isolated Metal Nanoparticle

The optical near fields of an SPN as recorded by a PSTM for an isolated metal nanoparticle result from the interference of the evanescent, exciting light field with the field scattered from the particle.[29] The optical near field of the excited particle shows two lobes (see Fig. 2.4(b)), whose intensities depend on the polarization of the exciting light. For ellipsoidal nanoparticles, this near field intensities can be qualitatively understood as an interference of the exciting plane light-wave with the field of a point dipole located at the center of the particle.

2.4.2. Optical Near Fields of Particle Arrays

Similar to the far field extinction, the optical near fields of particle arrays depend on the grating constants and one has to distinguish between different domains of the grating constant. If excited under total internal reflection condition, for sufficiently small grating constants no diffraction orders of a particle grating exist above the grating. Only the near fields of the particles are present. In case of the larger grating constants, additional to the near fields one or more grating order exist as well and lead to interferences with the exciting evanescent field.

The contributions of the SPN near field and the grating orders can be nicely observed in the case of one-dimensional particle arrays, i.e., particle chains (see Fig. 2.5).

As a result of the invariance of the chain against translation by a grating vector, the optical near field intensities always show the same periodicity as the particle chain and independent of the excitation (this can be considered as an implication of the "optical" Bloch theorem).

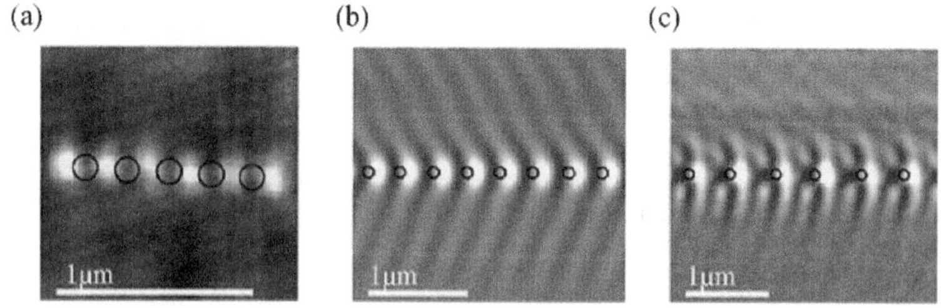

Figure 2.5. Optical near fields of nanoparticle chains with different grating constant; (a) 200 nm, (b) 400 nm, and (c) 500 nm. The chains are excited under total internal reflection from the right (see Fig. 2.4(a)) at 800 nm. The circles indicate the metal nanoparticles.

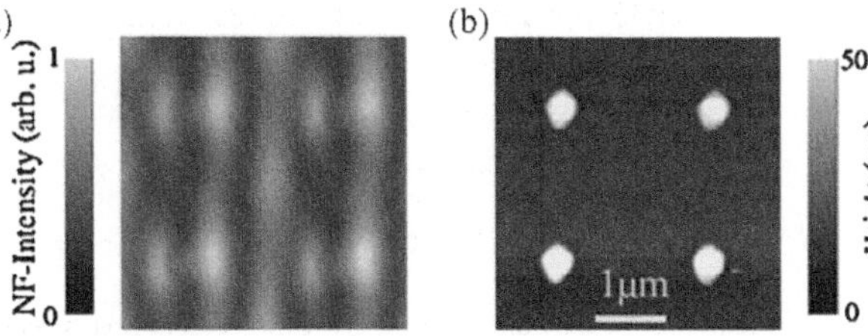

Figure 2.6. ptical near field of SPN (a) and topography (b) of a 2D particle array. The optical near fields exhibit the same periodicity as the array. The near field of the particles (lobes left and right of the particles) and grating orders interfering with the exciting evanescent wave (maxima in between the particles) determine the intensity pattern (unpublished results).

For small grating constants only the bright lobes due to the particles near field can be observed. The particles are so close that their enhanced near fields overlap and are strongly squeezed between the particles (Fig. 2.5(a)).[29] For larger particle spacing (Fig. 2.5(b) and (c)) grating orders exist above the substrate and additional to the bright lobes close to the particles also fringes further away from the particles can be seen. These fringes result from the interference of the exciting evanescent field with the grating orders scattered parallel to the substrate.[30]

For two-dimensional arrays of particles, the same behavior as for the particle chains is observed. Figure 2.6 shows exemplarily the measured optical near fields of metal nanoparticles arranged in a quadratic grating. In analogy to the chain, the invariance of the near field with respect to grating invariant translations can be recognized.

2.5. OPTICAL NONLINEARITIES

As described in Sect. 2.2.2, the surface plasmon resonance on metal nanoparticles can be considered to a first approximation in a simple oscillator model, where the quasi free electrons of the particle are shifted coherently with respect to the metal ion background by an applied electromagnetic field (light). The restoring force is due to the charging of the opposite surface elements by the coherent shift of the electron cloud. This restoring force responds linear to the displacement of the electron cloud only at low electromagnetic field strength. At higher field strength (some MW per cm^2) the restoring force (and therefore the polarization) depends nonlinearly on the exciting field and consequently nonlinear optical effects occur.

Qualitatively, one has to distinguish between centrosymmetric and non-centrosymmetric particle shapes and array configurations (see Fig. 2.7). For centrosymmetric particles (the particle is symmetric with respect to an inversion center) the restoring force and particle polarization as a function of the applied electromagnetic field is symmetric around the particle-center as well (see Fig. 2.7(a)). When a harmonic field excites such an oscillator, the anharmonic polarization will lead to scattering of the exciting wave (fundamental) and to the emission of waves with an odd-numbered multiple frequency of the fundamental (third, fifth, etc. harmonics).

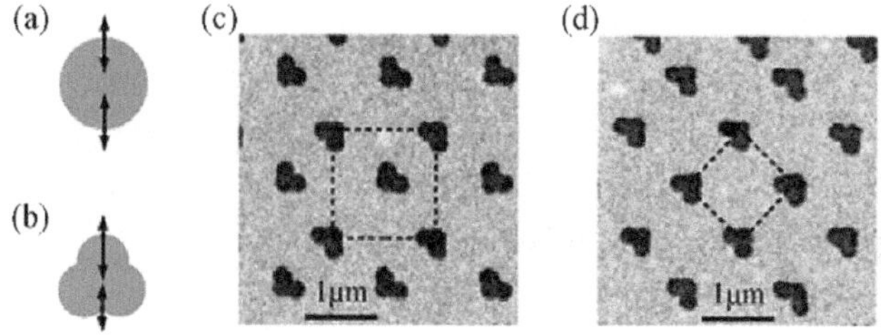

Figure 2.7. Optical nonlinearities of metal nanoparticles. For centrosymmetric particles (a), also the polarization is centrosymmetric and only odd order nonlinearities are allowed. For non-centrosymmetric particles (b), also even order nonlinearities exist. Part (c) depicts an electron micrograph of a centrosymmetric array of non-centrosymmetric particles (no even order nonlinearities), part (d) a non-centrosymmetric array of non-centrosymmetric particles (also even order nonlinearities). The unit-cells of the particle arrays are sketched by the dashed lines in (c) and (d).

Since the intensity of higher harmonics strongly decreases with the order of the nonlinearity, usually only the lowest (i.e., the third) harmonic wave has considerable strength.

For non-centrosymmtric particles (see Fig. 2.7(b)), also the polarization will not be non-symmetric with respect to the applied electromagnetic field and, additionally to the odd harmonics, also even harmonic waves (i.e., second, fourth etc harmonics, dominated by the second harmonic) will be emitted (scattered) from the particle.[31]

It is important to note, that not only the particles themselves, but also the array in which they are arranged has to maintain a non-centrosymmetry to produce even order nonlinear effects. If the particles are non-centrosymmetric, but are arranged in a centrosymmetric array (see Fig. 2.7(c)) the even order harmonics produced by the single particles of a unit cell interfere destructively. Therefore, in the far field no second-order nonlinear effect can be observed.

2.6. PARTICLE INTERACTIONS

Up to here, particle arrays were treated, in which the surface plasmons on the individual nanoparticles only showed some interaction in terms of a modified far field of the scattered light due to the coherent excitation and therefore, coherently scattered fields. This had influences on the spectral width of the surface plasmon resonance, but the fields scattered by one particle were so weak at the positions of the other particles compared to the exciting field, that the direct interaction of the particles could be neglected. However, if the particles get closer to each other, the electromagnetic near fields get comparable to the exciting field and the particles become electromagnetically coupled. The simplest case for the discussion of the effects expected in this distance domain is a system of only two interacting particles.

To gain qualitative insight, the coupled SPN system of a particle pair can be described as a system of two interacting dipolar oscillators. Depending on the

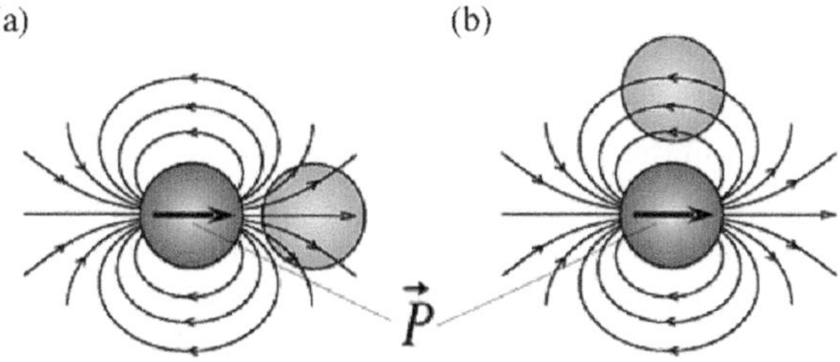

Figure 2.8. Two possible configurations of a pair of metal nanoparticles with respect to the polarization of the exciting light field. The thin lines sketch the electric field lines of the polarized particle.

polarization with respect to the particle pair orientation, one can distinguish between two cases of polarization either parallel (Fig. 2.8(a)) or perpendicular to the pair-axis (Fig. 2.8(b)). For parallel polarization, the near field of one particle dipole acting on the other points to the same direction, whereas for polarization perpendicular to the pair axis, it points to the opposite direction. Thus, if both particles are excited together, the dipole (SPN) field either weakens (perpendicular polarization) or strengthens (parallel polarization) the exciting field at the position of the other dipole (particle). In analogy to coupled harmonic oscillators, one therefore expects a shift of the resonance to higher frequencies for perpendicular polarization and to lower frequencies for parallel polarization.[32]

The experimentally measured positions of the SPN resonances are in agreement with these expectations: The degenerate modes of the two identical nanoparticles split up and shift to the red or the blue depending on the polarization of the exciting light (see Fig. 2.9).

It is important to note that, due to the retarded interaction of the polarized particles, the damping of particle pairs ("radiation-damping") is larger compared to single particles.

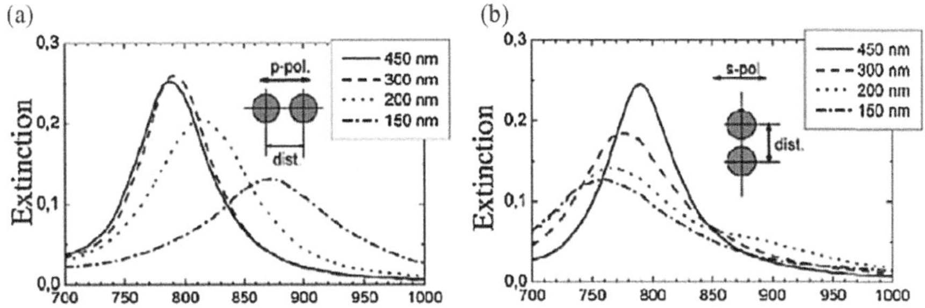

Figure 2.9. Experimental extinction spectra of arrays of particle pairs for polarization parallel (a) and perpendicular (b) to the pair axis. The oblate particles have a diameter of ~150 nm and a height of ~14 nm and varying spacing (center to center), see insets.

2.7. CONCLUSION

On hand of the results of different exemplarily chosen experiments, this chapter demonstrates that for the understanding and description of the near-field and far-field properties of metal nanoparticle arrays, the properties of the surface plasmons on the individual particles (Sect. 2.2.2) and their interaction to the surface plasmons on the other particles have to be taken into account. For arrays of closely spaced particles, this interaction is due to the optical near fields of the surface plasmons on neighboring particles (Sect. 2.2.6). For larger distances the particles still influence each other due to coherent grating effects and changed radiation patterns (see Quasistatic Approximation in Sect. 2.2.3.2).

Therefore, by controlling the geometry of the particles and their arrangement in the array, the arrays' optical properties can be controlled, in the near-field (near-field enhancement and -pattern (Sect. 2.2.4)) as well as in the far field (spectral resonance position (see Mie Theory 2.3.1), spectral width and decay time (see Quasistatic Approximation 2.3.2)). This controllability of the optical properties suggests the potential of nanoparticle arrays as thin layer materials with tailored optical properties in numerous micro- and nano-optical applications.

ACKNOWLEDGMENTS

We would like to thank former and present members of our group who have contributed to this chapter. For their financial support, the Austrian Government and the European Union (Projects FP6 NMP4-CT-2003-505699 and FP6 2002-IST-1-507879) are acknowledged.

REFERENCES

1. J.I. Gersten. Surface shape resonances. In: *Surface Enhanced Raman Scattering*, ed by R.K. Chang and T.E. Furtak (Plenum Press, New York, 1982), p. 89.
2. M. Moskovits: Surface-enhanced spectroscopy, Rev. Mod. Phys. **57**, 783 (1985).
3. A. Wokaun: Surface enhancement of optical fields. Mechanism and applications, Mol. Phys. **56**, 1 (1985).
4. U. Kreibig, M. Vollmer: *Optical Properties of Metal Clusters* (Springer-Verlag, Berlin, 1995).
5. T. Schalkhammer: Nanoclusters as transducers for molecular structure and recognitive binding. In: *Encyclopedia of Nanoscience and Nanotechnology*, ed by H.S. Nalva (American Scientific Publishers, New York, 2004).
6. J.L. West, N.J. Halas: Engineered nanomaterials for biophotonics applications: Improving sensing, imaging, and therapeutics, Ann. Rev. Biomed. Eng. **5**, 285 (2003).
7. K. Baba, Kazutaka, Miyagi, Mitsunobu: Optical polarizer using anisotropic metallic island films with a large aperture and a high extinction ratio, Opt. Lett. **16**, 964 (1991).
8. G. Bauer, J. Hassmann, H. Walter, J. Hagemüller, C. Maier, T. Schalkhammer: Resonant nanocluster technology—From optical coding and high quality security features to biochips, Nanotechnology **14**, 1289 (2003).
9. M. Quinten, A. Leitner, J.R. Krenn, F.R. Aussenegg: Electromagnetic energy transport via linear chains of silver nanoparticles, Opt. Lett. **23**, 1331 (1998).

10. S.A. Maier, M.L. Brongersma, P.G. Kik, S. Meltzer, A.A.G. Requicha, H.A. Atwater: Plasmonics—A route to nanoscale optical devices, Adv. Mater. **13**, 1501 (2001).
11. J.R. Lakovicz: Radiative decay engineering: Biophysical and biomedical applications, Anal. Biochem. **298**, 1 (2001).
12. J.R. Lakovicz, Y. Shen, S. D'Auria, J. Malicka, J. Fang, Z. Gryczynski, I. Gryczynski: Radiative decay engineering 2: Effects of silver island films on fluorescence intensity, lifetimes, and resonance energy transfer, Anal. Biochem. **301**, 261 (2002).
13. J. D. Jackson: *Classical Electrodynamics* (Wiley, New York, 1962).
14. C.F. Bohren, D.R. Huffman: *Absorption and Scattering by Small Particles* (Wiley, New York, 1983).
15. M. Kerker: *The Scattering of Light and Other Electromagnetic Radiation* (Academic Press, New York, 1969).
16. A. Wokaun, J.P. Gordon, P.F. Liao: Radiation damping in surface-enhanced Raman scattering, Phys. Rev. Lett. **48**, 957 (1982).
17. J. Kittel: *Introduction to Solid State Physics* (Wiley, New York, 1996).
18. Sönichsen, T. Franzl, T. Wilk, G. von Plessen, J. Feldmann, O. Wilson, P. Mulvaney: Drastic reduction of plasmon damping in gold nanorods, Phys. Rev. Lett. **88**(7), 77402 (2002).
19. W.Gotschy, K. Vonmetz, A. Leitner, F.R. Aussenegg: Thin films by regular patterns of metal nanoparticles: tailoring the optical properties by nanodesign, Appl. Phys. B **63**, 381 (1996).
20. H. Ditlbacher, J.R. Krenn, B. Lamprecht, A. Leitner, F.R. Aussenegg: Spectrally coded optical data storage by metal nanoparticles, Opt. Lett. **25**(8), 563 (2000).
21. E.D. Palik: *Handbook of Optical Constants of Solids* (Academic Press, New York, 1985).
22. B. Lamprecht, J.R. Krenn, A. Leitner, F.R. Aussenegg: Metal nanoparticle gratings: influence of dipolar particle interaction on the plasmon resonance, Phys. Rev. Lett. **84**(20), 4721–4 (2000).
23. B. Lamprecht, J.R. Krenn, A. Leitner, F.R. Aussenegg: SHG studies of plasmon dephasing in nanoparticles, Appl. Phys. B **69**, 223 (1999).
24. S. Zou, N. Janel, G. C. Schatz: Silver nanoparticle array structures that produce remarkably narrow plasmon lineshapes, J. Chem. Phys. **120**(23), 10871 (2004).
25. B. Lamprecht, A. Leitner, F.R. Aussenegg: SHG studies of plasmon dephasing in nanoparticles, Appl. Phys. B **68**, 419 (1999).
26. B. Lamprecht, J.R. Krenn, A. Leitner, F.R. Aussenegg: Resonant and off-resonant light-driven plasmons in metal nanoparticles studied by femtosecond-resolution third-harmonic generation, Phys. Rev. Lett. **83**, 4421 (1999).
27. P.F. Liao, A. Wokaun: Lightning rod effect in surface enhanced Raman scattering, J. Chem. Phys. **76**, 751 (1982).
28. Dereux, C. Girard, J.C. Weeber: Theoretical principles of near-field optical microscopies and spectroscopies, J. Chem. Phys. **112**(18), 7775 (2000).
29. J.R. Krenn, A. Dereux, J.C. Weeber, E. Bourillot, Y. Lacroute, J.P. Goudonnet, G. Schider, W. Gotschy, A. Leitner, F.R. Aussenegg, G. Girard: Squeezing the optical near-field zone by plasmon coupling of metallic nanoparticles, Phys. Rev. Lett. **82**(12), 2590 (1999).
30. M. Salerno, J.R. Krenn, A. Hohenau, H. Ditlbacher, G. Schider, A. Leitner, F.R. Aussenegg: The optical near-field of gold nanoparticle chains, Opt. Comm. **248**(4–6), 543–9 (2005).
31. B. Lamprecht, A. Leitner, F.R. Aussenegg: Femtosecond decay-time measurement of electron-plasma oscillation in nanolithographically designed silver particles, Appl. Phys. B **64**, 269 (1997).
32. W. Rechberger, A. Hohenau, A. Leitner, J.R. Krenn, B. Lamprecht, F.R. Aussenegg: Optical properties of two interacting gold nanoparticles, Opt. Comm. **220**, 137 (2003).

CHAPTER THREE

THEORY OF LIGHT TRANSMISSION THROUGH PERIODICALLY STRUCTURED NANO-APERTURES

F.J. GARCÍA-VIDAL[1], F. LÓPEZ-TEJEIRA[2], J. BRAVO-ABAD[1] AND L. MARTÍN-MORENO[2]

[1]Departamento de Fisica Teorica de la Materia Condensada, Universidad Autonoma de Madrid, E-28049, Spain

[2]Departamento de Fisica, de la Materia Condensada, ICMA-CSIC, Universidad de Zaragoza, E-50009 Zaragoza, Spain

3.1. INTRODUCTION

As illustrated throughout this book, surface plasmons (SPs) are well known for their capabilities of concentrating light in subwavelength volumes and also for guiding light along the surface of a metal. But this does not exhaust the phenomena related to SPs. Even the scientific community working on this subject was greatly surprised in 1998 by the suggestion that SPs could enhance the transmission of light through subwavelength holes.[1] That seminal paper reported that, when subwavelength holes are disposed on a metallic film forming a two-dimensional (2D) array, the transmission of light through this structure is greatly enhanced at some particular wavelengths. The locations of the transmission peaks appearing in the experimental spectra could be approximately found from the dispersion relation of SPs modes running on the metal surface. Then, from the beginning of the history of this subject, it was clear that there was a close connection between extraordinary optical transmission (EOT) and the excitation of SPs. Since 1998, several experimental and theoretical groups around the world have reproduced the main features found in the first set of experiments. The dependence of this phenomenon with the type of metal (noble metals show larger enhancements), type of lattice (square or triangular), shape of the holes (circular, elliptical, square or rectangular) and frequency regime (optical, THz or microwave) have been thoroughly analyzed.[2–13] Four years after the discovery of EOT in 2D hole arrays, it was also reported[14] that EOT phenomenon also appears in a single aperture (hole or slit) flanked by periodic corrugations in the side of the metal film the light is impinging on. Moreover, it was also found that very strong directional emission

M.L. Brongersma and P.G. Kik (eds.), Surface Plasmon Nanophotonics, 27–38.

(beaming) is possible through single apertures if the corrugation is placed on the exit side. In this chapter, we concentrate on the explanation of the fundamental physics behind the phenomenon of EOT both in 2D hole arrays and in single apertures and also behind the beaming effect observed in single apertures surrounded by periodic corrugations.

3.2. TWO-DIMENSIONAL SUBWAVELENGTH HOLE ARRAYS

Our goal in this section is to explain the physical origin of EOT in subwavelength hole arrays.

Let us briefly describe the basic ingredients of our theoretical formalism that was presented in Ref. 3 for the case of a 2D array of square holes and in Ref. 15 for circular holes. In our formalism, the dielectric constant of the metal is taken into account by considering surface impedance boundary conditions[16] (SIBC) on the metal-interfaces defining the metal film. However, in the metal walls defining the hole, the metal is treated as a perfect conductor. This approximation greatly simplifies the formalism as it allows the expression of the electromagnetic (EM) wavefield inside the hole in terms of the eigenmodes of the hole. For simple hole shapes (such as rectangular, triangular or circular), these eigenmodes are known analytically.[17] This approximation, therefore, neglects absorption by the metal walls surrounding the hole. It is expected that this approach is not too bad as the area of the "horizontal" metal-dielectric interfaces (in which absorption is properly taken into account) is much larger than the "vertical" ones, for the geometrical parameters typically analyzed in the experiments. However, assuming perfect conductor walls also neglects the penetration of the EM fields. This is an important deficiency, as it is well known that in the optical regime EM fields penetrate into the metal up to a distance mainly controlled by the skin depth of the metal (of the order of 10–20 nm for noble metals). We circumvent this deficiency by considering an (wavelength dependent) effective hole radius such that the propagation constant inside the hole is equal to the one extracted from an exact calculation. Within these approximations, the calculation of the transmission properties of 2D hole arrays amounts to expanding the EM fields in terms of the Bloch EM modes in each spatial region (plane waves in vacuum regions and hole waveguide modes inside the holes), and obtaining the expansion coefficients by just matching appropriately the parallel components of the E- and H-fields in the two metal-dielectric interfaces.

In Fig. 3.1 we plot the result of our numerical simulation corresponding to the geometrical values of the structure experimentally analyzed in Fig. 1 of Ref. 3, but this time the calculation considers circular holes. Clearly, our model is capturing the main features of the experimental spectrum; the position of the highest peak (located at around 780 nm) is in reasonable agreement with the experimental data. However, the experimental peak is lower and wider than the one obtained in the calculations. This could be indicative of the presence of disorder and/or finite size effects.

In order to gain physical insight into this phenomenon, it is important to look for the minimal model in which EOT is still present. In Fig. 3.2 we compare the result of the fully converged calculation (solid curve) displayed in Fig. 3.1 with the one

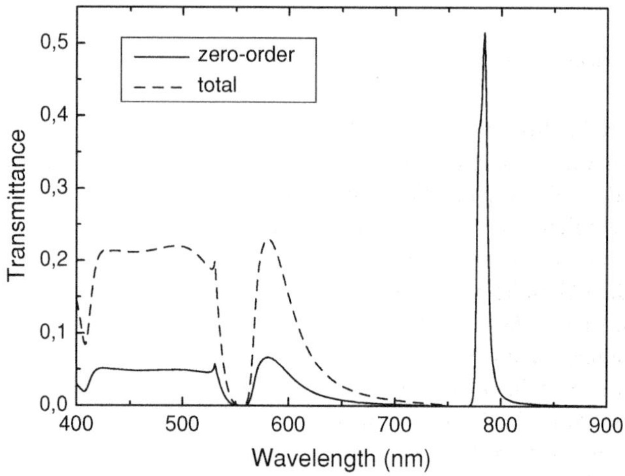

Figure 3.1. Zero-order transmittance (solid curve) and total transmittance (dashed curve) calculated for a 2D hole array (period of the array, $d = 750$ nm and the diameter of the circular holes, $a = 280$ nm) perforated in a silver film of thickness $h = 320$ nm.

(dashed curve) obtained by just considering one eigenmode inside the hole (the TE_{11} mode, the least decaying evanescent mode). As clearly seen in the figure, the inclusion of more evanescent modes inside the hole provokes a very small shift (2 nm) in the transmission peaks to shorter wavelengths but the overall picture of the spectrum

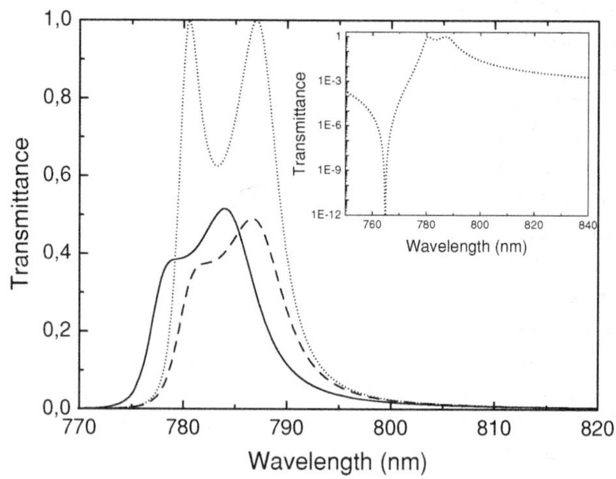

Figure 3.2. Zero-order transmittance for the structure analyzed in Fig. 3.1 obtained by using a fully converged calculation (solid curve), by only considering the TE_{11} mode inside the holes (dashed curve) and the same calculation as the dashed curve but assuming that no absorption is present in the metal (dotted curve); this case is also displayed in the inset but in a logarithmic scale.

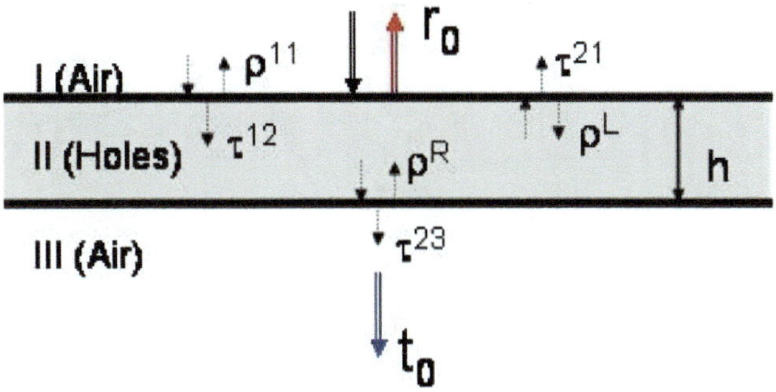

Figure 3.3. Schematic drawing of the different scattering magnitudes at interfaces I–II and II–III. See text for a detailed explanation of the different terms.

remains the same. Moreover, by neglecting the absorption in the metal film (in our calculations we can do it very easily by assuming that the imaginary part of the dielectric constant of silver is zero), we find that the spectrum has two peaks (dotted curve) that reach 100% transmittance and that the net effect of the absorption in the metal is to reduce the amount of light emerging from the structure without altering the physical picture. In the inset of the figure, we plot this last curve in logarithmic scale in order to better illustrate the presence of a zero (the so-called Wood's anomaly, see Ref. 1) in the transmittance spectrum. We will discuss the origin of this zero later on.

From now on in this section we are going to analyze the results of this minimal model (only TE_{11} considered and $\mathrm{Im}(\varepsilon(\omega)) = 0$). In order to unveil the physical mechanism responsible for EOT and to relate EOT with the SPs modes of the metal-dielectric interfaces, we calculate the transmittance through the structure within the multiple scattering formalism. Within this framework, transmission amplitudes for crossing the whole system can be obtained from the scattering amplitudes for crossing the two different individual metal-dielectric interfaces and the propagation constant of the fundamental (TE_{11}) mode inside the hole (see Fig. 3.3).

The zero-order transmission amplitude (t_0) can be expressed then as:

$$t_0 = \frac{\tau_{12} e^{ik_z h} \tau_{23}}{1 - \rho^R \rho^L e^{2ik_z h}}, \tag{3.1}$$

where $\tau_{12}$and τ_{23} are the transmission amplitudes for crossing the I-II and the II-III interfaces, respectively. $k_z = \sqrt{k_0^2 - (1,84/a)^2}$, k_0 is the EM wavenumber in vacuum, and ρ^R and ρ^L are the amplitudes for the TE_{11}mode to be reflected back into the hole at the II-III and II-I interfaces, respectively. In the symmetric system we are now considering where dielectric constants in reflection and transmission regions are equal, $\rho^R = \rho^L = g\rho$. In Fig. 3.4 we show the behavior of the modulus of τ_{12}, τ_{23} and ρ as a function of the wavelength for the case of a 2D square array with $d = 750$ nm of holes and $a = 280$ nm. There are several interesting features appearing in these scattering magnitudes. Firstly, the three quantities present a maximum at around

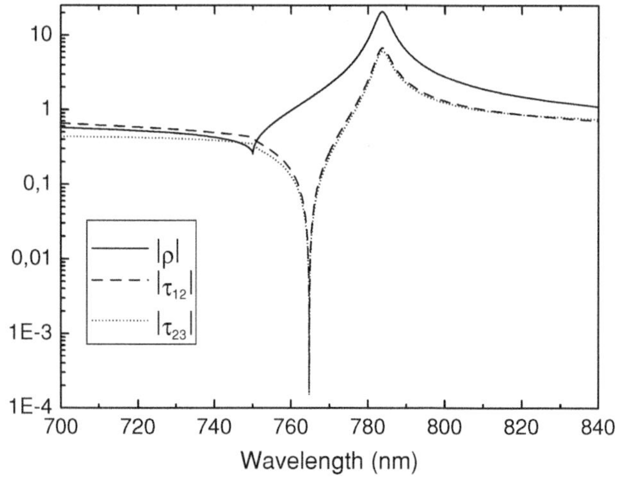

Figure 3.4. Modulus of ρ (solid), τ_{12} (dashed) and τ_{23} (dotted) as a function of the wavelength for a silver surface perforated with a 2D array of period $d = 750$ nm of circular holes with diameter $a = 280$ nm.

785 nm. Moreover, $|\rho| \gg 1$ at this resonant location. This counterintuitive result is due to the fact that the fundamental eigenmode inside the hole is evanescent, for which current conservation only restricts Im $\rho \geq 0$, with no restrictions applied to the real part of this scattering magnitude.

The strong peak in ρ (and in τ_{12}, and τ_{23}) signals the existence of a surface resonance (or surface leaky mode) **of the perforated metal surface**. Its spectral width is related to the time the EM field spends at the surface before it is either radiated or absorbed. This large reflection amplitude opens up the possibility of a resonant denominator in Eq. (3.1) even for metal thicknesses such that $e^{-2|k_z|h} \ll 1$.

Figure 3.5 illustrates graphically that the peaks appearing in the zero-order transmittance occur at the wavelengths for which the *distance* between $|\rho|$ and $e^{|k_z|h}$ is minimal. **This figure unambiguously shows that EOT in 2D hole arrays has a resonant nature and that the origin of this resonant behavior is the existence of SPs decorating the metal-dielectric interfaces**. For thin films ($h = 100 - 400$ nm in Fig. 3.5), the two curves intersect at two different wavelengths giving rise to the appearance of two transmission peaks in the spectrum. It can be shown that these two peaks correspond to the symmetric and antisymmetric combinations of the two SPs of the two interfaces that are coupled through the evanescent fields inside the holes. These two coupled surface modes are able to transfer energy very efficiently (100% if no absorption is present in the system) through the structure. When h is further increased, there is no crossing between the two curves and only one peak with associated transmittance less than 100 % remains in the spectrum. As commented above, the location of this peak coincides with the location of the SP at parallel momentum $2\pi/d$ of the silver surface perforated with a 2D array of holes. A detailed discussion of the formation of the coupled surface modes and the typical times in the transmission process can be found in Ref. 3.

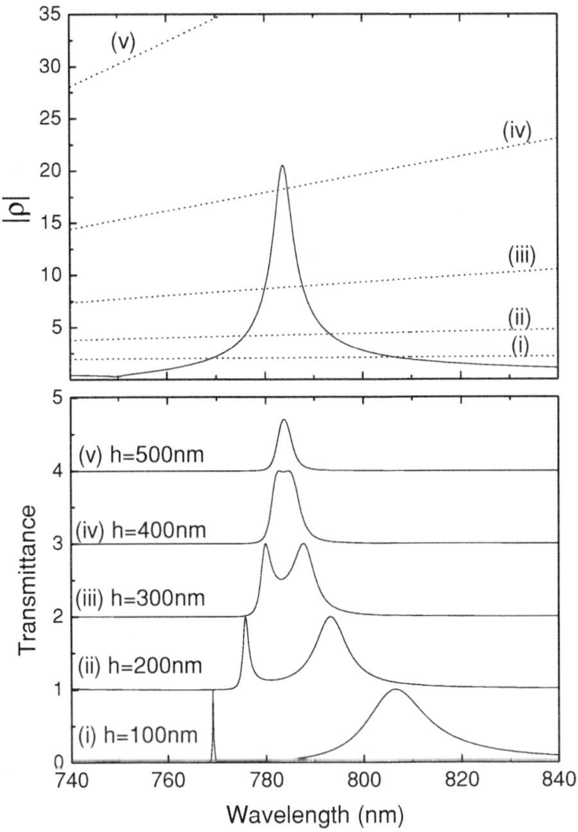

Figure 3.5. Upper panel: modulus of ρ and curves $\exp(|k_z|h)$ for different values of h (100, 200, 300, 400 and 500 nm) for the same geometrical parameters than in the previous figures. Bottom panel: zero-order transmittance versus wavelength for the silver thicknesses considered in the upper panel.

An additional feature appearing in Fig. 3.4 is that both $|\tau_{12}|$ and $|\tau_{23}|$ have a zero located at around 765 nm, that translates into a minimum in the zero-order transmittance theoretical spectrum (see Fig. 3.2), the so-called Wood's anomaly. It is worth commenting that the location of this zero does not coincide with the location of the Rayleigh minimum that occurs when a propagating diffracted wave becomes evanescent (in this particular case this should appear at 750 nm). On the contrary, it can be shown that the location of the minima in both $|\tau_{12}|$ and $|\tau_{23}|$ coincides with the location of the SP at parallel momentum $2\pi/d$ of the **plain** (without holes) silver surface. This is the origin of some criticism on the relation of EOT and excitation of SPs.[18] As we have stated here and already shown in Ref. 3, EOT is mediated by SPs but those corresponding to the structured metal surface.

Once EOT is explained in terms of the excitation of SPs in the optical regime, the question about the transferability of EOT to other frequency regimes naturally arises. In Ref. 3 we showed that the EOT phenomenon also appears even in a perfect conductor film perforated with a 2D array of holes. A more extensive theoretical analysis of the existence of EOT in perfect conductors can be found in Ref. 15. But, flat perfect conductor interfaces do not posses SP modes. This could imply that the origins of EOT for metals in the optical regime and for perfect conductors are different. Importantly, surface EM modes appear in **corrugated** perfect conductors and, in particular, in perfect conductors perforated with 2D hole arrays. Very recently, we have shown that these surface EM modes are responsible for the existence of EOT in perfect conductors.[19] Therefore, EOT seems to be a more general phenomenon that will appear in any electromagnetic structure in which surface EM modes are present and can couple to radiative modes. This hypothesis has been verified for metals in the THz[12] and microwave regimes[13] and for photonic crystal waveguides.[20]

3.3. EOT IN SINGLE APERTURES FLANKED BY CORRUGATIONS

As was discussed in the previous section, surface EM modes are at the origin of the EOT phenomenon. The two necessary ingredients for observing EOT are: (i) the existence of a surface EM mode and (ii) the presence of a grating coupler that allow the incident light to interact with the surface mode. Therefore, it was reasonable to expect that EOT phenomenon could also appear in a **single aperture** surrounded by a finite periodic array of indentations. This hypothesis was experimentally verified in Ref. 14 both for a 1D slit surrounded by a finite array of grooves and for a 2D circular hole flanked by circular trenches, the so-called *bull's eye* geometry. Here we present the theoretical foundation of this phenomenon for the 1D case.

We present a brief description of the theoretical formalism used to simulate the transmission of light through a single slit of width a symmetrically flanked by a finite array (with period d) of $2N$ grooves of width a and depth w (see Fig. 3.6). A normal incident p-polarized plane wave is impinging at the structure.

The theoretical formalism we have developed in order to describe the transmission properties of this type of structures is a non-trivial extension to finite structures of the framework previously used for analyzing 2D hole arrays. First we consider an artificial supercell with cell parameter L that includes the finite set of indentations we are

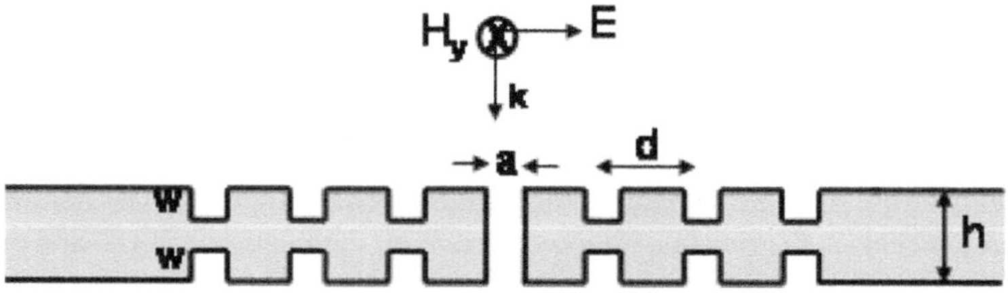

Figure 3.6. Schematic figure of the structure analyzed in this section: a single slit of width a, surrounded by grooves of width a and depth w in the input and output surfaces. The metal thickness is h and we analyze the transmission properties of this system for a normal incident p-polarized plane wave.

considering. Then we express the EM-fields in different regions in terms of their mode expansion. In vacuum we expand the fields by a set of plane waves whereas for the grooves and central slit only the fundamental propagating eigenmode is considered. That is, inside indentation α, E_x is a linear combination of $\phi_\alpha(x)e^{\pm ikz}$, where $2\pi/\lambda$ and $\phi_\alpha(x) = 1/\sqrt{a}$. Then we match the fields appropriately on all interfaces (as in the case of 2D hole arrays, in the horizontal interfaces we apply SIBC while perfect metal boundary conditions are assumed in the vertical ones). At the end the limit $L \to \infty$ is taken, leading to a set of linear equations for the unknowns $\{E_\alpha, E'_\gamma\}$:

$$\begin{aligned}
&[G_{\alpha\alpha} - \varepsilon_\alpha]E_\alpha + \sum_{\beta\neq\alpha} G_{\alpha\beta}E_\beta - \delta_{\alpha 0}G_V E'_\alpha = I_\alpha \\
&[G_{\gamma\gamma} - \varepsilon_\gamma]E'_\gamma + \sum_{\nu\neq\gamma} G_{\gamma\nu}E'_\nu - \delta_{\gamma 0}G_V E_\gamma = 0
\end{aligned} , \tag{3.2}$$

where α and γ runs over all indentations (slit or grooves). The set $\{E_\alpha\}$ gives the x-component of the electric field right at the indentations in the input surface: $E_x(z = 0^+) = \sum_\alpha E_\alpha\phi_\alpha(x)$ whereas the set $\{E'_\gamma\}$ describes the x-component of the electric field at the output surface: $E_x(z = h^-) = \sum_\gamma E'_\gamma\phi_\gamma(x)$. The different terms appearing in these *tight-binding* equations have a clear physical interpretation. I_α takes into account the direct initial illumination term over object α and it is basically the overlap integral between the incident p-polarized plane wave and wavefield ϕ_α. In this structure, the two metal interfaces are only connected through the central slit by the term $G_V = 1/\sin(kh)$. ε_α measures the back and forth bouncing of the EM fields inside indentation α:$\varepsilon_\alpha = \cot(kw)$ at the grooves ($\alpha \neq 0$) and $\varepsilon_0 = \cot(kh)$ for the slit. The term $G_{\alpha\beta}$ controls the EM coupling between indentations. It takes into account that each point in the indentation β emits radiation that can be collected by indentation α. Mathematically, $G_{\alpha\beta}$ is the projection onto wavefields ϕ_α and ϕ_β of the Green's function $G(\vec{r}, \vec{r}')$. It can be shown that this Green function contains the contribution of both diffraction modes and the SP channel. In general, it has to be computed numerically although in the case of perfect conductors its expression is known analytically as: $G = (i\pi/\lambda)\, H_0^{(1)}(k\,|\vec{r} - \vec{r}'|)$, $H_0^{(1)}$ being the 0-order Hankel function of the first kind. Once the values for $\{E_\alpha, E'_\gamma\}$ are calculated the normalized-to-area transmittance can be obtained from $T = G_V Im(E_0 E'_0)$.

In Fig. 3.7 we show the theoretical results for the normalized-to-area transmittance $T(\lambda)$ for a single slit surrounded by $2N$grooves in the input side and located symmetrically (N to the left and right) with respect to the central slit. The set of geometrical parameters used is typical for experimental studies of this phenomenon in the optical regime ($a = 100$ nm, $d = 600$ nm, $w = 100$ nm and $h = 400$ nm).

The curve for $N = 0$ (solid curve) corresponds to the single slit case; in this frequency range, the spectrum presents two broad peaks that correspond to the excitation of slit waveguide modes inside the central slit.[21] As the number of indentations increases, a maximum in $T(\lambda)$ develops at $\lambda_M = 755$ nm. For this set of geometrical values and for the metal considered, maximum in $T(\lambda)$ saturates at about $N = 5$–10, when T is enhanced by a factor close to 5. With respect to the output corrugation, we have demonstrated in previous works (see Ref. 22) that it has little effect on the

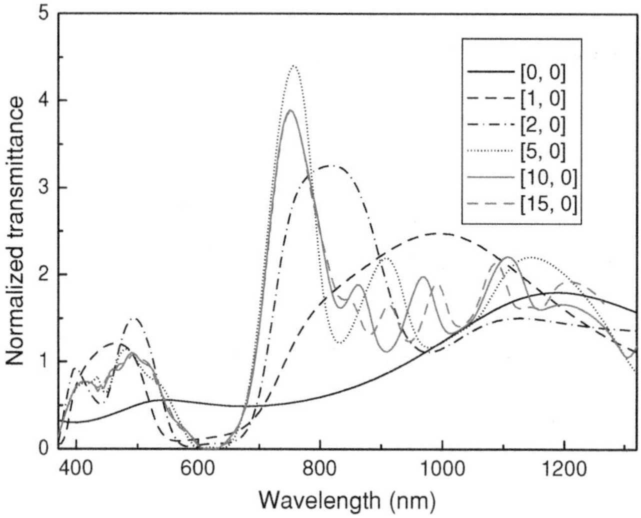

Figure 3.7. Normalized transmittance versus wavelength for a single slit of width $a = 100$ nm surrounded by 2N grooves (N ranging from 0 to 15) located symmetrically with respect to the central slit and only disposed in the input surface. In these calculations the output surface is not corrugated. The period of the array is $d = 600$ nm, the width of the grooves is also 100 nm and their depth is 100 nm. The silver film has a thickness of 400 nm.

total transmittance. From the set of Eqs. (3.2), it is possible to identify the different mechanisms that help to enhance the transmission of light through the central slit and are present in Fig. 3.7. The corresponding two equations for $\{E_0, E_0'\}$ are (assuming that the slit is flanked by the grooves only at the input surface):

$$\begin{aligned} &[G_{00} - \varepsilon_0]E_0 + \sum_{\alpha \neq 0} G_{0\alpha}E_\alpha - G_V E_0' = I_0 \\ &[G_{00} - \varepsilon_0]E_0' - G_V E_0 = 0 \end{aligned} , \qquad (3.3)$$

As commented before, one mechanism is already present in a single slit. For this particular case, $E_0 = 2(G_{00} - \varepsilon_0)/D$ and $E_0' = 2G_V/D$ where the denominator $D = (G_{00} - \varepsilon_0)^2 - G_V^2$. For some particular wavelengths, D is very close to zero. These resonances are essentially slit waveguide modes. Corrugating the input surface opens up the possibility of having a large E_0 by having a large E_α. If we have a look to the equation for E_α, this magnitude can be large if $G_{\alpha\alpha} - \varepsilon_\alpha \approx 0$, that is the condition of the excitation of a groove cavity mode. However, in order to have a large E_0, the illumination coming from the different grooves has to reach the central slit in phase. The phase in this re-illumination process is controlled by $G_{0\alpha}$. By looking at the asymptotic expression of $H_0^{(1)}(x) \approx e^{ikx}$, it is expected that all light re-emitted from the grooves reaches the other grooves and the central slit in phase for $\lambda \approx d$, although the presence of a SP channel in $G_{0\alpha}$ modifies slightly this condition as seen in Fig. 3.7. The combination of the two mechanisms described above (groove cavity mode and

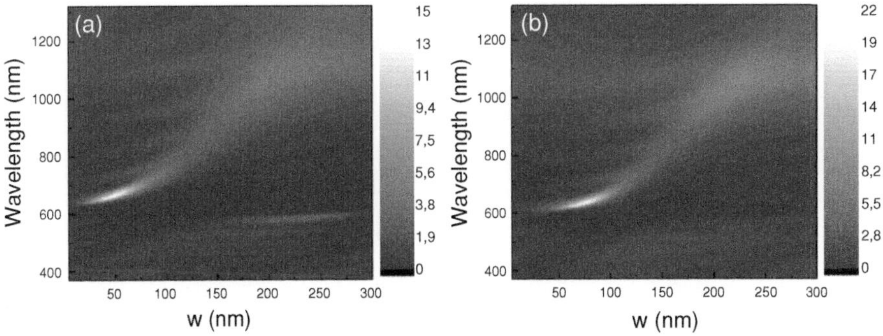

Figure 3.8. Normalized-to-area transmittance versus wavelength and depth of the grooves for a single slit of width 100 nm surrounded by 20 grooves located symmetrically in the input surface. The thickness of the metal film is 400 nm as in the previous figure. In panel (a), we are assuming SIBC in the horizontal interfaces of the structure whereas panel (b) shows the results for a perfect conductor film.

in-phase groove re-emission) is responsible for the peak located at around 755 nm appearing in Fig. 3.7.

In order to illustrate how the three different mechanisms influence the transmittance through the central slit, in Fig. 3.8(a) we show T versus λ and depth of the grooves, w, for $a = 100$ nm, $h = 400$ nm, $d = 600$ nm and $N = 10$. As clearly seen in this figure, when two mechanisms coincide, there is a boost in the transmittance. For small w, maximum transmittance appears close to the $\lambda \approx d$ condition. **It can be shown that this line corresponds to the excitation of a surface EM mode**, originated by the interplay between the groove cavity modes and the in-phase groove re-emission mechanisms. This surface mode has strong similarities with the one responsible for EOT in periodic apertures. It is quite interesting to compare the results we have presented in (a) with the ones obtained within the perfect conductor approximationand rendered in panel (b) of Fig. 3.8. The similarities between the results obtained in these two cases reinforces the conclusion that the main ingredients of the EOT phenomenon in 2D hole arrays and in single apertures is already present in corrugated perfect conductor surfaces.

3.4. BEAMING OF LIGHT IN SINGLE APERTURES

As commented before, it was found experimentally in Ref. 14 that the radiation pattern emerging from the structure (basically controlled by the output corrugation) presents a very small angular divergence at some resonant wavelengths. As a way of example, in Fig. 3.9 we show our calculated radial component of the Poynting vector, $S_r(\theta)$, in the far-field region and normalized to the total transmittance for a single slit surrounded symmetrically by $2N$ grooves in the output surface. The theoretical framework used is the same as the one described in previous section. Several N are presented (from 1 to 15) for the resonant wavelength $\lambda_M = 750$ nm.

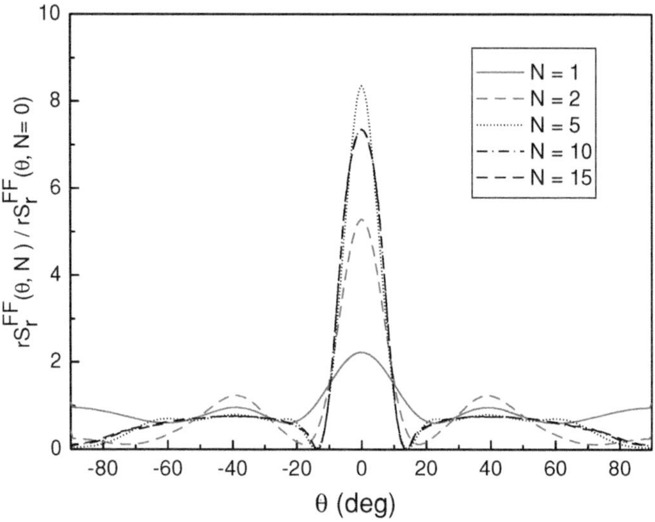

Figure 3.9. Radial component of the Poynting vector evaluated in the far field versus angle for a single slit of width a = 100 nm surrounded symmetrically by $2N$ grooves (N ranging from 1 to 15) of width 100 nm and depth 100 nm. The wavelength of the incident radiation is 750 nm.

Note that this resonant wavelength is the same as the one found for EOT in a single slit flanked by a finite array of grooves in the input surface for the same set of geometrical parameters. This fact clearly shows that the origin of the beaming effect is the same as the EOT in single apertures surrounded by periodic corrugations: the excitation of a surface EM mode in the output surface. Details about the formation of this surface mode and its relation with the radiation pattern mode can be found in Ref. 23.

REFERENCES

1. T.W. Ebbesen, H.J. Lezec, H.F. Ghaemi, T. Thio, P.A. Wolff: Extraordinary optical transmission through sub-wavelength hole arrays, Nature **391**(6668), 667–669 (1998).
2. H.F. Ghaemi, T. Thio, D.E. Grupp, T.W. Ebbesen, H.J. Lezec: Surface plasmons enhance optical transmission through subwavelength holes, Phys. Rev. B **58**(11), 6779–6782 (1998).
3. L. Martín-Moreno, F.J. García-Vidal, H.J. Lezec, K.M. Pellerin, T. Thio, J.B. Pendry, T.W. Ebbesen: Theory of extraordinary optical transmission through subwavelength hole arrays, Phys. Rev. Lett. **86**(6), 1114–1117 (2001).
4. L. Salomon, F.D. Grillot, A.V. Zayats, F. de Fornel, Near-field distribution of optical transmission of periodic subwavelength holes in a metal film, Phys. Rev. Lett. **86**(6), 1110–1113 (2001).
5. A. Krishnan, T. Thio, T.J. Kima, H.J. Lezec, T.W. Ebbesen, P.A. Wolff, J. Pendry, L. Martín-Moreno, F.J. García-Vidal: Evanescently coupled resonance in surface plasmon enhanced transmission, Opt. Commun. **200**(1–6), 1–7 (2001).
6. A. Degiron, H.J. Lezec, W.L. Barnes, T.W. Ebbesen: Effects of hole depth on enhanced light transmission through subwavelength hole arrays, Appl. Phys. Lett. **81**(23), 4327–4329 (2002).

7. N. Bonod, S. Enoch, L. Li, E. Popov, M. Nevière: Resonant optical transmission through thin metallic films with and without holes, Opt. Exp. **11**(5), 482–490 (2003).
8. C. Genet, M. P. van Exter, J.P. Woerdman: Fano type interpretation of red shifts and red tails in hole array transmission spectra, Opt. Comm. **225**(4-6), 331 (2003).
9. W.L. Barnes, W.A. Murray, J. Ditinger, E. Devaux, T.W. Ebbesen: Surface plasmon polaritons and their role in the enhanced transmission of light through periodic arrays of subwavelength holes in a metal film, Phys. Rev. Lett. **92**(10), 107401 (2004).
10. R. Gordon, A.G. Brolo, A. McKinnon, A. Rajora, B. Leathem, K.L. Kavanagh: Strong polarization in the optical transmission through elliptical nanohole arrays, Phys. Rev. Lett. **92**(3), 37401 (2004).
11. K.J. Klein Koerkamp, S. Enoch, F.B. Segerink, N.F. van Hulst, L Kuipers: Strong influence of hole shape on extraordinary transmission through periodic arrays of subwavelength holes, Phys. Rev. Lett. **92**(18), 183901 (2004).
12. J. Gómez-Rivas, C. Schotsch, P. Haring Bolivar, H. Kurz: Enhanced transmission of THz radiation through subwavelength holes, Phys. Rev. B **68**(20), 201306 (2003).
13. M. Beruete, M. Sorolla, M. Campillo, J.S. Dolado, L. Martín-Moreno, J. Bravo-Abad, F.J. García-Vidal: Enhanced millimeter-wave transmission through subwavelength hole arrays, Opt. Lett. **29**(21), 2500–2502 (2004).
14. H.J. Lezec, A. Degiron, E. Devaux, R.A. Linke, L. Martín-Moreno, F.J. García-Vidal, T.W. Ebbesen: Beaming light from a subwavelength aperture, Science **297**(5582), 820–822 (2002).
15. L. Martín-Moreno, F.J. García-Vidal: Optical transmission through circular hole arrays in optically thick metal films, Opt. Express **12**(16), 3619–3628 (2004).
16. J.D. Jackson: *Classical Electrodynamics*, 2nd ed. (Wiley, New York, 1975).
17. P.M. Morse, H. Feshbach: *Methods of Theoretical Physics* (McGraw-Hill, New York, 1953).
18. Q. Cao, P. Lalanne: Negative role of surface plasmons in the transmission of metallic gratings with very narrow slits, Phys. Rev. Lett. **88**(5), 57403 (2002).
19. J.B. Pendry, L. Martín-Moreno, F.J. García-Vidal: Mimicking surface plasmons with structured surfaces, Science **305**(5685), 847–848 (2004).
20. E. Moreno, F.J. García-Vidal, L. Martín-Moreno: Enhanced transmission and beaming of light via photonic crystal surface modes, Phys. Rev. B **69**(12), 121402 (2004).
21. J.A. Porto, F.J. García-Vidal, J.B. Pendry: Transmission resonances on metallic gratings with very narrow slits, Phys. Rev. Lett. **83**(14), 2845–2848 (1999).
22. F.J. García-Vidal, H.J. Lezec, T.W. Ebbesen, L. Martín-Moreno: Multiple paths to enhance optical transmission through a single subwavelength slit, Phys. Rev. Lett. **90**(21), 213901 (2003).
23. L. Martín-Moreno, F.J. García-Vidal, H.J. Lezec, A. Degiron, T.W. Ebbesen: Theory of highly directional emission from a single aperture surrounded by surface corrugations, Phys. Rev. Lett. **90**(16), 167401 (2003).

CHAPTER FOUR

DEVELOPMENT AND NEAR-FIELD CHARACTERIZATION OF SURFACE PLASMON WAVEGUIDES

J.-C. WEEBER[1], A.-L. BAUDRION[1], M. U. GONZÁLEZ[1], A. DEREUX[1], RASHID ZIA[2], AND MARK L. BRONGERSMA[2]

[1]Laboratoire de Physique de l'Université de Bourgogne, 9 Avenue A. Savary, BP 47870, F-21078 Dijon, France, jcweeber@u-bourgogne.fr
[2]Geballe Laboratory for Advanced Materials, Stanford University, Stanford, CA 94305

4.1. INTRODUCTION

A polariton is an electromagnetic mode related to the oscillation of polarization charge density. At the interface between two media with frequency dependent complex dielectric functions ε_1 and ε_2, surface polaritons with electromagnetic field exponentially decaying into both media may occur according to the well-known dispersion relation[1]: $k_{sp} = (\omega/c)\sqrt{(\varepsilon_1\varepsilon_2)/(\varepsilon_1 + \varepsilon_2)}$ (where k_{sp}, ω and c are respectively the in-plane wave-vector of the surface polariton, the angular frequency and the speed of light) provided the real part of the dielectric functions in the two media are of opposite sign. If the material with the negative real part dielectric function is a metal, the polarization charge density oscillation corresponds to the oscillation of the electron gas, and then the surface polariton is called a surface plasmon polariton (SPP). Unlike SPP excited on extended metal thin films, which have been studied for decades, SPP sustained by thin metal films of finite width (metal strips) have been considered only recently. These metal strips, that can be viewed as SPP waveguides, could play an important role in the development of surface wave based optical devices.

When embedded inside a dielectric medium, thin metal strips (MS) can support long-range type SPP modes with propagation distances of a few millimeters at telecommunication frequencies.[2–4] Based on these MS, passive and active devices such as couplers and modulators have been recently demonstrated.[5–7] If the metal strips are deposited onto a substrate with a refractive index different from that of the superstrate, the field of the SPP penetrates more deeply into the metal, leading

M.L. Brongersma and P.G. Kik (eds.), Surface Plasmon Nanophotonics, 39–54.

to much shorter propagation distances than in the symmetric case. Nevertheless, asymmetric MS are still promising for applications such as reduced size optical components[8] or bio-sensors where the propagation distances of practical interest are small. Additionally, the field confinement in the direction perpendicular to the surface along which the SPP propagates is much higher in the asymmetric case than in the symmetric case, making asymmetric MS appealing for the design of optical systems integrated into coplanar geometries.

In this work, we present a detailed study of the SPP propagation along gold strips, performed by means of a near-field optical microscope known as the photon scanning tunneling microscope (PSTM).[9] We first analyze the near-field intensity distributions obtained for MS of varying widths. Once the guiding properties of straight SPP waveguides have been described, we focus on the use of double-bent MS for SPP routing. Then with the aim of improving such routing, we consider structured waveguides equipped with micro-gratings formed by periodic surface defects (bumps, slits). We show that such micro-gratings act as Bragg mirrors on the SPP modes. Tilted Bragg mirrors integrated into metal strips are used to achieve highly efficient SPP guiding along 90° sharp bends. Finally, we demonstrate the fabrication of metal strip based SPP beamsplitters.

4.2. EXPERIMENTAL BACKGROUND

4.2.1. Near-Field Microscopy

The experimental PSTM setup used in this work is sketched in Fig. 4.1. The microscope consists of three parts: an optical stage, a piezo-tube scanner, and a data acquisition system. The optical stage is used to excite the SPP in the Kretschmann-Raether configuration. The samples are glued on the hypotenuse of a right angle glass prism using an index matching fluid that provides the refractive index continuity between the sample substrate and the prism. The samples are illuminated by means of

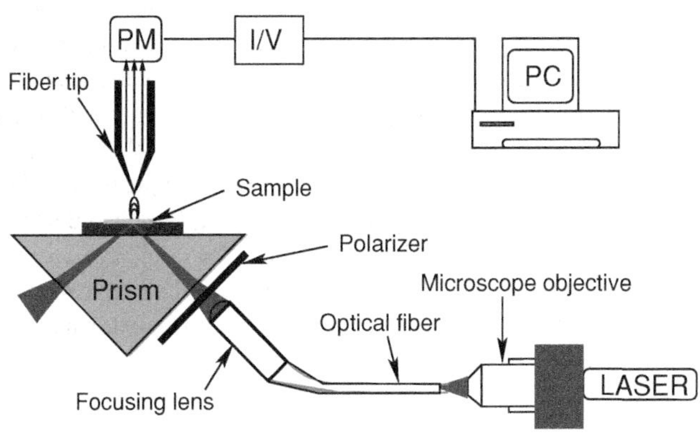

Figure 4.1. Schematic view of the PSTM setup.

either a collimated or a focused beam, obtained by injecting a Titanium–sapphire laser into a suitably lensed monomode optical fiber. The output end of the excitation fiber is mounted on a three axis micro-positionner, which controls the location and focal position of the spot on the sample. The excitation of the SPP is achieved by finely adjusting the angle of incidence with a rotation platform. The SPP excitation is effective for an angle of incidence θ given by: $n\ k_0 \sin(\theta) = k_{sp}$, where nk_0 is the incident light wave-vector into the prism and k_{sp} is the in-plane wave-vector of the SPP. All the experimental results shown in this work have been obtained using PSTM probes comprising tapered multi-mode fibers coated with a thin layer of chromium. The tapered fibers are fabricated by a standard heat–and–pull technique and are subsequently coated with chromium while rotating around their axis. The PSTM probes are mounted on the piezo-tube and scanned just above the sample surface. Because the refractive index of the PSTM probe is larger than that of the surrounding medium (air in our case), the evanescent field of the SPP is frustrated and converted into guided modes. The output end of the PSTM probe is connected to a photo-multiplier (PM) tube, where the near-field intensity detected by the tip is converted into a current signal and then amplified (typical gain 10^6) using a current/voltage amplifier. Thus, for each position of the tip relative to the sample, a voltage proportional to the near-field intensity is transmitted to the data acquisition system. Scanning the tip at a constant height (typical tip-sample distances are around 150 nm) allows us to obtain the electric near-field distribution in an observation plane parallel to the sample surface.

4.2.2. Sample Fabrication

The samples considered in this work are fabricated by electron beam lithography (EBL), combined with focused ion beam (FIB), milling when needed for the additional patterning. The EBL is a six-step fabrication process which starts by spin coating indium tin oxide (ITO) doped glass substrates with a polymethylmethacrylate (PMMA) layer (typical thickness of 200 nm). After baking the PMMA layer, the coated substrates are exposed in a scanning electron microscope equipped with a beam motion-control software. The subsequent development of the sample dissolves the exposed PMMA areas. The samples are then coated with metal, and finally lift-off of the metal film is performed by chemically removing the non-exposed PMMA areas. In our case, a gold film with a typical thickness of 50–60 nm is evaporated onto the samples. Thus, after the lift-off step, we end up with positive gold patterns lying on the glass substrate.

4.3. RESULTS

4.3.1. Field Distribution of Metal Strip Modes

By analogy to the different guided volume electromagnetic modes supported by channel waveguides and planar slab waveguides, the properties of a guided SPP are ex-

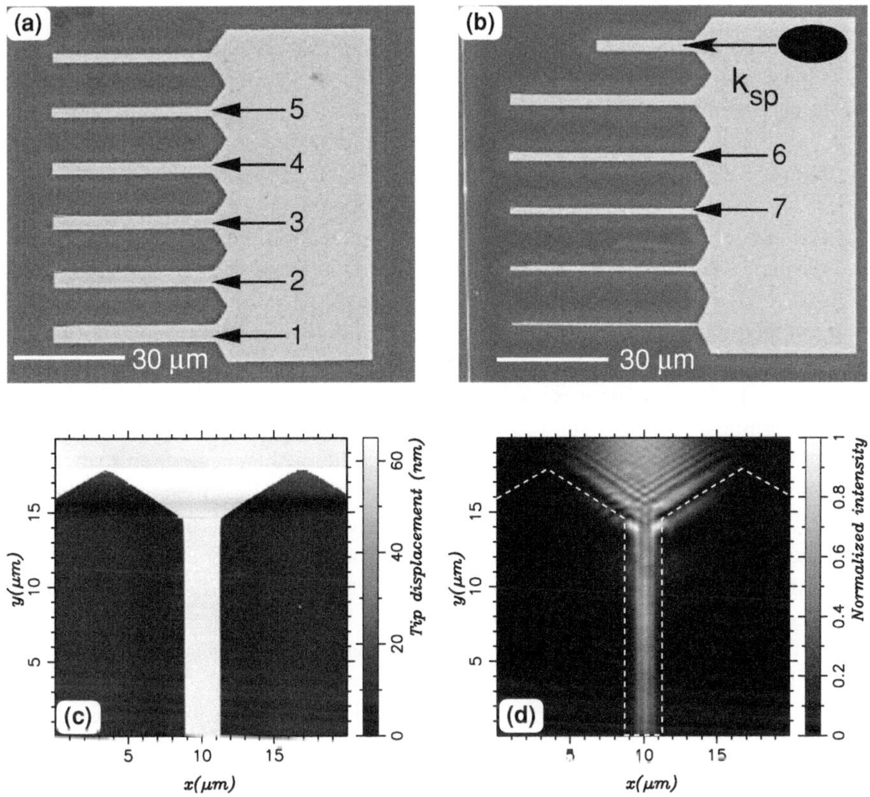

Figure 4.2. (a) and (b) Scanning electron microscope (SEM) images of the samples used to study the near-field response of metal strips of different widths. (c) Atomic force microscope (AFM) image of the taper connecting the extended thin film area and a strip with a width of 2.5 μm. (d) PSTM image of the 2.5 μm wide strip. A SPP launched on the extended thin film area couples with a SPP propagating along the strip.

pected to differ along a finite width metal thin film (metal strip) as compared to an infinitely wide, extended thin film. For example, it has been shown that SPP attenuation depends on the width of the MS along which it propagates.[10] Similarly, the field distribution of the SPP sustained by MS (MS-SPP modes) is also expected to depend on the waveguide width. In order to study the modal field distribution of MS, we have considered the structures shown in Figs. 4.2(a) and (b). The samples consist of various width MS connected to a larger thin film area. A surface plasmon is launched on the extended thin film area through a focused spot according to the method described in the experimental section. The locally launched SPP propagates along the surface of the extended thin film and couples with the SPP sustained by the MS. Figs. 4.2(c) and (d) show an atomic force microscope (AFM) image and the corresponding PSTM image of a 2.5 μm wide MS excited with incident light having a free-space wavelength of $\lambda_0 = 800$ nm. It can be seen that, above the waveguide, the field distribution exhibits a neat three-peak longitudinal structure. This multiple-peak intensity distribution is characteristic of the SPP mode for a fixed incident wavelength and a given MS width.[11,12]

The influence of MS width on the near-field distribution has been studied by taking PSTM images of seven waveguides with widths ranging from $W = 4.5$ to $W = 1.5$ microns. All of these images have been obtained for an incident free-space wavelength of 800 nm. The transverse cross sections (taken along the x axis) of these PSTM images, together with the topographic profiles of the strips measured by AFM, are displayed in Fig. 4.3. Note that the MS provide a very efficient lateral confinement of the SPP field. Indeed, except for $W = 1.5$ μm (which corresponds to roughly $2 \times \lambda_0$), it can be seen that the near-field intensity drops to zero within the width of MS for wide waveguides or at the location of the vertical walls for thinner waveguides.

The near-field optical profiles of the MS are found to depend upon strip width (W). For decreasing W, the number of peaks in the optical profiles and/or the distance p between these peaks decrease. Clearly, the sustained nature of these transverse intensity patterns is characteristic of optical waveguide modes. However, the physical origin of these peaks and, in particular, their connection to the guided modes of the MS are still the subject of debate.

It has been shown recently that finite width MS lying on a glass substrate, can sustain leaky surface plasmon modes with phase constants that differ only slightly from that of a SPP launched on an extended metal thin film.[13] The physical basis for these modes is the interference of counter-propagating SPPs, which undergo continuous total internal reflection at the edges of the MS. Thus, it is possible to understand these modes by analogy to those of conventional dielectric slab waveguides, where lateral confinement is provided by index guiding.[14] Just as the normalized phase constant for a dielectric slab, waveguide lies between that of a high index core and low index cladding. The normalized phase constant for the surface plasmon MS mode resides between that of a high index SPP core and a low index air cladding.

Considering our illumination method, the initial SPP excited along the extended thin film area, could couple to the leaky surface plasmon modes on the MS, which are well matched in terms of momentum and overlap. Thus, the observed field distribution may result from the multimode excitation of several such modes. This hypothesis is supported by the correspondence of the observed lateral peaks with the simulated electric field intensities above such MS.[13] With increasing stripe width, an increasing number of leaky MS modes are supported with additional intensity peaks. For example, 1.5 μm and 2.5 μm wide strips support only a single fundamental MS mode with three lateral intensity peaks, while a 3.5 μm wide strip also supports a higher order mode with four such peaks. (Note that in addition to intensity peaks, which result from the modal interference of counter-propagating SPPs, there are also two peaks associated with the severe dielectric discontinuities at the edge of the MS; hence, the fundamental MS mode has three peaks instead of one.) These simulated results are in good agreement with the experimental results shown in Fig. 4.3, but questions

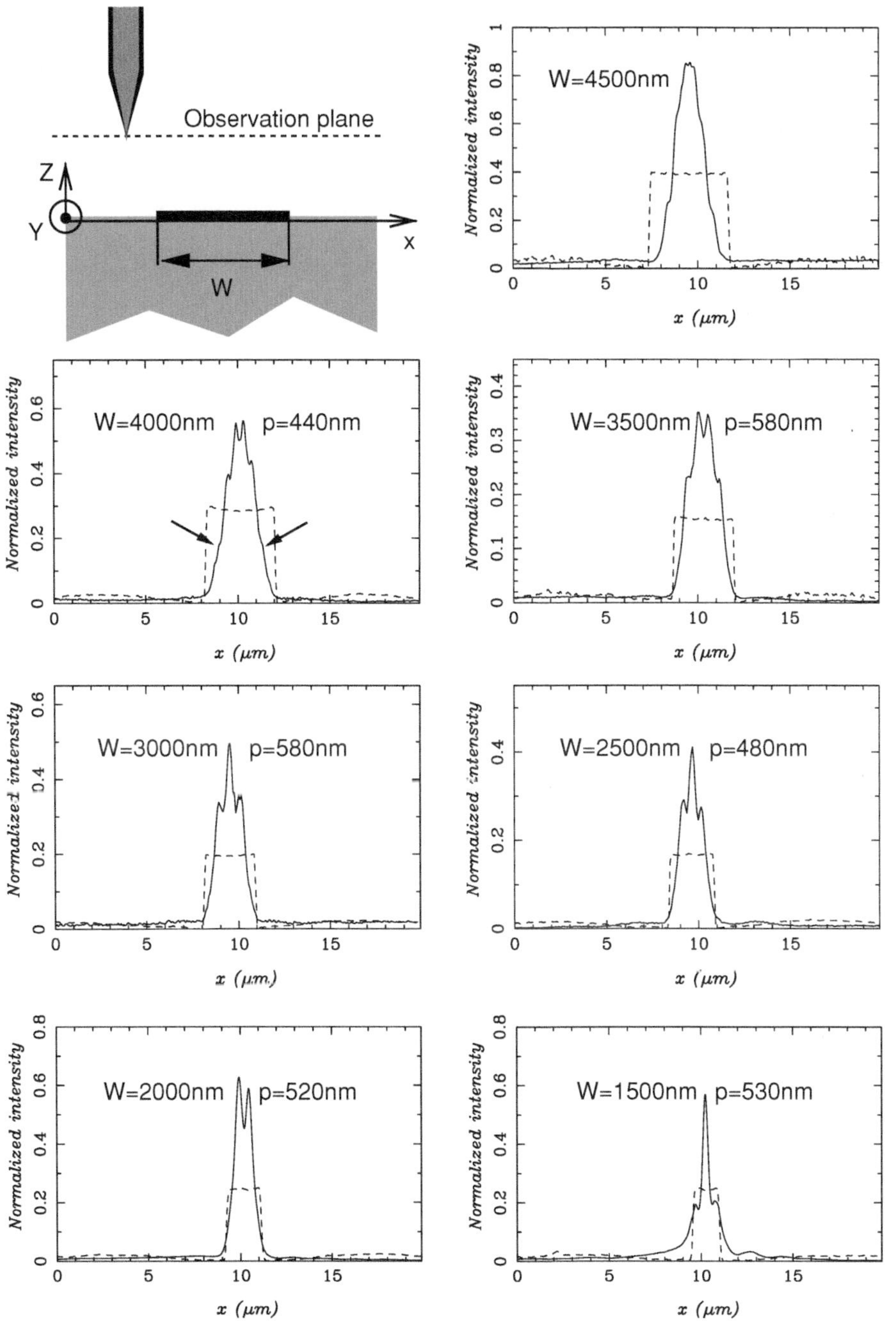

Figure 4.3. (solid lines) Transverse cross sections of the near-field images recorded at constant height over several strips. All these results have been obtained for an incident free-space wavelength of 800 nm. (Dashed lines) Topographic profiles of the considered strips.

remain regarding the experimental relevance of higher order modes which have odd lateral parity. Given that the initial SPP from the extended film area should be laterally symmetric (i.e. even parity), one would suspect that the overlap integral with odd modes would be small. However, it is unclear if potentially small asymmetries in illumination or structure may be sufficient to provide for the excitation of such modes.

On the other hand, it has been shown unambiguously that the edges of the strip play a key role in the near-field distribution of the MS-SPP modes. On the basis of these experimental results, the MS field distribution can also be attributed to the coupling of a finite interface SPP mode (which corresponds to the fundamental leaky mode sustained by a given MS) and eigenmodes supported by the MS edges.[12]

While not fully understood, the near-field intensity distributions of the MS-SPP modes are nevertheless of great interest for future applications. In particular, the field confinement achieved by MS is a key parameter for the design of integrated optical devices such as directional couplers. Because of their very efficient field confinement, MS can be viewed as SPP waveguides. However, to be useful for practical applications, MS should also allow for manipulation of the SPP propagation direction; in particular, MS should allow SPP to be guided along bent paths. From this point forward, we focus our analysis on two different methods to direct SPP along bent paths: (1) routing along bent strips and (2) integration of basic optical elements, such as mirrors, inside the strips.

4.3.2. Routing Metal Strip Modes

Bent Homogeneous Strips

The simplest strategy to guide MS-SPP modes along bent paths is to use bent strips. To assess the guiding efficiency of SPP along bent MS, the sample shown in Fig. 4.4(a), has been fabricated. This sample comprises dual MS connected to a large launching area. The separation between the straight and the bent MS has been chosen large enough (1 μm) to prevent significant cross-talk between the two waveguides. For continuity, the straight and the bent waveguides have a cross sectional dimension of 2.5 μm along the x axis, which results in the bent regions having a slightly smaller width ($W = 2.5\ \mu\text{m}/\cos(\theta)$). Likewise, the length of the bent region is 10 μm along the y axis, which results in a slightly longer path for increasing bend angle. The excitation of a dual MS is performed by focused beam illumination on the extended thin film area: the waist of the incident spot is large enough to achieve the excitation of both the straight and the bent MS at the same time. A typical PSTM image recorded at $\lambda_0 = 800$ nm over the dual MS with a bend angle of 10° is shown in Fig. 4.4(b). As expected from the previous discussion, a near-field intensity distribution with three peaks is observed throughout the entire length of the straight MS. For the bent MS, this modal structure is clearly visible at the input, but becomes more difficult to identify following the initial bend. Clearly, the SPP traveling along the doubly bent MS suffers additional losses (mainly scattering losses) as compared to the straight strip. In order to quantify these losses, we compare the near-field intensities at the output of the bent (I_o) and straight MS (I_o^r), normalized by their corresponding input levels (I_i and I_i^r

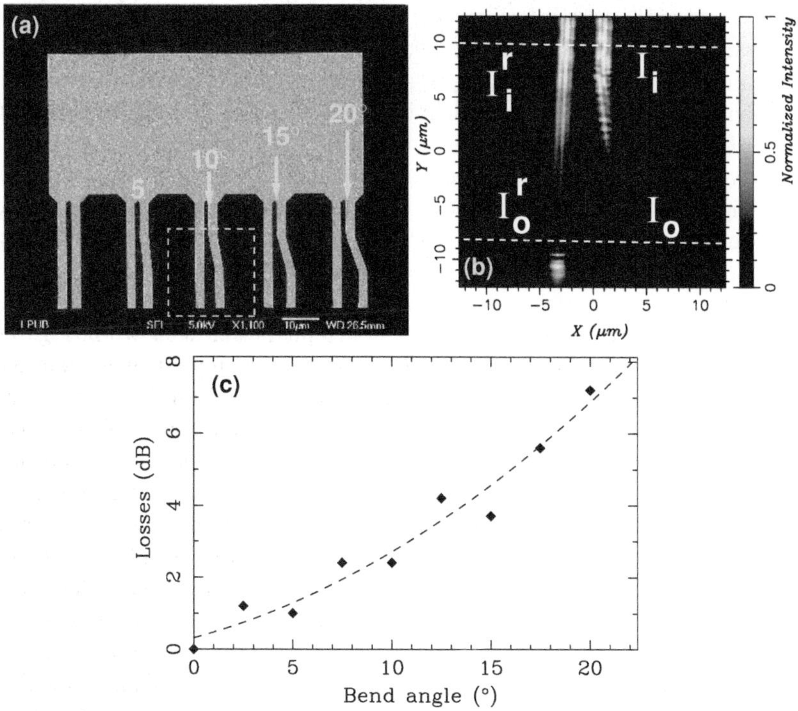

Figure 4.4. (a) SEM image of the sample used to assess bend losses. (b) PSTM image of dual MS consisting of a straight reference waveguide and a doubly bent waveguide (bend angle 10°). (c) Plot of the bend losses as a function of the bend angle.

for the bent and straight MS, respectively). The bend losses are then given by:

$$\mathcal{L}(\text{dB}) = -10 \times \log\left[\frac{I_o}{I_o^r} \times \frac{I_i^r}{I_i}\right]. \tag{4.1}$$

The input and output near-field levels are extracted from transverse cross-sections of the PSTM images (see Fig. 4.4(b)). Note that, by using these levels for the computation of the bend losses, we do not account for the longer propagation distance of the SPP along the bent MS as compared to the straight MS. However, this additional propagation distance is small, and the related damping is expected to be negligible as compared to the bend losses. Figure 4.4(c) shows the experimentally measured losses as a function of the bend angle. As expected, the losses increase with the bend angle in such a way that the signal at the end of the bent MS is roughly half that of the straight strip for an angle of about 12°. While such a loss level is reasonable for a guiding system with small damping, this is a limiting result for SPP waveguides. Indeed, owing to the short propagation distances for SPP, the distances needed to change direction cannot be of arbitrary length and thus, only modest lateral shifts of the SPP propagation direction can be achieved with small bend angles. For example, a bend angle of 10° corresponds to a lateral shift of 1.5 μm for the sample we have used. Note that this shift is smaller than the width of the MS; thus, a part of the incident SPP mode traveling along a straight trajectory can reach the output end of the bent MS without being deflected by the bends. For this reason, the losses obtained for a given bend angle are expected to increase with the length of the tilted strip until the lateral overlap between the input and output waveguides no longer exists. In any case, guiding SPP along sharply bent MS with low losses is not realistic, so alternative solutions have to be developed. It has been shown recently that a SPP traveling along an extended thin film can be efficiently deflected by a micro-grating of closely packed lines of gold nano-particles acting as a SPP Bragg mirror.[15] Thus, it is worth investigating the ability of tilted Bragg mirrors integrated inside MS for similar routing of MS-SPP.

Micro-structured metal strips

The field of a MS-SPP mode can be laterally confined over typical distances of a few SPP wavelengths. This tight confinement corresponds necessarily to a broad plane-wave spectrum and thus a large numerical aperture for the SPP waveguide. The sample shown in Fig. 4.5(a), has been designed to assess the numerical aperture of a MS with a width of 2.7 μm excited at $\lambda_0 = 800$ nm. Figure 4.5(b) shows the PSTM image recorded over the white dashed perimeter displayed in Fig. 4.5(a). It can be seen that the MS mode spreads out when it reaches the wider pad ($W = 8$ μm). By using the scattering spot at the horizontal edge of the pad, we measure a spreading angle ϕ of roughly 17°. On the other hand, an extended thin film SPP is Bragg reflected by a surface-defects grating, if its in-plane wave-vector verifies:

$$k_{sp} = \frac{k_g}{2 \times \cos\theta} \tag{4.2}$$

where $k_g = 2 \times \pi/d$ denote the grating Bragg vector and θ the SPP angle of incidence on the grating. Thus, at a given frequency, a SPP Bragg mirror with a period d is expected to be efficient for a fixed angle of incidence θ (or possibly for angles

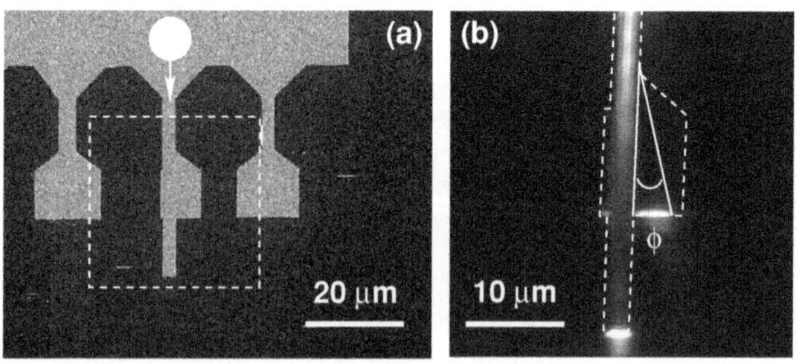

Figure 4.5. (a) SEM image of the sample fabricated to investigate the angular spreading of a MS mode. (b) PSTM image corresponding to the dashed perimeter shown in (a). The input strip has a with of 2.7 μm. When it reaches the wider area, the MS-SPP mode spreads out with an angle of 17°.

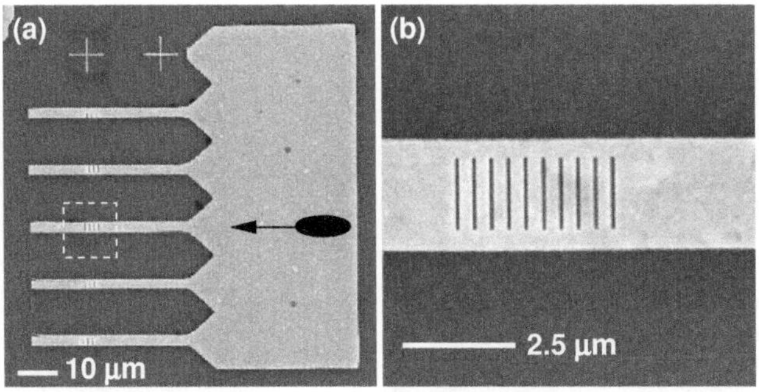

Figure 4.6. (a) and (b) SEM images of MS equipped with Bragg mirrors consisting of micro-gratings of slits. The period of the micro-grating is 400 nm and the width of the slits is 150 nm.

of incidence for a small range centered around θ). Combining this result with the significant spreading of the MS modes observed previously, we end up with the conclusion that the Bragg condition cannot be fulfiled for all components of the plane-wave expansion of a confined MS mode. In other words, the reflection efficiency of Bragg mirrors obtained for collimated SPP beams cannot be directly transposed to MS modes.

In order to gain insight into the interaction of MS modes with micro-gratings, a periodic arrangement of slits has been engraved by focused ion beam milling into a 2.5 μm wide strip (Figs. 4.6(a) and (b)). The period of the grating is 400 nm, and the width of each slit is 150 nm. The PSTM images obtained for a homogeneous strip and a micro-structured strip using an incident light with $\lambda_0 = 800$ nm are respectively displayed in Figs. 4.7(a) and 4.7(b).

The longitudinal cross sections taken along the center of each strip are shown in Figs. 4.7(c) and (d). While the dominant feature in the cross section of the homogeneous strip is the strong scattering spot at the end of the waveguide, the optical profile for the micro-structured strip clearly exhibits a standing-wave periodicity. This standing-wave pattern is created by the interference of the incident SPP mode with the mode that is back-reflected by the mirror.[16] By comparing the amplitude of the standing-wave pattern for the homogeneous and the structured MS, we conclude that the reflection efficiency of the micro-grating is significantly larger than that of the abrupt step at the strip end. Thus, the reflection efficiency of the micro-grating is due to an interaction of the incident SPP with several slits and not to the scattering from the very first slit of the grating. With the aim of proving that the micro-grating acts as a Bragg mirror, we have studied the influence of different structural parameters on reflection efficiency. The reflection efficiency of the mirrors is assessed by computing the modulation depth of the interference fringes visible on the experimental PSTM images. The modulation depth is obtained by averaging, for up to several tens of fringes, the contrast as defining by the following expression:

$$C = \frac{I_{max} - I_{min}}{I_{max} + I_{min}} \tag{4.3}$$

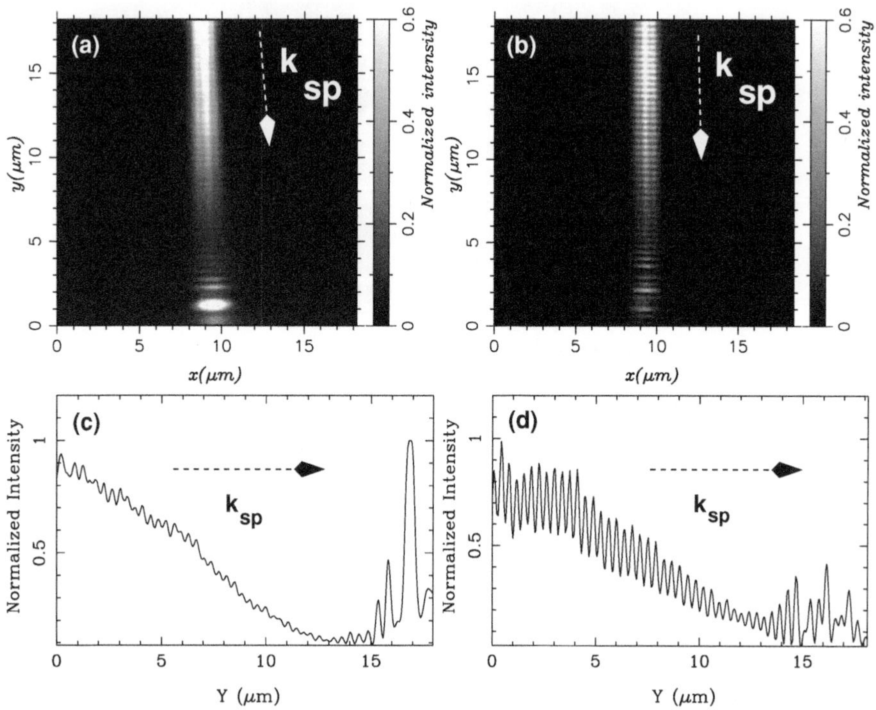

Figure 4.7. (a) PSTM image of an homogeneous MS (width 2.5 μm). (b) PSTM image of a micro-structured MS equipped with a Bragg mirror [see Fig. 4.6(b)]. The free-space incident wavelength exciting the propagating SPP is 800 nm. (c) (resp (d)) Longitudinal cross section taken along the center of the homogeneous (resp. textured) strip.

where I_{max} and I_{min} are respectively the near-field intensity at a given maximum and at the following minimum. Note that, by using this modulation depth we can compare the efficiency of various mirrors, but we cannot compute their respective reflectivity. Indeed, due to the convolution resulting from the finite size of the PSTM tip, the contrast we obtain experimentally is probably an underestimate and thus, prevents us from computing an accurate value for the mirror reflectivity. The modulation depth obtained for micro-gratings with an increasing number of slits is shown in Fig. 4.8(a). The reflectivity reaches an asymptotic value for a reduced number of $N_s = 10$ slits leading to the conclusion that large losses occur as the SPP propagates through the textured area. The interaction of the SPP with the micro-grating has been qualitatively simulated by using an equivalent Bragg mirror model (EBM) consisting a multilayer stack of two materials (A) and (B). The thicknesses of these two materials have been chosen equal to the width of the slits (150 nm) and the separation between two slits (250 nm), respectively. Considering an incident plane-wave normal incident upon the EBM, the refractive index of materials (A) and (B) have been obtained empirically in such a way that a saturation of the EBM reflectivity occurs for 10 layers

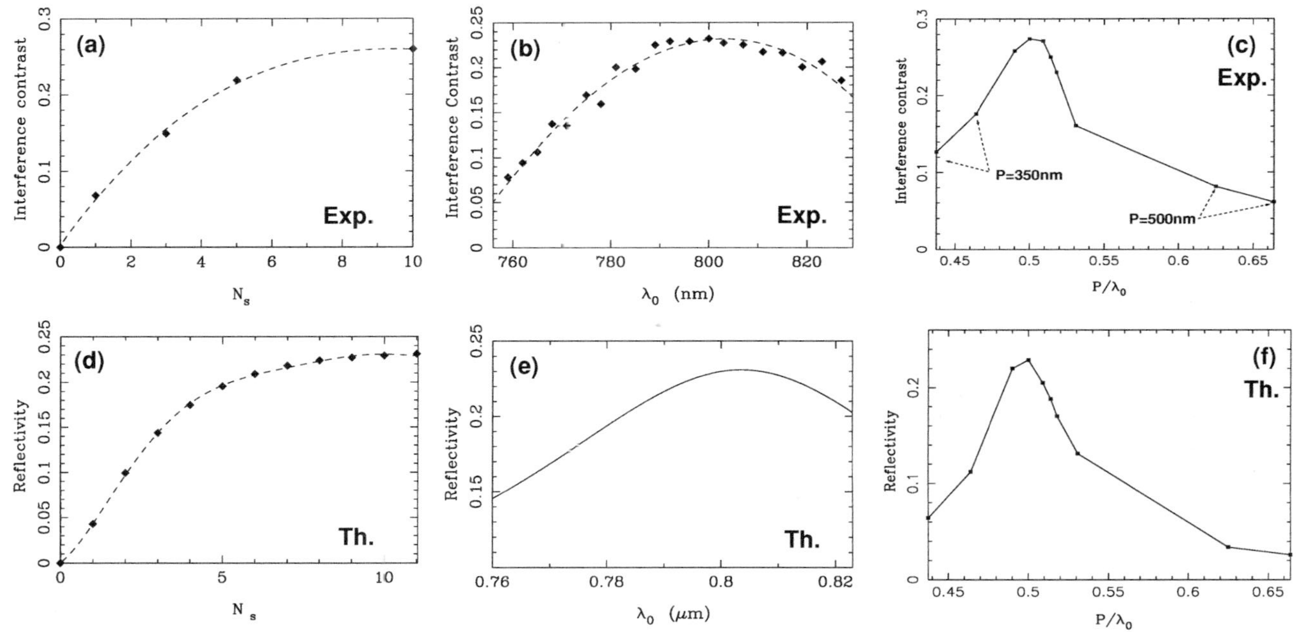

Figure 4.8. Figures (a), (b), (c): Modulation depth of the interference fringes as a function of (a) the number of slits N_s in the Bragg mirror (period = 400 nm, $\lambda_0 = 800$ nm), (b) the incident wavelength (period = 400 nm, $N_s = 10$), (c) the ratio period/λ_0. Figures (d), (e), (f): Computation of the reflectivity of an equivalent Bragg mirror (consisting of a stack of two materials (A) and (B), see text) as a function of various structural parameters (d) the number of layers (A) N_s in the Bragg mirror (period = 400 nm, $\lambda_0 = 800$ nm), (e) the incident wavelength (period = 400 nm, $N_s = 10$), (f) the ratio period/λ_0.

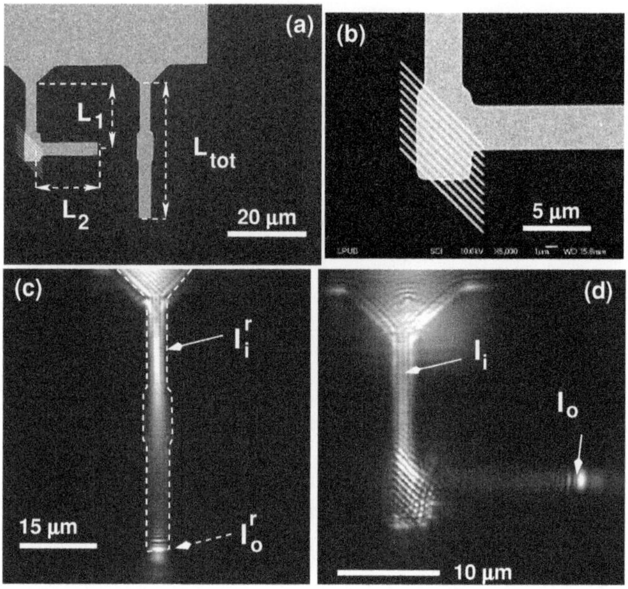

Figure 4.9. (a) and (b) SEM image of the 90° bent strip equipped with tilted Bragg mirrors. The mirror comprised gold ridges with a height of 60 nm. The losses related to the SPP guiding along the right angle bent is measured by comparison with a reference straight strip. (c) and (d) PSTM image of the reference strip and the right angle bent strip.

of material (A). A qualitative fair agreement with the experimental data has been obtained using refractive index $n_{(A)} = 1.0 + i0.3$ and $n_{(B)} = 1.01$ for materials (A) and (B) respectively (see Fig. 4.8(d)).[16] These refractive index values have been used to simulate the spectral response of a micro-grating with a period of 400 nm (Fig. 4.8(e)) and the reflectivity of gratings with different periodicities (Fig. 4.8(f)). We find a rather good agreement between the experimental results and the reflectivity computed using a simple Bragg mirror model. This allows us to conclude that a micro-grating integrated into a metal strip acts as a lossy Bragg mirror for the MS-SPP mode.

Tilted Bragg mirrors for routing MS-modes

Based on the conclusion of the previous study, micro-gratings tilted with respect to the strip axis have been used to route SPP along bent paths. In particular, tilted Bragg mirrors integrated into metal strips have been designed and fabricated to route MS-SPP modes along 90° bends (for an incident free-space wavelength of 800 nm, gratings with a period d of 550 nm have been used). The scanning electron microscope images of these integrated tilted Bragg mirrors are shown in Figs. 4.9(a) and (b). These samples have been fabricated by a double electron beam lithography process: The Bragg mirrors are defined first and then the metal strips. Prior to the second lithography step, an accurate alignment of the sample is necessary in order to superimpose the

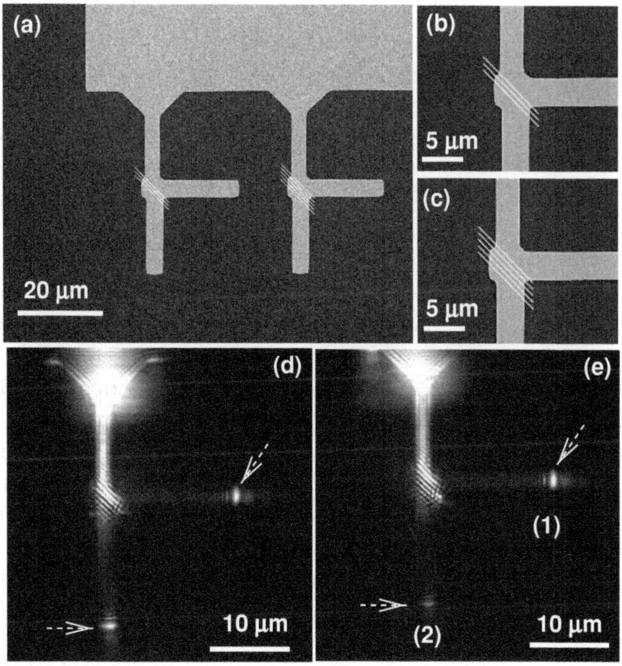

Figure 4.10. (a), (b) and (c) SEM images of MS based SPP splitters. Two configurations are considered: mirrors comprising three ridges (b) and five ridges (c). (d) (resp (e)) PSTM image of the splitter equipped with the three (resp. five) ridges mirror.

strips onto the mirrors. Unlike the previous micro-structured strips, the Bragg mirrors here do not comprise slits, but rather gold ridges with a width of about 150 nm and a height of 60 nm. In order to assess the bend losses related to the propagation of a MS-SPP mode along a 90° bend, we compare the intensity of the scattering spot at the end of the bent strip (I_o) with the intensity of the scattering spot at the end of a reference straight strip (I_o^r) with a length L_{tot} equal to the total length of the bent strip $L_1 + L_2$ (see Fig. 4.9(a)). Normalizing these output intensities by their corresponding input levels, the bend losses are computed according to Eq. 4.1. Figures 4.9(c) and (d) show typical PSTM images of respectively a reference strip and a bent strip equipped with a mirror consisting of 10 ridges. By taking small scan PSTM images of the input and output of both the reference and bent strip to accurately determine the relevant intensities, typical bend losses as low as 1.9 ± 0.6 dB have been observed.

A low-loss 90° bend is the elementary building block necessary for building many optical devices such as splitters. The samples shown in Figs. 4.10(a), (b) and (c) have been fabricated to demonstrate MS-SPP mode splitters. The splitters consist of Bragg mirrors with very few elements integrated at the crossing of two strips. For example, by using a mirror with only three ridges (see Fig. 4.10(b)), the PSTM image shown in Fig. 4.10(d) has been obtained. It can be seen that the scattering spots at the end of the two outputs are of same intensity leading to the conclusion that we have achieved a 50/50 MS-SPP mode splitter. The intensity level at the end of the two outputs of the splitter has been found to correspond to 40% of the intensity at the end of the corresponding reference strip. By increasing the number of ridges in the mirror from three to five (see Fig. 4.10(c)), the splitter shown in Fig. 4.10(e) has been obtained. In this case, the MS-SPP mode is mainly guided along the branch of the splitter perpendicular to the input strip. Indeed, we have found that the spot at the end of this branch (1) (resp. branch (2)) is 48% (resp. 19%) of the reference strip intensity, and therefore, this device corresponds to a 70/30 beam splitter.

4.4. SUMMARY AND OUTLOOK

In summary, the propagation of SPP along homogeneous and micro-structured metal strips has been characterized by means of near-field optical microscopy. We have shown that the field of a SPP traveling along a metal strip is tightly confined in the transverse direction. Owing to this confinement, one can anticipate weak cross-talks between strips placed in close proximity and, thus, an opportunity to integrate MS at high densities in coplanar geometries. Such prospects could be of great interest in the context of SPP bio-sensor miniaturization.

By texturing metal strips, we have shown that micro-gratings patterned from periodic surface defects act as lossy Bragg mirrors for the strip SPP modes. Tilted Bragg mirrors have been found to be efficient for routing SPP along sharp bends. In particular, we have demonstrated the fabrication of MS-based SPP beam splitters, opening the way toward the design of more complex devices such as interferometers. To this end, the next step in the development of SPP based optical devices is necessarily the investigation of dynamically controlled systems.

ACKNOWLEDGMENTS

The authors wish to thank Dr. E. Devaux and Pr. T. Ebbessen for the focused ion beam fabrication of the textured strips. This work has been supported by the European Network of Excellence “Plasmo-nano-devices” (Project FP6-2002-IST-1-507879) and the regional council of Burgundy.

REFERENCES

1. H. Raether: *Surface Plasmons*, Springer Tracts in Modern Physics, Vol. 111 (Springer, Berlin, 1988).
2. P. Berini: Plasmon-polariton waves guided by thin lossy metal films of finite width: Bound modes of symmetric structures, Phys. Rev. B **61**, 10484 (2000).
3. P. Berini: Plasmon-polariton waves guided by thin lossy metal films of finite width: Bound modes of asymmetric structures, Phys. Rev. B **63**, 125417 (2001).
4. R. Charbonneau, P. Berini, E. Berolo, E. Lisicka-Shrzek: Experimental observation of plasmon-polariton waves supported by a thin metal film of finite width, Opt. Lett. **25**, 844 (2000).
5. S.I. Bozhevolnyi, T. Boltasseva, T. Sondergaard, T. Nikolajsen, K. Leosson: Photonic bandgap structures for long-range surface plasmon polaritons, Opt. Commun. **250**, 328 (2005).

6. T. Nikolajsen, K. Leosson, S.I. Bozhevolnyi: In-line extinction modulator based on long-range surface plasmon polaritons, Opt. Commun. **244**, 455 (2005).
7. T. Nikolajsen, K. Leosson, S.I. Bozhevolnyi: Surface plasmon polariton based modulators and switches operating at telecom wavelengths, Appl. Phys. Lett. **85**, 5833 (2004).
8. A. Hohenau, J.R. Krenn, A.L. Stephanov, A. Drezet, H. Ditlblacher, B. Steiberger, A. Leitner, F.R. Aussenegg: Dielectric optical elements for surface plasmons, Opt. Lett. **30**, 893 (2005).
9. R. C. Reddick, R. J. Warmack, T. L. Ferrell: New form of scanning optical microscopy, Phys. Rev. B **39**, 767 (1989).
10. B. Lamprecht, J.R. Krenn, G. Schider, H. Ditlbacher, M. Salerno, N. Felidj, A. Leitner, F.R. Aussenegg, J-C. Weeber: Surface plasmon propagation in microscale metal stripes, Appl. Phys. Lett. **79**, 51 (2001).
11. J. C. Weeber, J. R. Krenn, A. Dereux, B. Lamprecht, Y. Lacroute, J.P. Goudonnet: Optical near-field distributions of surface plasmon waveguide modes, Phys. Rev. B **64**, 045411 (2001).
12. J. C. Weeber, Y. Lacroute, A. Dereux: Near-field observation of surface plasmon polariton propagation on thin metal stripes, Phys. Rev. B **68**, 115401 (2003).
13. R. Zia, M. D. Selker, M. L. Brongersma: Leaky and bound modes of surface plasmon waveguides, Phys. Rev. B **71**, 165431 (2005).
14. R. Zia, A. Chandran, M. L. Brongersma: Dielectric waveguide model for guided surface polaritons, Opt. Lett. **30**, 1473 (2005).
15. H. Ditlbacher, J. R. Krenn, G. Schider, A. Leitner, F. R. Aussenegg: Two-dimensional optics with surface plasmon polaritons, Appl. Phys. Lett. **81**, 1762 (2002).
16. J-C Weeber, Y. Lacroute, A. Dereux, E. Devaux, T. Ebbesen, C. Girard, M. U. González, A-L Baudrion: Near-field characterization of Bragg mirrors engraved in surface plasmon waveguides, Phys. Rev. B **70**, 235406 (2004).

CHAPTER FIVE

NUMERICAL SIMULATIONS OF LONG-RANGE PLASMONIC TRANSMISSION LINES

ALOYSE DEGIRON AND DAVID R. SMITH
Department of Electrical and Computer Engineering, Duke University, Durham, North Carolina, 27708, USA

5.1. INTRODUCTION

Structures that guide waves can be found in almost every optoelectronic or photonic device. Yet, the basic principles of guided waves in practical realizations have not evolved substantially over the past several decades. At microwave or radio frequencies (RF), waveguides typically comprise metal-enclosed volumes with or without a central conductor;[1] in the latter case, the lateral dimensions of the waveguide dictate the frequencies of operation. At optical wavelengths, metals are comparatively poor conductors and have traditionally been excluded as optical components. Instead, dielectric waveguides are employed in which the mismatch between a higher dielectric region and free space or a lower dielectric cladding constrains light in a plane perpendicular to propagation.[2] Because of the low losses in insulating dielectrics, optical waveguides (such as fiber optics) can support propagating modes with extraordinarily low absorption attenuation—often less than 1 dB per kilometer.

Although very powerful for current technologies, the minimum dimensions of optical waveguides are constrained to be on the order of or larger than the wavelength of light within the dielectric. As a consequence, waveguide based devices such as couplers, bends or interferometers occupy a relatively large footprint in photonic circuits, typically requiring many wavelengths for operation.

At RF and microwave frequencies, the use of metals allows for waveguides and transmission lines that are virtually unconstrained by the wavelengths of the electromagnetic waves they carry. Coaxial cables, for example, can transport electromagnetic energy at frequencies from zero (DC) up to over one hundred GHz. This transport can occur in metal structures because the energy of the wave is converted to surface currents and charge distributions relating to the inductance and capacitance of the designed metal structure.

M.L. Brongersma and P.G. Kik (eds.), Surface Plasmon Nanophotonics, 55–71.

In response to the need for miniaturization and multifunctionality in optical waveguides, there has been an emergence of wave guiding structures based on the controlled excitation of surface plasmon polaritons (SPPs).[3–12] SPPs are hybrid guided waves resulting from the resonant coupling between two-dimensional electromagnetic waves and collective electronic oscillations along a metal surface.[13] SPPs are characterized by large, localized electromagnetic fields confined within a subwavelength region near the interface, making them extremely sensitive to their immediate environment. Consequently, their properties can be tuned by an appropriate design of the interface. As a corollary, the manipulation of SPPs constitutes a powerful means to squeeze and control light at a scale smaller than the diffraction limit.[14] These unique properties suggest the pursuit of SPP structures as transmission lines in the optical regime: an increasing number of theoretical and experimental studies demonstrate that low-loss SPP modes can, in effect, be carried along metal strips with subwavelength cross-section.[6–12]

Because its localization is based on the inherent capacitance and inductance of the planar geometry, and not restricted by the wavelength, the SPP waveguide is in some sense the optical analog of the transmission line, common to RF and microwave frequency technology. However, the established techniques used to analyze transmission lines at lower frequencies or dielectric waveguides at optical frequencies cannot be applied to determine the properties of SPP transmission lines, which rely predominantly on the frequency dispersion inherent to metals at optical wavelengths. The frequency dispersive nature of the material and the subwavelength localization of the modes especially complicate the numerical analysis of SPP waveguide transmission lines. In both finite-difference and finite-element techniques, in which a discrete version of Maxwell's equations are solved, the number of volume elements required to achieve convergence is far greater than is needed for geometries involving only positive (dispersionless) dielectric materials or perfect conductors. The mesh requirements make all but the simplest geometries requiring the smallest computational domains unfeasible for solution on currently available personal computers.

Much of the analysis on SPP transmission lines has been performed using semi-analytic techniques. However, such techniques are restrictive in terms of the geometries that can be solved. For general engineering of SPP transmission lines, it is convenient to have a method capable of solving for the modes of SPP transmission lines of arbitrary geometry and material composition. For transmission lines whose properties do not vary in the propagation direction, the computational domain can be significantly reduced to being nearly a two-dimensional problem. In this chapter we describe a numerical method based on utilizing an eigenmode solver to simulate SPP transmission lines, which may be useful for the engineering of practical geometries and SPP transmission line applications. In Sect. 5.2, we will briefly review the theoretical and experimental framework for guiding SPP along metal strips; Sect. 5.3 will expose the principle of our simulations, based on using commercially available finite-element codes; and Sect 5.4 will provide several illustrative examples.

5.2. PHYSICAL BACKGROUND

The development of thin metal strips as plasmonic waveguides has been motivated by the short lifetime of SPPs supported by an isolated metal/dielectric interface. In this case, the propagation constant seldom exceeds a few tens of microns (at visible wavelengths) because the high field confinement near the interface leads to strong damping by absorption in the metal.[13] However, much longer propagation lengths can be achieved with a SPP mode whose field distribution extends significantly into the dielectric medium and is less confined to the metal. For such a mode—frequently referred to as a *long-range surface plasmon*—the losses due to absorption are minimized. Long-range surface plasmons are actually coupled modes that form when the SPPs of two or more interfaces overlap significantly.[15] This condition occurs when the individual modes have similar energies and are in close proximity–precisely the case for optically thin metal films and strips surrounded by a symmetric (or almost symmetric) environment.

The existence of long-range SPPs has been first reported in the case of freestanding metal films two decades ago.[15,16] At small thicknesses, the SPPs of both interfaces couple and form two transverse magnetic (TM) modes, whose field distributions are antisymmetric and symmetric with respect to the center plane of the film. Following a well-established convention, we call antisymmetric (resp. symmetric) mode the solution whose electric field component normal to the surface is antisymmetric (resp. symmetric) with respect to the center symmetry plane of the structure. While the antisymmetric mode is characterized by a field distribution mainly confined within the metal, thus yielding to high absorption and short propagation ranges, the symmetric mode can be made long-ranging for small enough thicknesses. In fact, the thinner the film, the less (most) attenuated the symmetric (asymmetric) mode so that at vanishing thicknesses, propagations over centimeters are theoretically possible in the symmetric case. However, such performances require such a low confinement of the fields that the modes are actually much closer to freely propagating waves in the surrounding dielectric than genuine SPPs bounded to the structure. Therefore the practical design of a plasmonic waveguide involves a trade-off between low-loss guiding and strong electromagnetic confinement.

Similar conclusions have been reached recently for thin metal films of finite width, although the modes are more complex and in greater number owing to the lower degree of symmetry of these strip-like structures. As an example, the field distribution of the modes supported by a straight rectangular strip embedded in a homogeneous dielectric medium (Fig. 5.1(a)) must be either symmetric or antisymmetric with respect to the two symmetry planes perpendicular to the propagation direction. For each of the four different field symmetries that are thus possible, a series of a fundamental and several higher-order modes can be excited along the strip in much the same way as electromagnetic modes supported by hollow waveguides. Among them, several long-ranging SPPs have been identified. In particular, one of these modes is of greater practical interest because it has neither a cutoff thickness nor a cutoff frequency

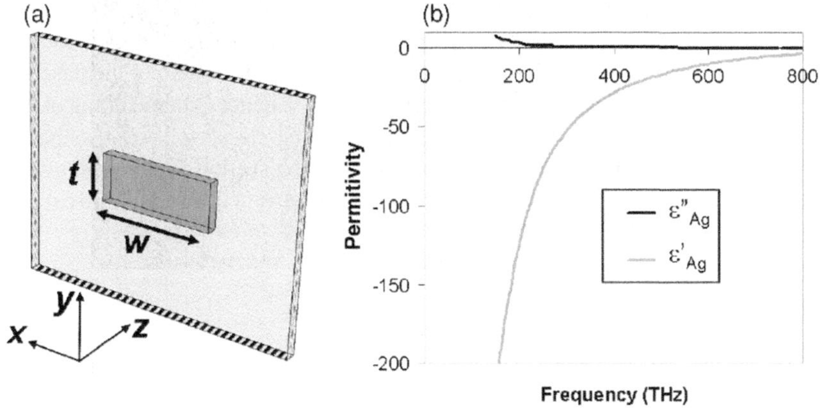

Figure 5.1. (a) Unit cell for simulating a straight waveguide of rectangular cross-section. The horizontal boundaries of the problem domain are perfect electric walls; the vertical patterned boundaries are perfect magnetic walls. (b) Real, $\varepsilon''_{\mathrm{Ag}}$, and imaginary, $\varepsilon''_{\mathrm{Ag}}$, parts of the Ag dielectric function $\varepsilon_{\mathrm{Ag}} = \varepsilon'_{\mathrm{Ag}} + i\varepsilon''_{\mathrm{Ag}}$.

and also because its field pattern allows excitation via a simple end-fire technique (e.g., focusing light on the input of the strip).[6] In fact, as the thickness decreases, this interesting mode evolves toward the transverse electromagnetic (TEM) wave supported by the adjacent dielectric medium. It is thus similar in many regards to the (long-ranging) symmetric mode of a freestanding thin metal film. A thorough description of modes corresponding to the rectangular strip has been made by Berini who used the semi-analytical method of lines to solve Maxwell's equations in this particular geometry.[6,7] Very recently, Zia *et al.* proposed an original approach to simplify the problem by modeling the modes with the solutions of two-dimensional dielectric slab waveguides.[12] Although such a model is highly idealized, the results are in good agreement to the rigorous calculations presented by Berini.[6] Other theoretical studies can be found in the literature but they mainly discuss very particular issues such as the near-field properties of short range SPPs.[3]

The SPP modes propagating along thin metal strips have also been studied experimentally. On one hand, experiments on short-range SPPs have been performed with asymmetric structures consisting of metal strips lying on a glass substrate. For example near-field measurements were used to map the fields around the air/metal interface of an asymmetric SPP mode.[4] Far-field measurements have further established that the propagation length for such asymmetric waveguides hardly exceeds 50 nm at optical frequencies.[5] On the other hand, the existence of long-ranging modes has been confirmed in the case of symmetric structures consisting of millimeter-long metal strips embedded in an isotropic dielectric medium.[8–11] These studies also show that single symmetric waveguides can be combined to make passive optical elements[9] (such as junctions or couplers) and even active components as demonstrated with thermo-optic interferometers and switches.[11]

All of these contributions successfully demonstrate the exciting potential of metal strips as plasmonic transmission lines. However, it should be noted that this topic is still at an early stage of development since existing theoretical works only apply for highly idealized structures, i.e. straight metal strips with perfectly smooth surfaces and sharp corners. As engineered SPP waveguides are becoming of increased interest, there is a need for novel simulation tools that allow a better understanding of the actual samples (including material dispersion, surface roughness, etc.) and also that could be used for designing innovative structures. In the next section, we present a simulation approach that may help to bridge the gap between current theoretical and experimental studies.

5.3. NUMERICAL METHODS

The structures considered in this study are infinitely long Ag strips embedded in a homogeneous dielectric medium, typically glass or SiO_2. Their properties are investigated by numerically solving a suitably defined eigenvalue problem with HFSS (Ansoft), a finite-element commercial code. The principle of the finite-element method is to decompose the computational domain into smaller volumes in which the fields can be solved separately. The global solution, obtained by assembling the solutions obtained in each element, converges to the true result by successive refinement of the calculation mesh. In our case, the computational domain is a box enclosing a lateral section of the waveguide, as shown in Fig. 5.1(a) for the case of a rectangular strip. Periodic boundary conditions are applied in the direction parallel to propagation in order to simulate an infinite structure whereas the remaining boundaries are perfect electric and magnetic walls placed far enough from the strip in order to prevent any perturbation of the SPP modes. The periodic boundary conditions ensure that the electromagnetic field on one surface of the computational domain matches the field on the opposite face to within a phase delay φ. A field component therefore satisfies $f(z+d) = \exp(i\varphi)f(z)$, where z is the space coordinate parallel to the propagation direction and d is the distance between the two planes of periodicity. Once φ is fixed, then an eigenvalue problem is specified by the materials within the unit cell and the boundary conditions. At every value of φ, an infinite number of eigenmodes can, in principle, be obtained, each with a different complex frequency.

A major difficulty in simulating plasmonic structures, common to most numerical methods, is that the mode density can often become very large, with modes spaced closely in frequency. For numerical solvers that rely on solving a discrete version of Maxwell's equations, the closely spaced plasmon modes translate effectively to a poorly conditioned matrix (in the frequency domain) whose inversion can be inaccurate. In the time domain, the extreme modal density results in slow convergence. We do not present a specific solution to this difficulty here, other than to note that we make use of symmetry (both translational and reflection) to reduce the size of the computational domain as much as possible. Even our restricted computational domain can support a wide variety of cavity modes in addition to the SPP modes of interest supported by the waveguide. Therefore a careful review of the field distributions is generally required in order to identify the solutions of interest, i.e. the modes having their field bounded to the strip.

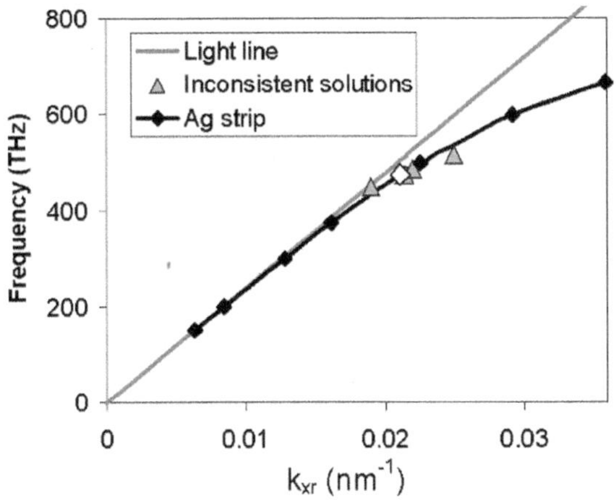

Figure 5.2. Dispersion curve for a rectangular Ag strip ($w = 1$ μm, $t = 40$ μm) embedded in an isotropic medium ($\varepsilon_d = 4$). The gray triangles illustrate the iteration process for obtaining a self-consistent solution at $\nu_0 = 473.61$ THz (which corresponds to the free-space wavelength $\lambda_{vac} = 633$ nm). The iteration process converges to the true result represented by the white diamond on the curve.

The definition of the material parameters within the unit cell presents an additional problem of self-consistency that is particular to our eigensolver approach. We address this issue here via a non-optimized iterative approach to converge at a mode whose material parameters, frequency and phase advance are in agreement. For example, in the following simulations we assume a lossless dielectric host with constant permittivity $\varepsilon_d = 4$ and a complex dielectric function ε_{Ag} of the silver strip as determined from a fit to the experimental data of Johnson and Christy.[17] As shown in Fig. 5.1 (b), ε_{Ag} strongly varies with frequency so that once a specific value ε_{const} is assigned to the strip, the frequency υ_0 of the mode must be consistent with $\varepsilon_{Ag}(\upsilon_0) = \varepsilon_{const}$ in order to be a valid solution. For each mode, we therefore adjust the initial phase delay φ between the two planes of periodicity to satisfy the previous equality. We start this iteration process with an initial guess for φ and complete the process once a self-consistent solution is achieved. After having determined φ, the real part of the propagation constant k_{xr} can be found from the definition $\exp(i\varphi) \equiv \exp(i \cdot k_{xr} \cdot d)$.

The dispersion curve of the mode $\upsilon(k_{xr})$ is obtained by repeating the simulations for different values of ε_{Ag}, as shown for instance in Fig. 5.2 for a 1 μm wide and 40 nm thick Ag strip. Then k_{xi}, the imaginary part of the propagation constant parallel to the surface, can be calculated at any frequency according to the relation $k_{xI} = 2\pi\upsilon''/\upsilon_g$, where υ'' is the imaginary part of the frequency and υ_g is the group velocity i.e. the derivative of $\upsilon(k_{xr})$.

It should be noted that our approach for modeling SP waveguides can be transposed to simulate conventional planar transmission lines at RF or microwave frequencies. This case is less challenging because such structures typically involve two conductors to guide electromagnetic waves so it is possible to directly focus on the mode of interest by specifying the polarization, the voltage, and the impedance between the two conductors.

5.4. RESULTS

We begin by reproducing some of the key results reported by Berini[6] for Ag strips having a rectangular cross-section and surrounded by a homogeneous medium of permittivity $\varepsilon_d = 4$ (Fig. 5.1(a)). As mentioned earlier, four families of modes are expected in this case owing to the geometrical symmetry. Each family is characterized by a specific field distribution with respect to the two symmetry planes of the structure and can thus be generated separately by placing the proper combination of electric and/or magnetic walls halfway through the structure. Consequently only one quarter of the structure need be simulated, which significantly reduces the calculation time. Still, simulating a rectangular strip with a finite element code is not straightforward because the 90° corners of the structure generate strong field singularities that cannot be solved properly without using an extremely large finite-element mesh. In the following, we avoid this problem and shorten the calculation time by slightly rounding the corners of the strips. As we will see, this modification does not alter the results significantly.

Figure 5.3 shows the real and imaginary parts of the wave vector as a function of the strip thickness for the four fundamental modes of the structure. The modes are computed at the free-space wavelength $\lambda_{vac} = 633$ nm for strips of width $w = 1$ μm; the radius of the rounded corners is $r = 5$ nm. We have labeled the modes according to the nomenclature proposed by Berini[6] in which a pair of letters indicates whether E_y, the dominant electric field component for a rectangular shape, is symmetric or antisymmetric with respect to the horizontal and vertical plane of symmetry, respectively. The subscript b means that the modes are bounded to the surface and the

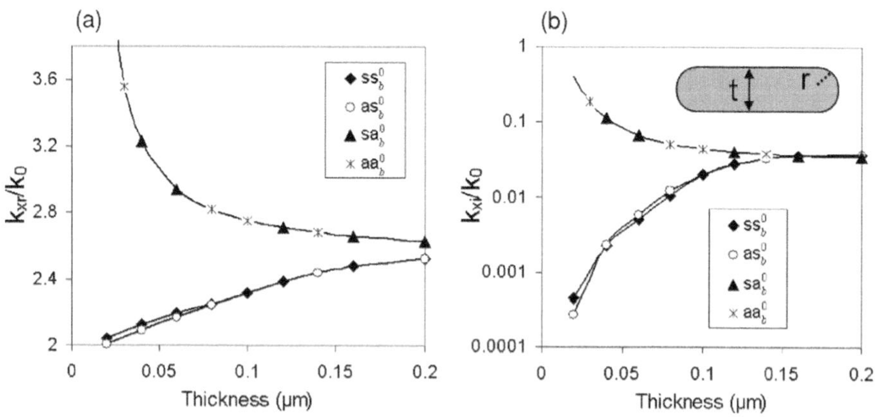

Figure 5.3. Dispersion curves for the real (a) and imaginary (b) parts of the four fundamental modes supported by a straight Ag strip ($w = 1$ μm, $r = 5$ nm) at the free-space wavelength $\lambda_{vac} = 633$ nm. The data are normalized against the free-space wave vector $k_0 = 2\pi/\lambda_{vac}$. Insert: geometry of the system.

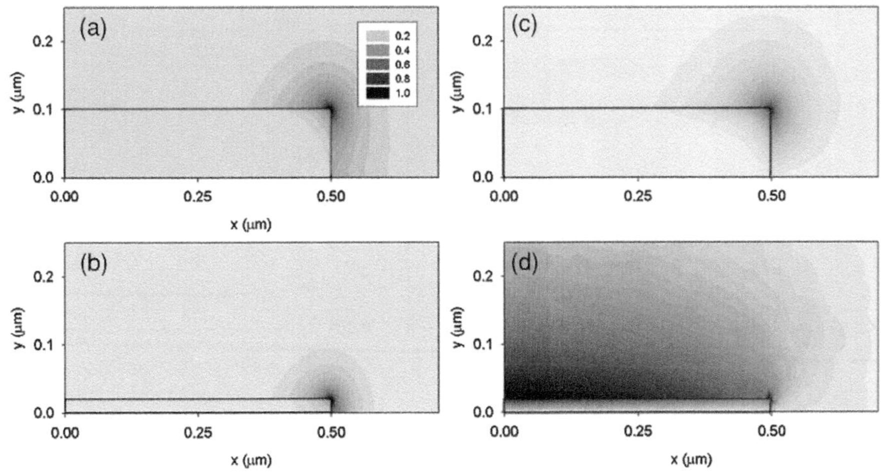

Figure 5.4. Electric field distributions of the sa_b^0 mode (a,b) and the ss_b^0 mode (c,d) for different thicknesses at the free-space wavelength $\lambda_{vac} = 633$ nm.

superscript indicates the number of maxima in the spatial distribution of E_y along the largest dimension.

The modes essentially follow two opposite behaviors depending whether their field component E_y is antisymmetric or symmetric with respect to the horizontal symmetry plane of the structure. While the modes following the upper branches of Figs 5.3 (a) and 5.3(b) remain degenerate for all studied thicknesses, the lower branches characterizing the two other modes eventually split as the strip gets thinner. In particular, the dispersion curve of the ss_b^0 mode becomes nearly identical to that of the symmetric mode of an infinitely wide Ag film.[6] The curvature of the branches is correlated with the field distributions of the corresponding modes. As shown in Figs 5.4 (a) and 5.4 (b), the fields of the upper branch modes remain localized near the corners and penetrate deeper into the strip with decreasing strip thickness. The localized modes possess smaller group velocities and consequently higher damping by absorption, thus increasing both the real and imaginary parts of the wave vector. Conversely, Figs 5.4 (c) and 5.4 (d) show that for decreasing thicknesses, the fields of the lower branch modes spread around the strip structure and extend farther into the dielectric region. These modes bear similarity to (unbounded) TEM modes that propagate freely within the dielectric. Being weakly bound, nearly TEM-like modes, the imaginary part of their wave vector vanishes while its real part tends to the value of a plane wave in the dielectric material–a behavior that corresponds to the expected signature of long-ranging SPPs. It should be noted that a compromise between propagation length and field confinement will typically be required for practical applications, as one is always achieved at the expense of the other.

While the previous results are in excellent agreement with Berini's calculations,[6] the branches of Figs 5.3 (a) and 5.3 (b) reach slightly different asymptotic values at large

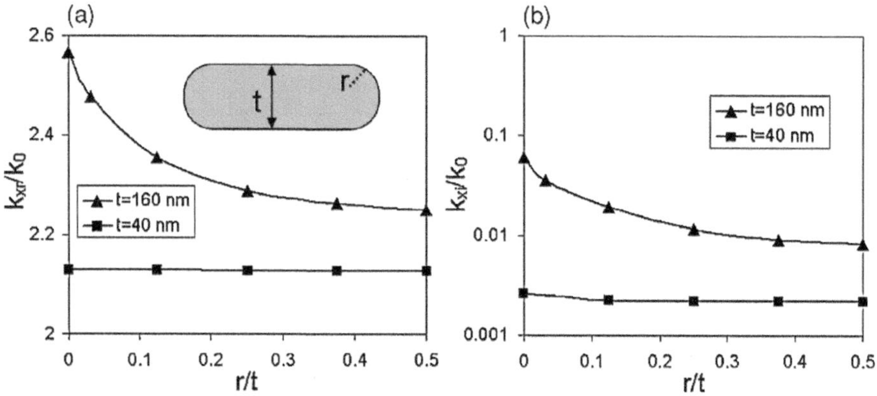

Figure 5.5. Influence of the corner curvature on the ss_b^0 mode for two different thicknesses (a,b) at $\lambda_{vac} = 633$ nm. Insert: geometry of the strip.

thicknesses. This discrepancy, on the order of 2–3% for both the real and imaginary parts of the wave vector, arises from the strong localization of the fields around the strip corners. This localization makes the modes very sensitive to the specific corner geometry, so it is not surprising that the results obtained here for rounded corners are somewhat different from those calculated by Berini, in which 90° corners are assumed. The same remarks apply for the remaining part of the upper branches, where the field remains mostly confined in the vicinity of the corners. In contrast, the lower branch modes supported by structures with sharp and rounded corners become nearly identical at small thicknesses. This limit is as expected because the fields surround the strip and therefore the general shape of the strip has more influence on the SPPs than its local details.

This qualitative discussion can be confirmed quantitatively by carefully analyzing the influence of the corner curvature on the fundamental mode field distributions and frequencies. Figure 5.5 plots the wave vector of the ss_b^0 mode as a function of the curvature for a 40 nm and 160 nm-thick strip, respectively. As expected from our previous considerations, the wave vector remains almost constant in the case of the thinnest strip because the field distribution is not localized near the corners. The converse holds true for the thickest strip which reveals a decrease in both the real and imaginary part of the wave vector. Interestingly, this result implies that the attenuation length can be increased in the case of thick strips by simply rounding the corners. Mode field distributions are mapped in Figs 5.6 (a) and 5.6 (c) for the cases of 90 degree corners and for completely rounded corners. The latter modes are not confined near the corner anymore but rather stretch along the long edge of the structure, extending deeper into the dielectric region and thus diminishing the damping by absorption in the metal.

Similar conclusions can be drawn for the other modes. However, it should be noted that the field distribution of the fundamental short-range modes (i.e. the upper branch

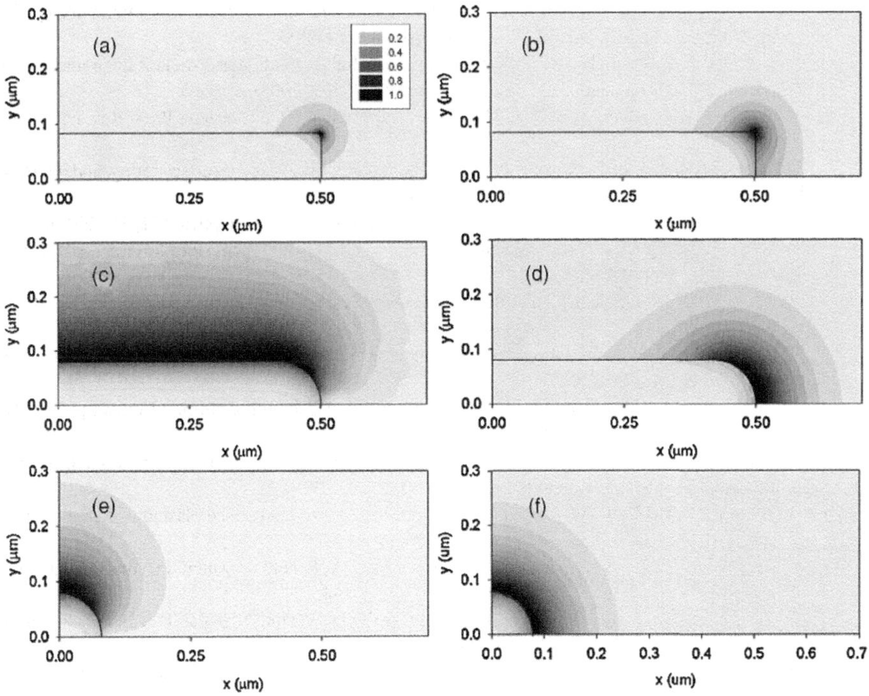

Figure 5.6. Electric field distributions of the ss_b^0 mode (a,c,e) and the sa_b^0 mode (b,d,f) for different geometries at $\lambda_{vac} = 633$ nm.

modes of Fig. 5.3) stretches along the smaller rather than the longer edge of the strip when the curvature of the corner increases (see Figs 5.6 (b) and 5.6 (d)). In order to explain this difference, we compared these results with the modes supported by a cylindrical wire whose cross-section has the same radius as the radius of curvature of the strips shown in Figs 5.6 (c) and 5.6 (d). Figures 5.6 (e) and 5.6 (f) display the electric field of the two modes computed by simulating a quarter of the wire with the same perfect walls as those used to generate the ss_b^0 and sa_b^0 modes of the strip, respectively. It turns out that the latter modes are similar to the field patterns of the ss_b^0 and sa_b^0 modes around the corner of the strip. In other words, rounding the corners of a strip of rectangular cross-section changes the nature of the SPPs as the latter tend to the modes of a cylindrical wire having the same field symmetries. Again, this conclusion is only valid if the modes of the strip are localized near the edges. For thin strips, the lower-branch modes of Fig. 5.3 are long-ranging and are rather insensitive to the shape of the corner. Also, the same modes are not much affected by a small surface roughness, as will be explained below.

Although a complete study on the effects of the surface roughness is beyond the scope of this chapter, we show here that some insight into the effect of surface disorder can be gained by the eigensolver techniques presented here. We consider a

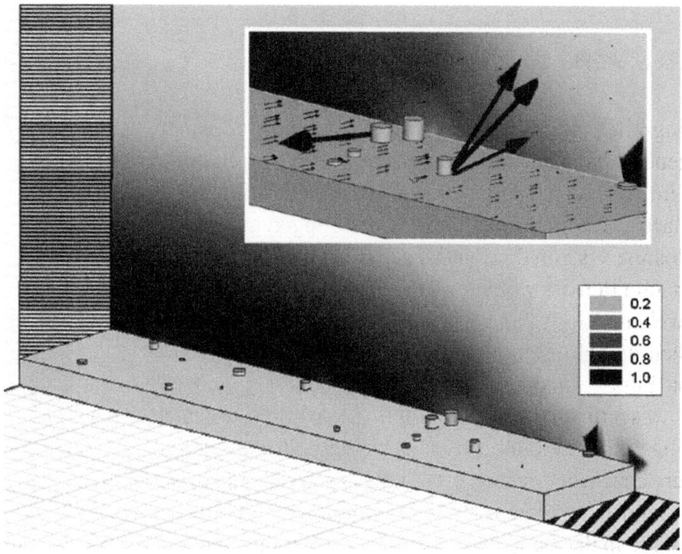

Figure 5.7. A unit cell for simulating a straight waveguide with subwavelength corrugations. The dashed zones are the boundaries of the computational domain. Here they consist in a horizontal electric wall and in a vertical magnetic wall in order to simulate the ss_b^0 mode. Also shown is the electric field pattern of the mode. Insert: detail of the Poynting vector above the surface.

20 nm thick Ag strip of rectangular cross-section surrounded by an isotropic medium ($\varepsilon_d = 4$) with a fictitious surface roughness modeled by the random addition of subwavelength Ag cylinders of various sizes on the surface of the strip. The height of the cylinders is treated as a random variable with values uniformly distributed between 0 and 5 nm, with the cylinder radius always smaller than half its diameter. This small roughness constitutes a small perturbation to the plasmon modes, as shown in Fig. 5.7 for the ss_b^0 mode: the field distribution is globally not very different than for the case of perfectly smooth surfaces except that localized "hot spots" can be observed around the cylinders. The randomly placed corrugations act as subwavelength scatterers, as indicated by the local Poynting vector, which generally assumes all directions in the vicinity of a given perturbation. The surface roughness thus presents an additional decay channel through which the damping of the SPP is increased due to local scattering. Radiation losses, however, remain small, as can be determined by comparing the complex wave vector for the rough strip with that of the perfectly smooth strip. This study provides evidence that long-ranging modes are rather insensitive to small perturbations of the strip shape because they are only weakly bound. This result is also consistent with the fact that long-ranging modes have been measured experimentally despite the unavoidable surface roughness of real samples.[8–11]

Although the results of Fig. 5.7 indicate that a small surface roughness can be analyzed by the eigensolver method, we should emphasize three following points. First, the solutions of a given problem are not unique. The corrugations strongly increase the mode density within the unit cell so that a large number of modes, each having very similar wave vector and energy, are generated. Secondly, it is expected that increasing the surface roughness will significantly enhance the radiation losses by scattering. As a result, the boundaries of the unit cell would need to be reconsidered in order to allow the light emitted by the strip to radiate freely. This important issue will be discussed later in the context of leaky modes. Last, but not least, it should be noted that the surface roughness modeled in the above example actually follows a periodic pattern due to the periodic boundary conditions in the direction of propagation and the use of perfectly conducting electric or magnetic walls to reduce the unit cell to a quarter of the actual structure. In other words, the numerical analysis of SPP transmission lines based on solving an eigenvalue problem enables only a qualitative analysis of a truly disordered surface roughness. However, and for the same reasons, this method can be intrinsically adapted to investigate metal strips with artificial periodic roughness.

The SPP modes traveling along a periodically patterned surface can be significantly altered by the periodicity. In particular, the periodic scatterers result in the formation of standing waves when the in-plane SPP wave vector k_{xr} reaches the edge of the first Brillouin zone $\pm\ \pi/P$, P being the lattice constant of the structure. For symmetry reasons, two standing waves actually exist when $k_{xr} = \pm\ \pi/P$ because the nodes of the electric field can be either centered at the minima or at the maxima of the periodic surface modulation.[18] These two field patterns correspond to two distinct surface charge distributions and thus to two different energy values. As a consequence, a gap in frequency opens up in the SPP dispersion relation, in direct analogy with Bragg scattering of electron waves in a periodic potential. These so-called photonic surfaces[19] have attracted much attention in recent years, mainly because they can be used as Bragg reflectors for SPP modes. For instance, they provide another means for guiding (short-range) SPP along a metal surface by creating a line defect in a two-dimensional lattice of metallic dots. If the period is tuned so as to produce a bandgap at the SPP mode frequency, then propagation is prevented except in the direction of the line defect[20]. Photonic structures have also begun to be investigated for their ability to exhibit a negative effective index of refraction, related to the zone-folding of certain SPP bands in the first (or higher) Brillouin zones.[21–22]

To illustrate the effect of periodicity on the SPP dispersion we compute the band structure of a SPP transmission line with periodic modulation. The structure consists in an Ag strip of rectangular cross-section with a crenel-shaped modulation of the top and bottom surface (Fig. 5.8 (a)). Strictly speaking, such a surface profile contains an infinite number of spatial Fourier components which each may influence the SPP modes. However, it has been shown that the lowest harmonic $2\pi/P$ dominates the scattering of the SPP modes so that the role of the higher Fourier components can be safely ignored for sake of clarity.[18] As in the previous calculations, the computational volume is filled with a lossless medium with $\varepsilon_d = 4$ and containing one quarter of the strip. The length of the unit cell in the direction of propagation is set to one period of the surface modulation, so that the phase advance between the two planes of periodicity is such that the corresponding propagation constant (k_{xr}) always lies within the first Brillouin zone.

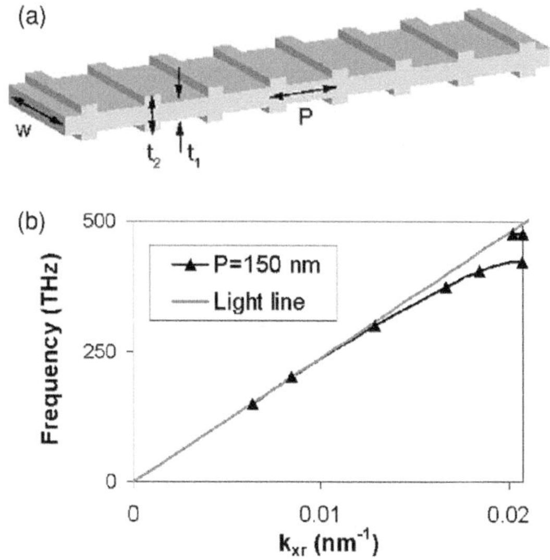

Figure 5.8. (a) Metal strip with periodic corrugations ($P = 150$ nm, $w = 1\,\mu$m, $t_1 = 20$ nm, $t_2 = 40$ nm). (b) Dispersion relation of the ss_b^0 mode.

Figure 5.8 (b) shows the dispersion curve of the ss_b^0 mode. As expected, a gap opens up at $k_{xr} = \pi/P$ owing to the formation of standing waves. More importantly, the group and phase velocities, $d\omega/dk$ and ω/k, respectively, of the upper branch have opposite signs, indicating that the energy and wavefront propagation velocities are antiparallel. In this sense, periodic SPP waveguides operating in this regime might be considered as the optical counterparts of the left-handed transmission line metamaterials that have been demonstrated at RF and microwave frequencies.[23–24] The property of "left-handedness," or negative refractive index, is usually associated with bulk materials for which the electric permittivity and magnetic permeability are simultaneously negative.[25–26] The "backward wave" transmission line modes available in periodic structures, however, are the planar analogs of left-handed materials and can exhibit many of the interesting and unusual properties initially proposed for bulk materials. For both the bulk material and the planar transmission line the modes of interest propagate as if in a medium with a negative index of refraction (or negative propagation constant). An experiment demonstrating the existence of backward modes in a hybrid plasmonic/photonic crystal has been recently reported, in which excitation of the modes was achieved via evanescent transfer of energy from a tapered optical fiber.[22] This coupling required a phase matching between the conventional forward modes in the fiber and the backward modes of the periodic waveguide. But since the group velocity of the former is parallel to the phase velocity and the converse holds true for the latter, a contra-directional power coupling can be achieved.

It should be noted that the upper band of Fig. 5.8(b) actually crosses the light line; however the mode density within the computational domain is so large at those

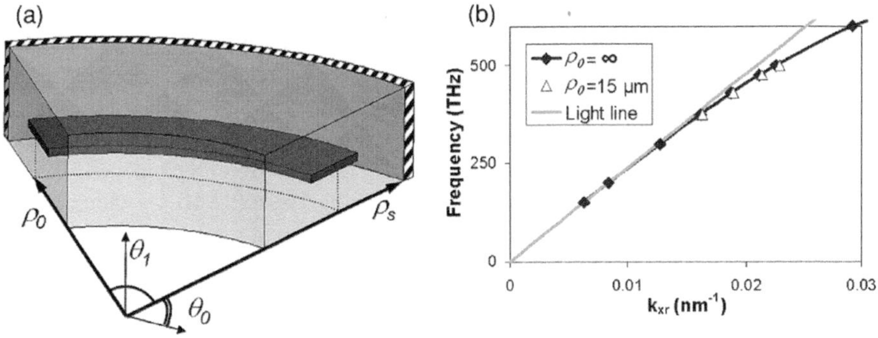

Figure 5.9. (a) Unit cell for simulating a bent waveguide. The dashed volume behind the boundary $\rho = \rho_s$ represents the PMLs. The remaining boundary conditions are periodic boundary conditions in the direction of propagation and perfect walls elsewhere. (b) Dispersion relation of the ss_b^0 mode for a straight and a bent structure.

frequencies that the solutions of interest could not easily be identified with our computational resources. We note that the dispersion curve for these modes lies within the light cone of Fig. 5.8(b), such that a coupling can occur between the guided modes and photons in free space satisfying the momentum conservation law:$\hbar k_{xr} = \hbar_0 \sin\alpha$, where α is the angle between the SPP wave vector k_{xr} and the free-space wave vector k_0. In other words, these states are leaky modes which spontaneously decay into the far field away from the strip.[27] Consequently, they cannot be generated in the same manner as were the bounded modes we have examined so far, because the light emitted from the strip would be artificially reflected from the external boundaries of the computational domain. The solution consists in replacing the problematic boundaries by surfaces having "open" boundary conditions; that is, surfaces that effectively fully absorb the electromagnetic fields impinging upon them. The efficiency of this method will be demonstrated in the following paragraphs by the leaky SPP modes sustained by waveguides swept along a circular curve.

Similar to the case of bent dielectric waveguides,[28] SPP transmission lines having a radius of curvature lose power by radiation, as an electromagnetic flux away from the structure and parallel to the plane of the strip develops. For this reason, and also because of the geometry of the system, the numerical analysis of bent metal strips imposes significant changes on the computational domain. An illustration of the unit cell can be seen in Fig. 5.9(a): it consists in an arc of waveguide, described by the cylindrical coordinates (ρ, θ, z). The curved transmission line is described by the parameters ρ_0, the mean radius of curvature (defined as the distance between the origin of the cylindrical coordinates and the middle of the strip), and ρ_s the position of the outer surface of the unit cell; we maintain the previous notations for the width and thickness of the strip. The periodic boundary conditions applied at the edges of the unit cell are now that a field component satisfies $f(\theta_0 + \theta_1) = \exp(i\varphi) f(\theta_0)$, where θ_0 and $(\theta_0 + \theta_1)$ are the position of the two planes of periodicity in cylindrical coordinates, and φ is the phase delay (Fig. 5.9 (a)). Note that when $\rho_0 \gg w$, the field of the guided modes should have almost the same form as those supported by a straight waveguide. The previous equality can thus be approximated by $f(\theta_0 + \theta_1) \approx \exp(i\theta_1 \cdot \rho_0 \cdot k_{xr}) f(\theta_0)$, which provides a qualitative relationship between φ and the in-plane component of the SPP wave vector k_{xr}.

An important point is that the light emitted by the SPP modes should be prevented from being numerically reflected by the outer face $\rho = \rho_s$ of the unit cell. To this end, we surround this surface with Perfect Matching Layers (PMLs), which have been specifically developed to avoid unwanted reflections at the boundaries of truncated computational domains.[29–30] PMLs are fictitious anisotropic materials that fully absorb the electromagnetic fields impinging upon them, thus essentially pushing the boundary infinitely far away from the structure. As already mentioned earlier, there is no need to apply PMLs on the remaining boundaries because the electromagnetic flux arising from the radiation losses is expected to cross the sole surface $\rho = \rho_s$. For this reason we place perfect conducting walls along the inner surface and the horizontal faces of the problem domain.

It should be noted that Fig. 5.9(a) is in fact a schematized representation of the actual unit cell. In order to reduce the calculation effort, we have made two modifications to this picture. First, we decrease the angle θ_1 between the two planes of periodicity to a few milliradians. Secondly, we need only simulate the upper half of the structure because of its symmetry along the center plane of the strip. It is thus necessary to generate the symmetric and antisymmetric modes separately, by successively placing a perfect electric wall and a perfect magnetic wall along the plane of symmetry.

We focus on the ss_b^0 mode of thin Ag strips ($t = 40$ nm and $w = 1$ μm), bent with several different radii of curvature. Experimental investigations[9] have demonstrated that this mode can propagate between two straight strips linked by an S-shaped segment without experiencing significant losses. These observations are confirmed by Fig. 5.9(b) showing that the dispersion relation computed for a strip with a radius of curvature $\rho_0 = 15$ μm is almost identical to that of a straight waveguide. However, both experimental and theoretical results are somewhat surprising if we keep in mind that the ss_b^0 mode supported by a thin strip is supposed to be only poorly bounded to the metal surface, which is likely to hamper its ability to be guided by bent structures.

To gain further insight into the behavior of long-range SPPs on bent strip transmission lines, we map the electric field of the mode for different radii of curvature. Figure 5.10 summarizes these simulations performed at the free-space wavelength $\lambda_{vac} = 633$ nm. As ρ_0 decreases, the field evolves from a symmetric pattern surrounding the strip to a localized and highly asymmetric distribution around the outer corner. In other words, the curvature improves the field confinement and hence allows the mode to round the bend without losing too much power. Consequently the radiation losses in real structures (which are typically composed of several straight and curved elements)[9,11] are probably dominated by the transition losses between the different parts of the strip rather than by the curvature losses themselves, as further suggested by the significant mismatch between the field patterns of Fig. 5.10. It would be thus interesting to analyze in detail the coupling between structures of different

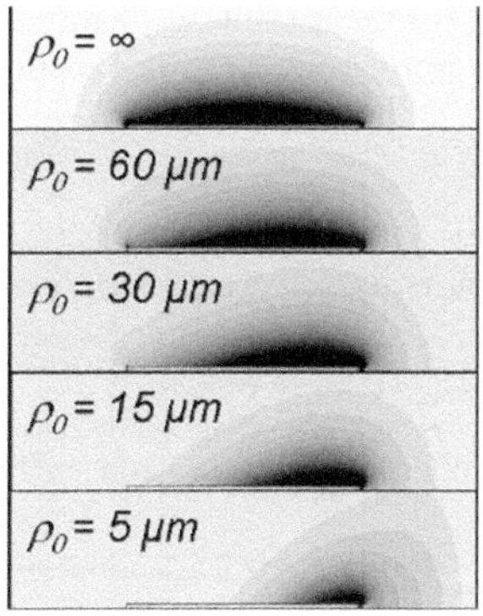

Figure 5.10. Electric field distribution of the ss_b^0 mode for different radii of curvature ρ_0.

curvatures. This would require simulating the junction and a significant length of strip, since the geometry of the problem does not allow the efficient use of periodic boundary conditions. Unfortunately, the size of the computational domain required would be too large to perform the simulations using the eigensolver as outlined.

Although the eigensolver method significantly limits the simulation domain, and thus the size and type of structures that can be analyzed, nevertheless, the variety of examples examined in this chapter illustrates the versatility of the method as well as its potential for use in designing new types of plasmonic waveguides and transmission lines. Through the application of eigensolutions we have been able to explore the rich propagation behavior of SPP transmission lines. Because of the wide diversity of modes supported, including modes localized to much less than the free-space wavelength, plasmonic transmission lines may be exploited in numerous situations where waveguides and other optical components are now typically used.

REFERENCES

1. D. M. Pozar: *Microwave Engineering* (John Wiley & Sons, New York, 1998).
2. A. Yariv: *Optical Electronics in Modern Communications* (Oxford University Press, New York, 1997).
3. J.-C. Weeber, A. Dereux, C. Girard, J. R. Krenn, J.-P. Goudonnet: Plasmon polaritons of metallic nanowires for controlling submicron propagation of light, Phys. Rev. B **60**(12), 9061–9068 (1999).
4. J.-C. Weeber, J. R. Krenn, A. Dereux, B. Lamprecht, Y. Lacroute, J.-P. Goudonnet: Near-field observation of surface plasmon polariton propagation on thin metal stripes, Phys. Rev. B **64**(4), 045411 (2001).
5. B. Lamprecht, J.R. Krenn, G. Schider, H. Ditlbacher, M. Salerno, N. Felidj, A. Leitner, F.R. Aussenegg: Surface plasmon propagation in microscale metal stripes, Appl. Phys. Lett. **79**(1), 51–53 (2001).
6. P. Berini, Plasmon-polariton waves guided by thin lossy metal films of finite width: bound modes of symmetric structures, Phys. Rev. B **61**(15) 10484–10503 (2001).
7. P. Berini: Plasmon-polariton waves guided by thin lossy metal films of finite width: bound modes of asymmetric structures, Phys. Rev. B **63,** 125417 (2001).
8. R. Charbonneau, P. Berini, E. Berolo, E. Lisicka-Shrzek: Experimental observation of plasmon-polariton waves supported by a thin metal film of finite width, Optics Lett. **52**(11), 844–846 (2000).

9. R. Charbonneau, N. Lahoud, G. Mattiussi, P. Berini: Demonstration of integrated optics elements based on long-ranging surface plasmon polaritons, Optics. Express **13**(3), 977–984 (2005).
10. T. Nikolajsen, K. Leosson, I. Salakhutdinov, S.I. Bozhevolnyi: Polymer-based surface-plasmon-polariton stripe waveguides at telecommunication wavelengths, Appl. Phys. Lett. **82**(5), 668–670 (2003).
11. T. Nikolajsen, K. Leosson, S.I. Bozhevolnyi: Surface plasmon polariton based modulators and switches operating at telecom wavelengths, Appl. Phys. Lett. **82**(5), 668–670 (2003).
12. Rashid Zia, Anu Chandran, Mark L. Brongersma: Dielectric waveguide model for guided surface polaritons, Optics lett. **30**(12), 1473–1475 (2005).
13. H. Raether: *Surface Plasmons* (Springer-Verlag, Berlin, 1988).
14. W.L. Barnes, A. Dereux, T.W. Ebbesen: Surface plasmon subwavelength optics, Nature **424**, 824–830 (2003).
15. D. Sarid: Long-range surface-plasma waves on very thin metal films, Phys. Rev. Lett. **47**(26), 1927–1930 (1981).
16. J.J. Burke, G.I. Stegeman, T. Tamir: Surface-polariton-like waves guided by thin, lossy metal films, Phys. Rev. B **33**(8), 5286–5201 (1986).
17. P.B. Johnson, R.W. Christy: Optical constants of the noble metals, Phys. Rev. B **6**(12), 4370–4379 (1972).
18. W.L. Barnes, T.W. Preist, S.C. Kitson, J.R. Sambles: Physical origin of photonic energy gaps in the propagation of surface plasmons on gratings, Phys. Rev. B **54**(9), 6227–6244 (1996).
19. W.L. Barnes, S.C. Kitson, T.W. Preist, J.R. Sambles: Photonic surfaces for surface-plasmon polaritons, J. Opt. Soc. Am. A **14**(7), 1654–1661 (1997).
20. S.I. Bozhevolnyi, V.S. Volkov, K. Leosson, J. Erland: Observation of propagation of surface plasmon polaritons along line defects in a periodically corrugated metal surface, Opt. Lett. **26**(10), 734–736 (2001).
21. P.E. Barclay, K. Srinivasan, M. Borselli, O. Painter: Probing the dispersive and spatial properties of photonic crystal waveguides via highly efficient coupling from fiber tapers, Appl. Phys. Lett. **85**, 4–6 (2004).
22. S.A. Maier, M.D. Friedman, P.E. Barclay, O. Painter: Experimental demonstration of fiber-accessible metal nanoparticle plasmon waveguides for planar energy guiding and sensing, Appl. Phys. Lett. **86**, 071103 (2005).
23. A. Lai, C. Caloz, T. Itoh: Composite right/left-handed transmission line metamaterials, IEEE Microwave Mag. **5**(3), 34–50 (2004).
24. R. Islam, F. Elek, G.V. Eleftheriades: Coupled-line metamaterial coupler having co-directional phase but contra-directiona power flow, Electronics Lett. **40**(5), 315–317 (2004).
25. V.G. Veselago: The electrodynamics of substances with simultaneously negative values of ε and μ, Sov. Phys. Usp. **10**, 509–514 (1968).
26. D.R. Smith, W.J. Padilla, D.C. Vier, S.C. Nemat-Nasser, S. Schultz: Composite medium with simultaneously negative permeability and permittivity, Phys. Rev. Lett. **84**(18), 4184–4187 (2000).
27. A. Christ, T. Zentgraf, J. Kuhl, S.G. Tikhodeev, N.A. Gippius, H. Giessen: Optical properties of planar metallic photonic crystal structures: experiment and theory, Phys. Rev. B **70**, 125113 (2004).
28. D. Marcuse: Curvature loss formula for optical fibers, J. Opt. Soc. Amer. **66**, 216–220 (1976).
29. J.-P. Berenger: A perfectly matched layer for the absorption of electromagnetic waves, J. Comput. Phys. **114**, 185–200 (1994).
30. R. Mittra, U. Pekel: A new look at the perfectly matched layer (PML) concept for the reflectionless absorption of electromagnetic waves, IEEE Microwave Guided Wave Lett. **5**, 84–86 (1995).

CHAPTER SIX

SURFACE PLASMON POLARITON GUIDING IN PHOTONIC BANDGAP STRUCTURES

THOMAS SØNDERGAARD[1] AND SERGEY I. BOZHEVOLNYI[1,2]
[1]Micro Managed Photons A/S, Ryttermarken 15, DK-3520 Farum, Denmark
[2]Institute of Physics and Nanotechnology, Aalborg University, Pontoppidanstræde 103, DK-9220 Aalborg Øst, Denmark

6.1. INTRODUCTION

Surface plasmon polaritons (SPPs) are quasi-two-dimensional electromagnetic excitations, propagating along a dielectric-metal interface and having the field components decaying exponentially into both neighboring media.[1] The field of a plane SPP comprises a magnetic field component, which is parallel to the interface plane and perpendicular to the SPP propagation direction, and two electric field components, of which the main one is perpendicular to the interface (Fig. 6.1(a)). SPPs can be tightly bound to the metal surface, penetrating on the order of 100 nm into the dielectric and ~10 nm into the metal. This feature implies the possibility of using SPPs for miniature photonic circuits and optical interconnects and has attracted a great deal of attention to SPPs.[2] It has been shown using numerical simulations that nanometer-sized metal rods can support extremely confined SPP modes, though only propagating over hundreds of nanometers.[3] Similar properties were expected[4] and indeed found[5] for the electromagnetic excitations supported by chains of metal nano-spheres. Metal stripes of finite width can also be employed to laterally confine the SPP propagation along the stripes.[6–8]

In conventional integrated optics based on dielectric waveguides, the problem of miniaturization is approached by making use of the photonic band gap (BG) effect that is essentially a manifestation of Bragg reflection of waves propagating (in any direction) because of periodic modulation of the refractive index.[9] Indeed, efficient waveguiding along straight and sharply bent line defects in two-dimensional (2D) BG structures (i.e., planar waveguide structures with periodic arrays of holes to control the light propagation in the waveguide plane) has been demonstrated for light wavelengths inside the BG.[10] It became clear that these BG structures, when properly designed and

M.L. Brongersma and P.G. Kik (eds.), Surface Plasmon Nanophotonics, 73–86.

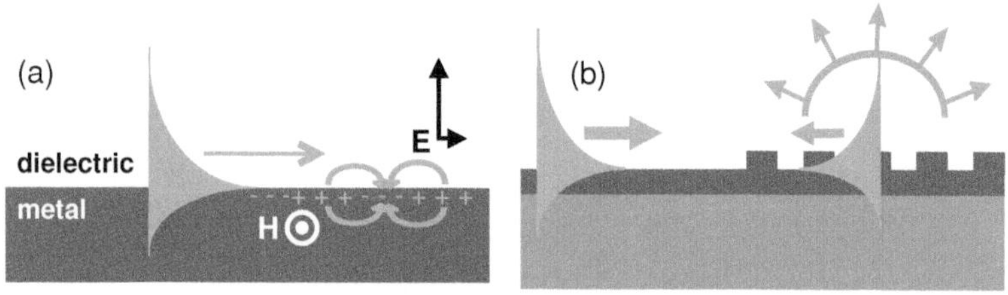

Figure 6.1. (a) Schematic representation of a SPP propagating along a metal-dielectric interface and orientations of electric and magnetic field components. (b) Illustration of the SPP reflection by a periodic array of surface scatterers, showing also scattered field components propagating away from the metal surface.

realized, might be advantageously used for miniature photonic circuits allowing for an unprecedented level of integration.[11] Furthermore, one can conjecture that other (quasi) 2D waves, e.g., SPPs, might be employed for the same purpose. In fact, the BG effect for SPPs excited along a specific direction was first observed more than 30 years ago in the experiments on light diffraction by periodically corrugated metal surfaces.[12] The SPPBG effect for all directions in the surface plane of a silver film having a 2D periodic surface profile has also been reported.[13] It should be emphasized that the SPP interaction with a periodic surface corrugation, similarly to the interaction of a waveguide mode with a periodic array of holes, produces inevitably scattered waves propagating *away* from the surface (Fig. 6.1(b)). This (unwanted) process results in the additional propagation loss and has to be taken into account when considering the SPPBG structures.

SPP guiding along line defects in SPPBG structures was first demonstrated with ~45 nm-high and 200 nm-wide gold bumps arranged in a 400 nm-period triangular lattice on the surface of a 45 nm-thick gold film.[14] The efficient SPP reflection by such an area and SPP guiding along channels free from scatterers was observed, as well as significant deterioration of these effects at 815 nm, indicating the occurrence of the SPPBG effect in these structures. The observations were made with a stand-alone scanning near-field optical microscope (SNOM) used to collect the radiation scattered by an uncoated sharp fiber tip into fiber modes and an arrangement for SPP excitation in the usual Kretschmann configuration (Fig. 6.2(a)). Similar experiments have followed revealing that the SPP guiding is strongly damped in narrow channels[15] contrary to what was expected from the experience in (conventional) photonic BG structures.[11] It was also found that the SPP guiding around sharp corners exhibited a bend loss increasing quadratically with the bend angle.[16] In fact the bend loss was found significant already at the bend angle of $>15^\circ$ (Fig. 6.2), an observation that was again somewhat unexpected. It became clear that further progress in the development of SPPBG structures for compact photonic circuits requires detailed theoretical studies.

In general, the process of SPP scattering by surface features is very complicated and yet to be understood in all its complexity, and the SPP scattering by *periodic* arrays of surface scatterers is not an exception. A similar scattering geometry was first

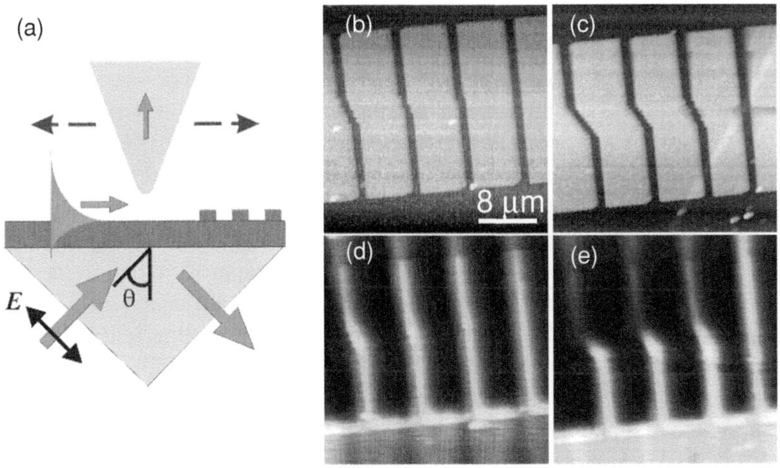

Figure 6.2. (a) Schematic representation of the excitation of SPPs and their detection with a fiber tip of a collection SNOM. The polarization of the incident laser beam is also indicated. Gray-scale (b, c) topographical and (d, e) near-field optical images (32×32 μm^2) for SPPBG structures (made of gold) at wavelength 737 nm with (b, d) and (c,e) taken simultaneously. Periodic surface structures are of ΓM orientation with the following parameters: the period is 410 nm; the bump height and width are 45 and 200 nm, respectively. The structures contain ~ 2 μm-wide straight defects followed (from right to left) by double-bent defects with (b,d) 5°, 10°, and 15° and (c, e) 20°, 25°, and 30°.[16]

considered by making use of a scalar two-dimensional multiple-scattering approach[17] accounting for some features observed experimentally, e.g., influence of the defect width on the SPP propagation loss. This scalar model, which considers only elastic (in the surface plane) and isotropic SPP scattering, being rather simple and efficient possess however important drawbacks that limit severely its applicability. The effective polarizability of a scatterer introduced to relate the amplitude of the scattered SPP to the incident one is only a phenomenological quantity that is difficult to connect to the scatterer parameters. Furthermore, in a rigorous consideration of SPP scattering from surface scatterers,[18] isotropic SPP scattering is only a limiting case of SPP scattering by a small particle and only for the component of the exciting field being incident on the scatterer which is perpendicular to the surface. Finally, the out-of-plane SPP scattering was completely ignored, thereby excluding the possibility of analyzing and taking into account this (very important) scattering process. Usage of the reduced Rayleigh equation allowed for the first time to relate characteristics of the SPPBG to the geometrical and material parameters of periodical arrays of hemiellipsoids.[19] However, it is questionable whether this approach can be adapted to consider line or point defects in finite-size scattering arrays, since the periodicity of a surface profile is a rather crucial assumption in the formulation used.

Recently, we have extended the scalar multiple-scattering approach into a vector dipolar multiple-scattering theory based on a dyadic Green's tensor description and applied it to finite-size SPPBG structures with and without line defects.[20] Our simulations have reproduced not only the influence of the defect width on the SPP propagation loss but also the threshold in sizes of scatterers with respect to the realization of efficient SPPBG. On the other hand, from the point of view of the point-dipole approximation used in this model, the threshold radius (75 nm) of spherical scatterers is rather large for the considered wavelength range (~800 nm). Our latest simulations of an SPP beam splitter using the same model have indicated that the sphere radii larger than 60 nm resulted in the violation of energy conservation and phase relationships for transmitted and reflected SPP beams.[21] We arrived thereby to the conclusion that it is imperative to go beyond the point-dipole approximation in the modeling of efficient SPPBG structures (i.e., made of relatively large scatterers).

In this Chap. we model the SPPBG phenomena, including SPP guiding and bending, treating surface scatterers as *finite*-size cylindrical bumps. Modeling results on transmission and reflection, and field magnitude images, are presented along with the consideration of waveguides and bends based on introducing channels in SPPBG structures. The idea is to provide a reader with practical guidelines on appropriate scatterer sizes and lattice constants for making SPPBG structures exhibiting the BG for both main orientations of the lattice. We investigate various, though not all possible, waveguide and bend designs, so that one can expect further improvements being made to the structure designs that we present.

As is the case with photonic crystals based on dielectric materials, SPPBG structures can also be tried out for creating components that are small and compact with the aim of integrating a large number of functionalities on a single optical chip. Due to the absorption by metal in the SPPBG structures it is not only desirable but also important to realize components that are small and compact since otherwise the propagation loss may become a major problem.

6.2. NUMERICAL METHODS

The method that we have applied for modeling of SPPBG structures, including waveguides and bends, is based on the Lippmann-Schwinger integral equation

$$\mathbf{E}(\mathbf{r}) = \mathbf{E}_0(\mathbf{r}) + \int \mathbf{G}(\mathbf{r}, \mathbf{r}')k_0^2(\varepsilon(\mathbf{r}') - \varepsilon_{\text{ref}}(\mathbf{r}')) \cdot \mathbf{E}(\mathbf{r}')d^3r'. \tag{6.1}$$

In this method the starting point is a reference structure with dielectric constant $\varepsilon_{\text{ref}}(\mathbf{r})$ as a function of the position $\mathbf{r}$, which in our case is a planar air-gold interface. The incident field $\mathbf{E}_0$ is a field solution in the reference structure. We have chosen $\mathbf{E}_0$ as the field of a Gaussian SPP beam propagating along the air–gold interface. The actual structure of interest has the dielectric constant $\varepsilon(\mathbf{r})$ so that the term $(\varepsilon(\mathbf{r}) - \varepsilon_{\text{ref}}(\mathbf{r}))$ in (6.1) represents the modification made to the reference structure, which in our case corresponds to placing an array of gold scatterers on the gold surface. This modification results in scattering of the incident beam $\mathbf{E}_0$ so that the total electric field becomes $\mathbf{E}$. In (6.1), the term $\mathbf{G}(\mathbf{r},\mathbf{r}')$ is the Green's tensor of the reference structure that here represents the field generated by a dipole as a function of the dipole position

r′ in the air–gold interface structure. The Green's tensor can be calculated numerically by use of Sommerfeld integrals.[22]

Since modeling of a single scatterer is already a demanding task,[18] and because we aimed at modeling structures with many ($\sim 10^3$) scatterers, we had to employ an approximation when treating an individual scatterer. In this approximation, the field *incident* on a scatterer ($\mathbf{E}_0$ + field originating from other scatterers) is assumed constant across the scatterer. This approximation is reasonable for small scatterers, and is better than treating the scatterers as (spherical) point particles (dipoles),[20,22] in which it is the *total* field inside the scatterers that is assumed constant, because the total field inside a small scatterer may vary much faster than the incident field. Having calculated the scattering in the case of a single scatterer with cylindrical symmetry,[23] being exposed to various orientations of the incident field, it is possible to use this result to significantly reduce the numerical task for scattering systems consisting of several thousand scatterers.

6.3. NUMERICAL RESULTS

The first task when considering the SPPBG structures is to determine the range of useful design parameters of the SPPBG structure, such as size and shape of scatterers and lattice constant. Our approach to this task is to calculate transmission and reflection (as a function of light wavelength) for a Gaussian SPP beam incident on rectangular SPPBG structures with various choices of the lattice constant, various scatterer heights and scatterer radii. From the transmission and reflection spectra, one can get an impression of BG properties (rigorous consideration of the BG effect requires an infinitely extended periodic scattering system) and try to find parameters where the structure exhibits the BG for both main orientations of the structure, ΓM and ΓK, with respect to the incident beam. Examples of this type of calculation for the lattice constant $\Lambda = 450$ nm, scatterer height h $= 50$ nm and two scatterer radii r $= 100$ nm and r $= 125$ nm are presented in Fig. 6.3. The structures considered consist of a finite array of cylindrical gold scatterers (30×30 or 26×34) arranged on a triangular lattice on the gold-air interface. The insets show the orientations ΓM and ΓK with respect to the incident beam.

In these particular calculations, the reflection and transmission are evaluated far from the regions occupied by scatterers along the axis of the Gaussian incident beam (beam waist 2.8 μm) but close to the air-gold interface. Considering first the structure with the scatterer radius r $= 100$ nm and ΓK orientation, one notices a reflection peak and the related transmission dip from $\sim$785 nm to $\sim$850 nm. For the corresponding structure with ΓM orientation, we also notice a reflection peak and a transmission dip. The reflection is somewhat lower in this case because most of the reflected light does not propagate directly backwards, where the reflection is evaluated, but propagates at 60° with respect to the incident direction (i.e., in the direction of Bragg reflection). The transmission dip at 750 nm is stable regarding changes in the dimensions of the SPPBG structure, and field calculations show that the incident SPP cannot propagate through the SPPBG structure at this wavelength. The dip at 785 nm is, however, not

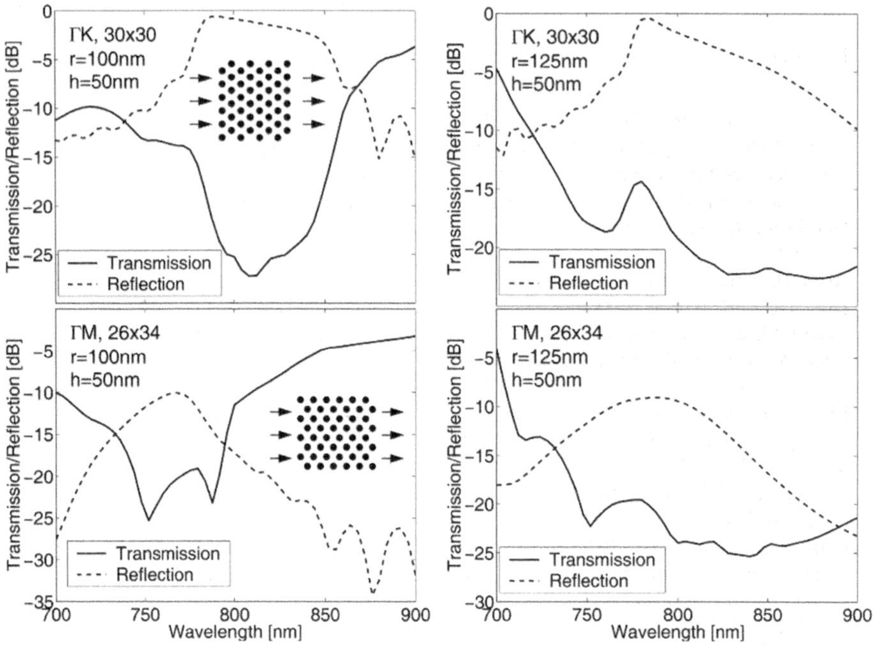

Figure 6.3. Transmission and reflection spectra for 450 nm-period SPPBG structures.

stable, and is here due to a resonance effect caused by the finite number of scatterers in the array resulting in light being transmitted in directions other than directly forward. For this particular choice of scatterers (r = 100 nm) there is no overlap between the reflection peaks for the two orientations. As the scatterer radius is increased to r = 125 nm, one may notice from the reflection spectrum for ΓK orientation that the short-wavelength BG edge remains at nearly the same wavelength, whereas the long-wavelength edge shifts toward longer wavelengths. For ΓM orientation, the reflection peak broadens and shifts to longer wavelengths. Similar trends were observed with changes in scatterer size when using the point-dipole approximation in our previous work.[20] It is no longer easy to extract the BG-related information from the transmission spectra, but the transmission is low for both orientations. It should also be borne in mind that the transmission spectra contain features associated with the coupling between SPPs propagating at flat and corrugated (with scatterers) surface regions. For the radius 125 nm, there appears to be an overlap in reflection peaks for the two main orientations at wavelengths close to 800 nm, a feature that can be considered as the indication of the complete (for all in-plane directions) SPPBG effect. In the following, we model straight and 30°—bent waveguides based on this SPPBG structure design.

A waveguide can be formed, e.g., by starting out with 60×30 periodically arranged scatterers having ΓK orientation relative to the incident beam and removing 7 adjacent center rows (W7) of scatterers. A waveguide can also be created by starting out with

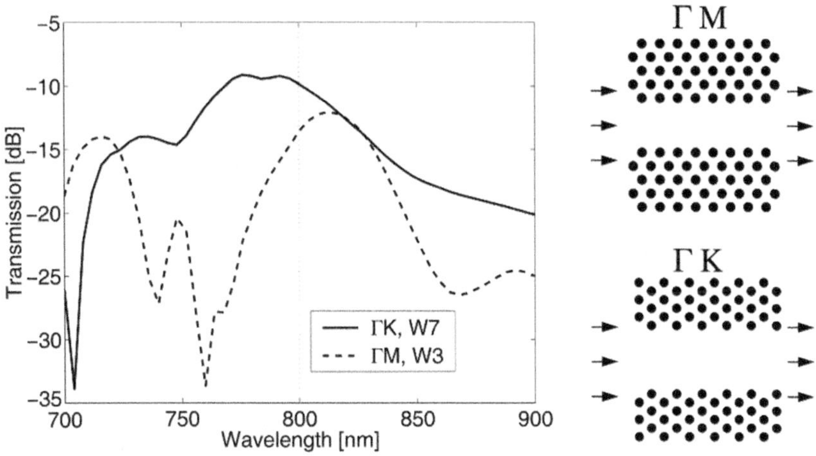

Figure 6.4. Transmission spectra for ΓK- and ΓM-oriented waveguides. These waveguides were formed by removing 3(W3) and 7(W7) adjacent center rows, respectively, from periodic arrays of scatterers.

52 × 34 scatterers and ΓM orientation by removing 3 adjacent center rows (W3) of scatterers. Note that the distance between rows is larger for ΓM orientation. For these two waveguides, the transmission spectra for an incident Gaussian beam with beam waist 2.8 μm are shown in Fig. 6.4.

From Fig. 6.4, it can be noticed that there is a peak in transmission near 800 nm for both waveguide orientations. The peak for ΓK orientation is, however, shifted towards shorter wavelengths, and the peak for ΓM orientation is shifted towards longer wavelengths. These shifts are opposite to the shifts of reflection peaks observed in Fig. 6.3 for ΓK and ΓM orientation. Note that efficient waveguiding implies efficient suppression of both SPP propagation through the front edge of SPPBG structures and SPP penetration through the sides of a waveguide channel. Since the arrangement of scatterers at the channel sides for the ΓK waveguide is similar to that at the front edge in the case of ΓM orientation (and vice versa), it seems reasonable to find waveguide transmission spectra with features associated simultaneously with both orientations. The wavelength of ~800 nm appears to be a reasonable compromise where both waveguides are efficient. Note that the transmission in Fig. 6.4 includes losses due to the SPP coupling in and out of the waveguide.

The electric field magnitude distributions at the height 300 nm above the air-gold interface for the two considered waveguides and wavelength 800 nm are shown in Fig. 6.5. The SPPBG structures and channel regions are indicated with dashed white lines. It is seen that the Gaussian beam (incident from the left) gradually transforms into a well-confined field propagating in the waveguide channel, which in turn transforms into a divergent beam radiating from the output end of the waveguide. Notice also that the backward scattering of light is rather different in the two cases. As was also mentioned above for SPPBG structures without channel waveguides, a large part of the backward scattered light for ΓM orientation is propagating at 60° relative to the direct backscattering.

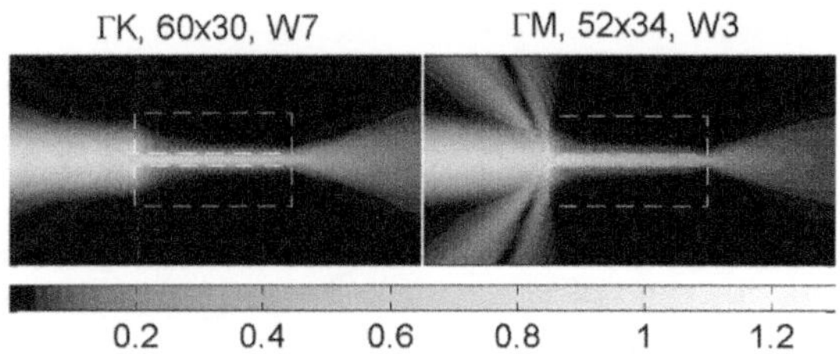

Figure 6.5. Electric field magnitude at the height 300 nm above the gold surface for ΓK- and ΓM-orientated waveguides corresponding to the wavelength (800 nm) marked in Fig. 6.4.

Since both waveguides are reasonably efficient at the same wavelength of 800 nm, one may expect that it is possible to combine them into an efficient 30°-bent waveguide. The corresponding transmission is shown in Fig. 6.6 Three different choices for the bend region are considered, namely a sharp bend, and smoothened versions of the sharp bend obtained by removing 1 or 3 scatterers in the bend region. The insets show the arrangements of scatterers in the bend region for these three cases. It is seen that, by moving three scatterers in the bend region, an increase in the transmission by a few dB can be achieved for wavelengths close to 800 nm, where the BGs for both main orientations overlap. It should be kept in mind that, although the transmission is only −20 dB, this value includes not only the bend loss and propagation loss (due to absorption by the metal) but also losses incurred when coupling into and out of the SPP waveguide mode. For comparison, the transmission of a straight ΓM oriented waveguide with width W3 is ∼ −12 dB, implying that the bend loss is at the level of ∼8 dB. One should also be aware of the following intricate detail of our simulations. The transmission was evaluated by calculating the average field intensity in a box

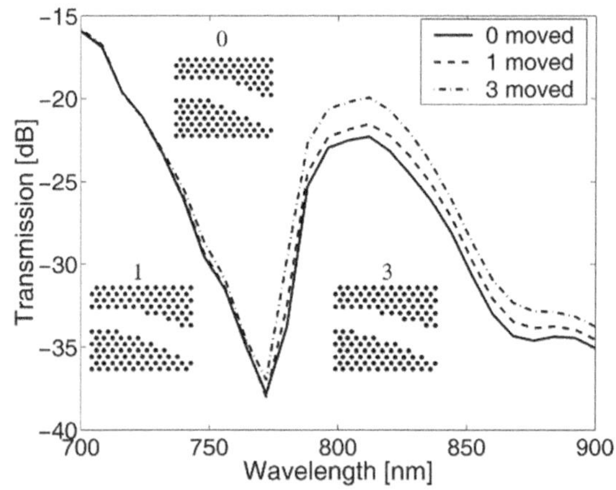

Figure 6.6. Transmission spectra for three designs of the 30° bend formed by the waveguides considered in Fig. 6.4.

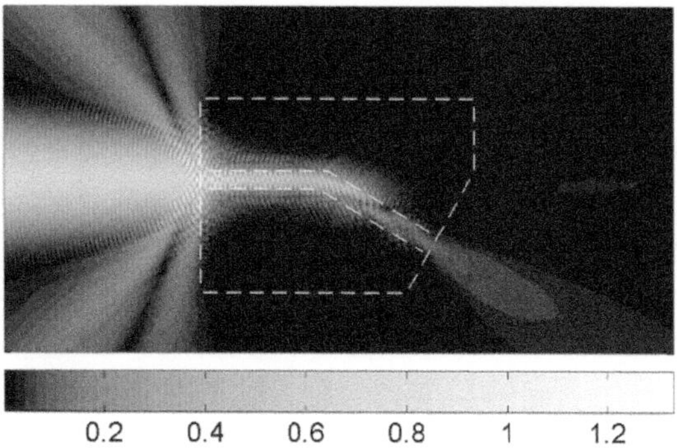

Figure 6.7. Electric field magnitude (wavelength = 800 nm) at the height 300 nm above the gold surface for the 30° bend (smoothened by removing 3 scatterers) considered in Fig. 6.6.

region placed near the exit of the bend. One may notice that it is not the wavelengths at ~800 nm that result in the largest transmission but that at ~700 nm. However, for the wavelengths of ~700 nm, field calculations reveal that the field intensity at the box position originates rather from scattered SPP fields than from the guided (around the bend) SPP mode. This can be implicitly confirmed by noticing that moving scatterers in the bend region does not lead to an improvement in the transmission at ~700 nm, but it does at ~800 nm. Furthermore, for wavelengths at ~800 nm, field calculations do show that the SPP mode is guided by the channel around a 30° bend.

An example of the electric field magnitude distribution at the height 300 nm above the air-gold interface for the 30° bend with 3 scatterers removed and wavelength 800 nm is shown in Fig. 6.7. One observes that, at the input of the bend, only a fraction of the incident beam power is coupled into the waveguide channel. The SPPBG structure and (bent) channel region are indicated with dashed white lines. It is clearly seen that an appreciable part of the SPP field does follow the channel being guided around the bend, and that a well-defined beam comes out of the exit waveguide propagating at the angle of 30° with respect to the incident beam direction. Although the loss is not negligible, this example still shows that the SPPBG structures can guide the radiation around sharp 30° bends.

Another method for changing the SPP propagation direction (using the SPPBG waveguides) is to gradually rotate the crystal lattice containing a line defect, an approach that was reported for conventional photonic crystal waveguides.[25] Transmission spectra for 45° bends corresponding to gradual bending of the ΓM waveguide (W3) were calculated for three different bending radii (23, 35 and 46 μm) and are shown in Fig. 6.8.

We may expect that the losses due to leakage out of the waveguide will decrease as the bending radius increases from 23 to 46 μm. However, at the same time, the bend will become longer, and the SPP mode has to propagate a longer distance inside the channel resulting in an increased propagation loss. For a given dielectric constant for gold[26] at wavelengths around 800 nm, the SPP propagation length (intensity reduced

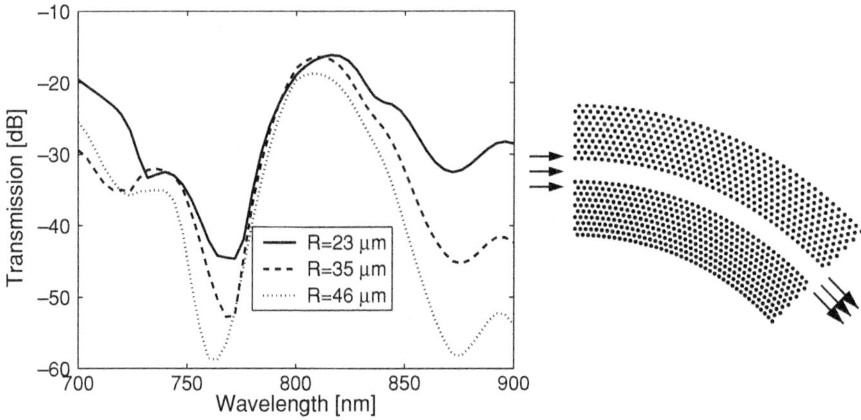

Figure 6.8. Transmission spectra for 45° bends obtained by gradual rotation (with different radii) of a straight waveguide. The structure for the smallest bending radius (23 μm) is shown to the right.

by a factor of $1e^{-1}$) is ~45 μm for a flat air-gold interface. The SPP scattering at the waveguide sides reduces the propagation length further (depending on the channel width[15]). All in all, it turned out that, when increasing the bending radius from 23 to 35 μm, the decrease in the leakage loss and the increase in the propagation loss cancel each other to some extent, so that the peak transmission does not significantly change (Fig. 6.8). Note that, in this case, the bend loss is ~5 dB lower compared to that for the sharp bends considered in Fig. 6.6. Further increase of the bending radius to R = 46 μm results in the decrease of the peak transmission (due to the increase in the propagation loss). The SPP propagation through a 45° bend with the radius 35 μm is shown in Fig. 6.9.

There are at least two ways in which the SPP propagation loss can be reduced. The first is to increase the wavelength moving further away from the plasma resonance and thereby decreasing the SPP propagation loss. The second approach is to replace a metal-air interface structure with a symmetric structure of a thin metal film (10–20 nm thickness) embedded in a dielectric. The metal film supports both long-range and short-range SPPs, which differ in the symmetry of field components across the center of the film. Compared to the SPPs of the metal-air interface structure, the long-range SPPs have smaller propagation loss, whereas the short-range SPPs have larger propagation loss by proper symmetric design with scatterers on both sides of the film, it is (in theory, and for perfect structures) possible to avoid coupling between the long-range and short-range SPPs. However, in both cases (i.e., for SPPs at longer wavelengths and for long-range SPPs) a reduction in the SPP propagation loss is partly related to the fact that the SPP modes in question are weaker bound to the metal and penetrate farther in the dielectric. Consequently, it may also become necessary to use larger scatterers in order for the scatterers to cover the same fraction of the mode and to achieve the same efficiency in the SPP scattering. On the other hand, the efficiency of unwanted out-of-plane SPP scattering (into modes propagating away

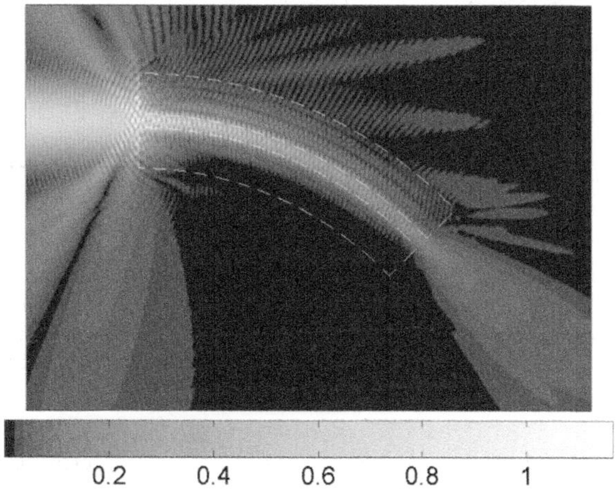

Figure 6.9. Electric field magnitude (wavelength = 810 nm) at the height 300 nm above the gold surface for a 45° bend obtained by gradual rotation of a straight waveguide (bend radius = 35 μm).

from the metal surface) increases as SPPs become less strongly bound to the metal.[24] These general considerations imply that the reduction in propagation loss obtained by changing the wavelength or by using long-range SPPs can only be obtained at a price, namely, that larger scatterers may be needed, and the SPP out-of-plane scattering might increase (whereas the SPP in-plane scattering would decrease accordingly). The latter circumstance may even jeopardize the possibility of realizing the SPPBG effect, which is originated in efficient multiple (in-plane) SPP scattering.

The effect of increasing the operating wavelength is illustrated with the reflection spectra calculated for SPPBG structures at telecom wavelengths are shown in Fig. 6.10. The arrangement and number of scatterers on the air-gold interface are the same as in Fig. 6.3. Two sizes of scatterers and four choices of lattice constants are considered. Notice that, for ΓK orientation, the reflection does not increase much above −3 dB (50%), whereas for shorter wavelengths it could approach 100% (Fig. 6.3). This difference we attribute to higher losses due to stronger out-of-plane SPP scattering related, as elaborated above, to weaker SPP localization at the air-gold interface.[24] Note that longer (ΓK-oriented) structures do not significantly increase the SPP reflection. For example, with SPPBG a structure consisting of 60×30 scatterers instead of 30×30 scatterers it is only possible to increase the reflection by ~0.5 dB.

For the scatterers with radius 200 nm and height 100nm, there is no overlap between the reflection maxima for ΓM and ΓK orientations for lattice constants Λ ranging from 800 nm to 950 nm, although it is close to being the case for $\Lambda = 800$ nm. When increasing the radius to 248 nm, we observe tendencies similar to those seen previously for shorter wavelengths (Fig. 6.3). Namely, for ΓK orientation, the short-wavelength edges of the reflection maxima are almost unaffected, and the long-wavelength edges shift to longer wavelengths, whereas, for ΓM orientation, the reflection maxima broaden and shift (as a whole) to longer wavelengths. These changes result in an

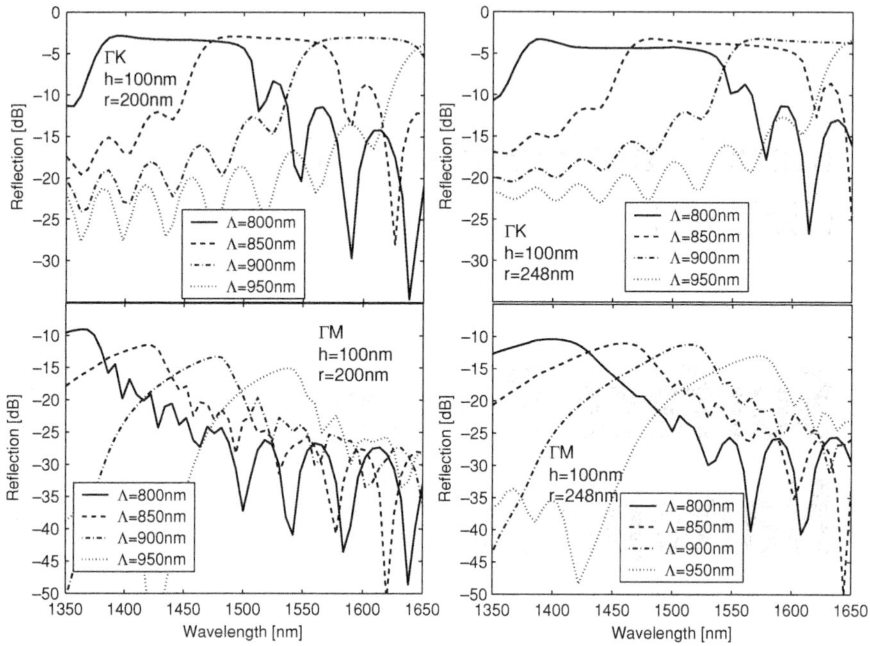

Figure 6.10. Reflection spectra for SPPBG structures at telecom wavelengths.

overlap between reflection peaks for the lattice constant 800 nm but not for the other lattice constants. It is interesting that, compared to the scatterers that resulted in the reflection overlap for wavelengths of ~800 nm (Fig. 6.3), the radius and height are twice as large. However, the corresponding wavelength and lattice constant have not been doubled, so that, effectively, the dimensions of scatterers were increased more than the wavelength and lattice constant. This is in agreement with the above general consideration conjecturing that weaker SPP localization requires the use of larger scatterers. Note that, for wavelengths of ~1430 nm, the SPP propagation length along a flat gold-air interface is approximately 4 times longer compared to that at ~800 nm.

6.4. CONCLUSIONS

SPP scattering phenomena associated with the SPPBG effect is analyzed theoretically in this Chap. We considered the SPP propagation in SPPBG waveguides and bends by making use of the Lippmann-Schwinger integral equation method. The SPPBG structures consisted of a gold-air interface with gold scatterers arranged on a triangular lattice. From calculations of reflection spectra we found appropriate sizes of the scatterers and lattice constants that, for certain wavelengths, result in high reflection for both main orientations of the SPPBG structure with respect to the incident SPP beam. Our results indicate that, for longer wavelengths (e.g., for 1400–1600 nm instead of 700–800 nm), it is necessary to use progressively larger scatterers in order to obtain the overlap in reflection peaks for the two main SPPBG structure orientations. Furthermore, the out-of-plane SPP scattering increases for longer wavelengths as well, decreasing the efficiency of SPPBG effects. We argue that these features stem from the SPP being less confined to the metal surface for longer wavelengths, a circumstance, which is in agreement with our recent analytical investigations of SPP scattering from a single scatterer.[24]

We present transmission spectra for two waveguide designs based on removing a number of rows of scatterers in SPPBG structures. It is observed that the waveguide created by removing rows in the SPPBG structure of ΓK orientation is most efficient for the BG seen along ΓM orientation and vice versa. This feature can be accounted for by noticing that the channel sides for a ΓK-oriented waveguide resemble the reflection interface for a ΓM-oriented structure (and vice versa). We show that both waveguides are reasonably efficient for the wavelengths of ~800 nm and considered their combination forming a 30° bend. Transmission calculations indicate an appreciable efficiency of the bend transmission that could be further increased by moving 3 scatterers in the bend region and thereby smoothing the bend. Another type of bend obtained by gradual rotation of one of the straight SPPBG waveguides is found even more efficient, e.g. the improvement of ~5 dB is expected for a 45° bend with the bending radii of ~30 μm. Finally, it is worth noting that simulations of the field intensity distributions above the SPPBG structures show for the parameters considered, the SPP mode field can be reasonably well confined to the waveguide channels and bend regions.

The experimental and theoretical results reported previously[14–17,19,20] and in this Chap. indicate that SPPBG structures having carefully chosen parameters can be employed for efficient SPP guiding and bending in plasmonic circuits, though the losses incurred are not negligible and would limit the number of SPPBG components that could be integrated. It is important to note that further improvements in the designs of SPPBG structures, waveguides, and bends, can be expected, since there still exists many unexplored possibilities for the design of individual scatterers and scatterer configurations.

ACKNOWLEDGMENTS

The research has been carried out in the framework of the European Network of Excellence, PLASMO-NANO-DEVICES (FP6-2002-IST-1-507879).

REFERENCES

1. H. Raether: *Surface Plasmons* (Springer-Verlag, Berlin, 1988).
2. W.L. Barnes, A. Dereux, T.W. Ebbesen: Surface plasmon subwavelength optics, Nature **424**, 824 (2003).
3. J. Takahara, S. Yamagishi, H. Taki, A. Morimoto, T. Kobayashi: Guiding of a one-dimensional optical beam with nanometer diameter, Opt. Lett. **22**, 475 (1997).

4. M. Quinten, A. Leitner, J.R. Krenn, F.R. Aussenegg: Electromagnetic energy transport via linear chains of silver nanoparticles, Opt. Lett. **23**, 1331 (1998).
5. S.A. Maier, P.G. Kik, H.A. Atwater, S. Meltzer, E. Harel, B.E. Koel, A.A.G. Requicha: Local detection of electromagnetic energy transport below the diffraction limit in metal nanoparticle plasmon waveguides, Nature Mater. **2**, 229 (2003).
6. J.R. Krenn, J.C. Weeber: Surface plasmon polaritons in metal stripes and wires, Philos. Trans. Roy. Soc. A **326**, 739 (2004).
7. P. Berini: Plasmon-polariton waves guided by thin lossy metal films of finite width: Bound modes of symmetric structures, Phys. Rev. B **61**, 10484 (2000).
8. A. Boltasseva, T. Nikolajsen, K. Leosson, K. Kjaer, M.S. Larsen, and S.I. Bozhevolnyi: Integrated optical components utilizing long-range surface plasmon polaritons, J. Lightwave Technol. **23**, 413 (2005).
9. J.D. Joannopoulos, R.D. Meade, J.N. Winn: *Photonic Crystals* (Princeton University Press, Princeton, 1995).
10. T.F. Krauss, R.M. De La Rue: Photonic crystals in the optical regime—past, present, and future, Prog. Quant. Elect. **23**, 51 (1999).
11. C.M. Soukoulis, ed.: *Photonic Crystals and Light Localization in the 21st Century*, (Kluwer, Dordrecht, 2001).
12. R.H. Ritchie, E.T. Arakawa, J.J. Cowan, R.N. Hamm: Surface-plasmon resonance effect in grating diffraction, Phys. Rev. Lett. **21**, 1530 (1968).
13. S.C. Kitson, W.L. Barnes, J.R. Sambles: Full photonic band gap for surface modes in the visible, Phys. Rev. Lett. 77, 2670 (1996).
14. S.I. Bozhevolnyi, J. Erland, K. Leosson, P.M.W. Skovgaard, J.M. Hvam: Waveguiding in surface plasmon polariton band gap structures, Phys. Rev. Lett. **86**, 3008 (2001).
15. S.I. Bozhevolnyi, V.S. Volkov, K. Leosson, J. Erland: Observation of propagation of surface plasmon polaritons along line defects in a periodically corrugated metal surface, Opt. Lett. **26**, 734 (2001).
16. S.I. Bozhevolnyi, V.S. Volkov, K. Leosson, A. Boltasseva: Bend loss in plasmon polariton band-gap structures, Appl. Phys. Lett. **79**, 1076 (2001).
17. S.I. Bozhevolnyi, V.S. Volkov: Multiple-scattering dipole approach to modeling of surface plasmon polariton band gap structures, Opt. Comm. **198**, 241 (2001).
18. A.V. Shchegrov, I.V. Novikov, A.A. Maradudin: Scattering of surface plasmon polaritons by a circularly symmetric surface defect, Phys. Rev. Lett. **78**, 4269 (1997).
19. M. Kretschmann: Phase diagrams of surface plasmon polaritonic crystals, Phys. Rev. B **68**, 125419 (2003).
20. T. Søndergaard, S.I. Bozhevolnyi: Vectorial model for multiple scattering by surface nanoparticles via surface polariton-polariton interactions, Phys. Rev. B **67**, 165405–1–8 (2003).
21. V. Coello, T. Søndergaard, S.I. Bozhevolnyi: Modeling of a surface plasmon polariton interferometer, Opt. Commun. **240**, 345 (2004).
22. L. Novotny, B. Hecht, D. Pohl: Interference of locally excited surface plasmons, J. Appl. Phys. **81**, 1798 (1997).
23. T. Søndergaard, B. Tromborg: Lippmann-Schwinger integral equation approach to the emission of radiation by sources located inside finite-sized dielectric structures, Phys. Rev. B **66**, 155309 (2002).
24. T. Søndergaard, S.I. Bozhevolnyi: Surface plasmon polariton scattering by a small particle placed near a metal surface: An analytical study, Phys. Rev. B **69**, 045422 (2004).
25. J. Arentoft, T. Søndergaard, M. Kristensen, A. Boltasseva, M. Thorhauge, L. Frandsen: Low-loss silicon-on-insulator photonic crystal waveguides, Electron. Lett. **38**, 274 (2002).
26. E. Palik: *Handbook of Optical Constants of Solids* (Academic, San Diego, CA, 1985).

CHAPTER SEVEN

SUBWAVELENGTH-SCALE PLASMON WAVEGUIDES

HARRY A. ATWATER, JENNIFER A. DIONNE AND LUKE A. SWEATLOCK
Thomas J. Watson Laboratory of Applied Physics, California Institute of Technology, Pasadena, California 91125

7.1. INTRODUCTION

By the mid-17th century, numerous scientists—notably including Hooke and Gallileo—had developed transparent ground lenses and applied them in the construction of compound optical microscopes. This development revolutionized the contemporary understanding of the natural world by, for example, enabling the imaging of blood cells and microbes. Ever since this now bygone era of fantastic development of optics principles and instrumentation, the size and performance of photonic devices has been largely limited by diffraction. Photonic devices of today are generally composed of dielectric materials with modest dielectric constants, and are much bigger than the smallest electronic devices (e.g., transistors in silicon integrated circuits) for this reason.

By contrast, subwavelength spatial confinement of light at dimensions down to less than 10% of the free-space wavelength is possible using plasmonic components, owing to the large and highly tunable dielectric permittivities for metals, especially near their plasmon resonances. Ultimately it may be possible to employ plasmonic components to form the building blocks of a chip-based nanophotonic device technology that is scaleable essentially to molecular dimensions, with potential imaging, spectroscopy and interconnection applications in computing, communication and chemical/biological detection. The basic component required to demonstrate feasibility of this approach through realization of nanophotonic networks is a subwavelength scale waveguide structure, several designs for which are outlined in this chapter.

7.2. NANOPARTICLE CHAIN PLASMONIC WAVEGUIDES

Photonics is in transition toward a world of sub-100 nm designs and components. This is an evolution to a subwavelength world, and actual electromagnetic wavelengths in

M.L. Brongersma and P.G. Kik (eds.), Surface Plasmon Nanophotonics, 87–104.

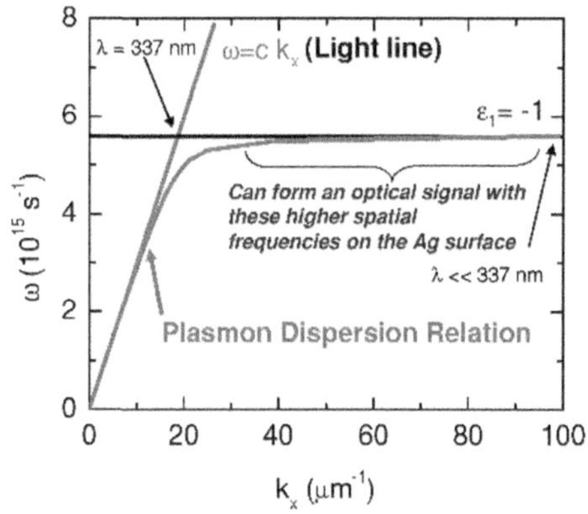

Figure 7.1. A typical planar interface plasmon dispersion relation. Surface plasmons are collective excitations that have non-radiating modes bound to the surface of a metal or thin metallic film. Light can be coupled from free space into the surface plasmon mode only by matching the momentum of the plasmon; this can be done, e.g., via index matching or grating coupling. Near the surface plasmon frequency, the real part of the dielectric constant ε_1 changes sign from positive to negative (while ε_2 is positive) and thus the accessible in-plane wavevector values k_x can be very large. High wavevectors are just what is needed to produce strongly localized optical modes on surfaces, potentially giving rise to imaging and light manipulation at optical frequencies and soft X-ray wavelengths (10's of nm).

some media, such as metals, can be much shorter than free-space optical wavelengths. Photonic devices and systems that are conventionally said to operate at certain wavelengths (e.g., 1300–1600 nm telecommunication wavelengths), but the dispersion curve in Fig. 7.1 amply illustrates that it is more appropriate to describe device operation in terms of frequencies, since for plasmonic materials[1] and devices, frequency is invariant, but wavelength is highly variable, depending on the situation.

Advances in particle synthesis and fabrication techniques have recently allowed for studies of ordered arrays of noble metal nanoparticles. Recently, the use of metal nanostructure arrays as waveguides has been considered.[2–5] In such arrays, each nanoparticle with a diameter much smaller than the wavelength of the exciting light acts as an electric dipole.[6–8] Thus, two types of electromagnetic interactions between the particles can be distinguished, depending on the spacing d between adjacent nanoparticles. For particle spacing on the order of the excitation wavelength, far-field dipolar interactions with a d^{-1} dependence dominate.[4–9] For particle spacing much smaller than the wavelength of light, near-field dipolar interactions between adjacent particles with a distance dependence of d^{-3} dominate. These strongly distance-dependent interactions lead to the formation of collective eigenmodes of the chain, which are manifested by a splitting of the plasmon dipolar peak for regular, one-dimensional chain arrays of metal nanoparticles,[10–12] implying the possibility of using such structures as waveguides.

Individual noble metal nanoparticles strongly interact with visible light at their dipole surface plasmon frequency due to the excitation of a collective electron motion giving rise to a plasmon excitation inside the metal particle.[7] The surface of the nanoparticle confines the conduction electrons inside the particle and sets up an effective restoring force leading to resonant behavior at the dipole surface plasmon frequency. When the frequency of the light is far from the intrinsic plasmon resonance of the metal nanoparticle, energy flow around the particle is only slightly perturbed. At the plasmon resonance frequency, the strong polarization of the particle effectively draws energy into the particle, giving rise to a strong enhancement in the scattering cross-section in optical extinction measurements.[7] The dipole surface plasmon resonance is most pronounced for particles much smaller than the wavelength of the exciting light, since in this case all conduction electrons of the particle are excited in phase. The resonance frequency is determined by the particle material, the shape of the particle and by the refractive index of the surrounding host. Surface plasmons can be efficiently excited in noble metals, such as gold, silver, and copper, due to their free electron like behavior. For these metals the plasmon resonance occurs in the visible range of light in a variety of hosts.

The strong interaction of individual metal nanoparticles with light can be used to fabricate waveguides if energy can be transferred between nanoparticles. The dipole field resulting from a plasmon oscillation in a single metal nanoparticle can induce a plasmon oscillation in a nearby particle due to near-field electrodynamic interactions. The finding that ordered arrays of closely spaced noble metal particles exhibit a collective behavior under far field illumination confirms the existence of this near-field coupling.[8] In 1999, we proposed the name *plasmon waveguides* apply to structures operating on this principle and the name *plasmonics* for the field of study, to draw attention to the energy guiding mechanism via surface plasmons.[9]

When metal nanoparticles are spaced closely together (separation a few tens of nanometers), as depicted in the inset of Fig. 7.2, the strongly distance-dependent, near-field term in the expansion of the electric dipole interaction dominates. The interaction strength and the relative phase of the electric field in neighboring particles are both polarization and frequency dependent. This interaction leads to coherent modes with a wavevector **k** along the nanoparticle array. One can calculate a dispersion relation for energy propagation along the nanoparticle chain by taking into account nearest-neighbor interactions, and non-nearest-neighbor interactions via a polarization dependent interaction frequency ω_1 derived from the electromagnetic interaction term and the plasmon dipole resonance ω_0. Internal and radiation damping are also accounted for in the model.[3] Fig. 7.2 shows the results of such a calculation for modes with the electric field polarized along the chain (longitudinal modes) and for modes polarized perpendicular to the chain (transverse modes). Both analytic calculations and full-field electromagnetic simulations illustrate the dispersion relations that arise from the near-field coupling.

Calculations that include nearest-neighbor coupling only (solid lines) and including up to five nearest neighbors in the coupling term (dotted lines) for an infinite linear array of metal nanoparticles. The inclusion of up to five nearest-neighbors has little effect on the dispersion curves, confirming that the interaction is dominated by nearest-neighbor coupling. For both polarizations, the propagation velocity of the guided

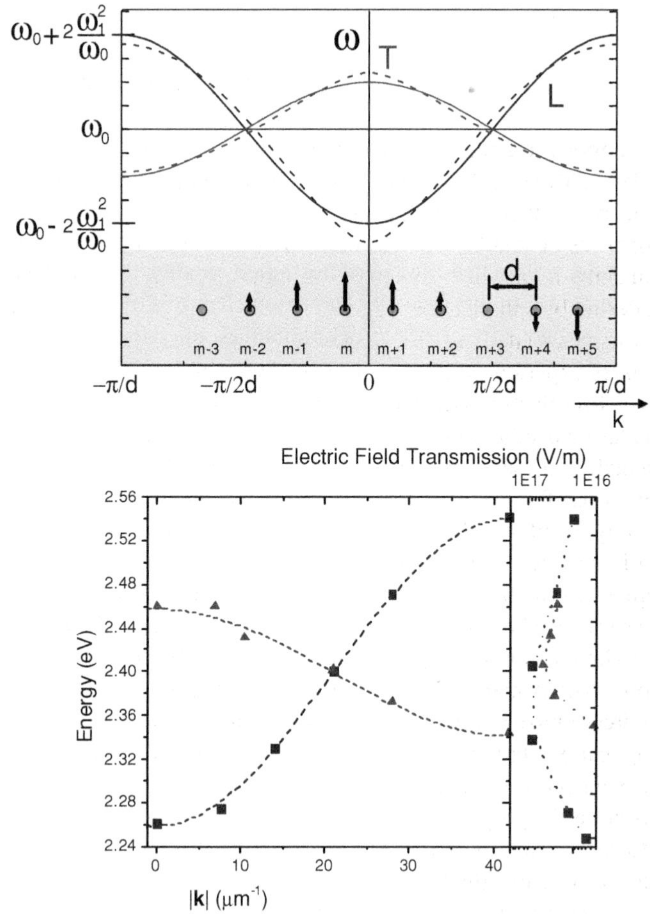

Figure 7.2. At top, dispersion relation for plasmon modes in a linear chain of metal nanoparticles (see inset), showing the twofold degenerate branch corresponding to transverse modes (T) and the branch corresponding to longitudinal modes (L). Results are shown for calculations incorporating nearest-neighbor interactions only (solid curve), and including up to fifth nearest-neighbor interactions (dotted curve). The small difference in the dispersion curves illustrates that mode propagation in plasmon waveguides is dominated by near-field interactions. At bottom, comparison of full-field electromagnetic simulations for the longitudes (squares) and transverse (triangles) modes of a chain of 50 nm diameter spherical Au particles separated by a 75 nm center-to-center spacing. The left panel gives the energy-wavevector dispersion relation whereas the right panel gives the electric field strength associated with modes of different energies.

energy, given by the slope $d\omega/d\boldsymbol{k}$ of the dispersion relation, is highest at the resonance frequency ω_0. Calculations for 50 nm diameter silver spheres with a center-to-center distance of 75 nm show energy propagation velocities of about 10% of the speed of light. This is ten times faster than the velocity of electrons in typical semiconductor devices.

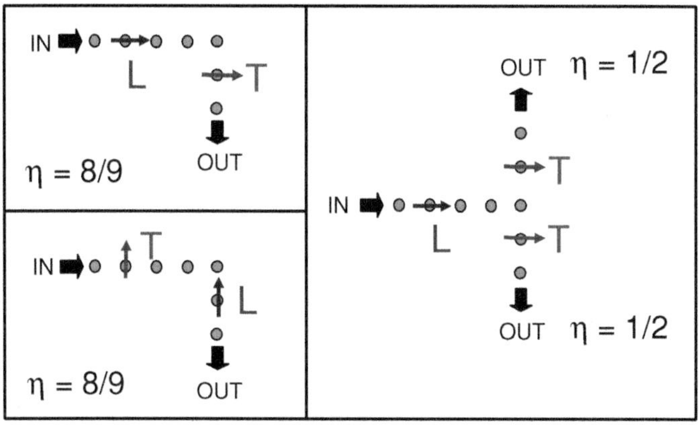

Figure 7.3. Calculated power transmission coefficient η in nanoparticle chain arrays for a 90 degree corner and a tee structure. The thick arrows indicate the direction of the energy flow, the thin arrows indicate the direction of polarization corresponding to either the longitudinal mode (L) or the transverse mode (T).

The dispersion relations for Au nanoparticle chain arrays have been derived from full-field electromagnetic simulations. The waveguides were excited with an oscillating point-dipole placed at a distance of 75 nm to the center of the first Au nanoparticle of the waveguide. In consecutive runs, the point-dipole source was driven continuously at various frequencies in the vicinity of E_0. The field distribution along the nanoparticle chain structure was then analyzed to determine the wave vector **k** of the propagating waves. The bottom part of Fig. 7.4 shows the dispersion relation for both longitudinal (squares) and transverse (triangles) excitations. The dotted lines indicate the dispersion relation calculated using a simple point-dipole model. The obtained $\omega(\boldsymbol{k})$ data are in excellent agreement with the predictions from the point-dipole model despite the limitations of the latter. The right panel of this figure shows the electric field amplitude at the center of the last nanoparticle for longitudinal excitations (squares). Note that the waveguide loss is a minimum at the center of the dispersion band, as expected since the group velocity is a maximum at this point.

Aside from the dispersion relation, another important parameter in waveguide design is loss. In plasmon waveguides, the loss can be due to radiation into the far field and due to internal damping. Radiation losses into the far field are expected to be negligible due to the dominance of near-field coupling. Internal damping of the surface plasmon mode is due to resistive heating. This damping was shown to induce transmission losses of about 6 dB μm^{-1}.

Waveguides can be concatenated to form circuit elements such as corners and tee structures (Fig. 7.3). Due to the near-field nature of the coupling, signals can be guided around 90 degree corners and split via tee structures without radiation losses into the far field at the discontinuity. Power transmission coefficients for the guiding of energy around corners and for signal splitting in tee structures were calculated by requiring continuity of the plasmon amplitude and of the energy flux at the corner where the wave gets partly transmitted and reflected. The transmission coefficients are a strong function of the frequency of the guided wave and of its polarization and show a maximum at the dipole plasmon frequency. Transmission coefficients close to 100% are possible for propagation around 90 degree corners for certain polarizations, and lossless signal splitting was shown in tee structures. Over length scales in which plasmon excitations remain coherent, it is also possible to design switches that rely on interference effects, such as Mach-Zehnder interferometers.[3]

We have also investigated optical pulse propagation through these plasmon waveguides driving local dipole sources.[14] In Fig. 7.4(a) a pulse centered at the resonance energy $E_0 = 2.4$ eV corresponding to the $\mathbf{k} = \pi/2d^{-1}$ modes with the highest group velocities is launched from the dipole source located on the left-hand side of the array. The upper inset of Fig. 7.4(a) shows a snapshot of the x-component of the electric field in the xy-plane for longitudinal polarization on a linear color scale. The periodicity of the field distribution along the chain confirms that the pulse is centered on the $\mathbf{k} = \pi/2d^{-1}$ mode corresponding to a wavelength of 4 particle spacings. An analogous snapshot for the y-component of the electric field for transverse polarization is shown in the lower inset. The main part of Fig. 7.4(a) shows the pulse position, defined as the location of maximum field amplitude, over time for both longitudinal (squares) and transverse (triangles) excitation. Linear fits of the datasets yield values for the group velocities of transverse and longitudinal modes. These full-field electromagnetic simulations quantitatively confirm the possibility of using plasmon waveguides for information transport with $v_g = 0.01c$. The calculations also show the occurrence of negative phase velocities in these plasmon waveguides when excited in a transverse mode. For the transverse pulse, the group velocity and thus the direction of energy propagation and the phase velocity of the individual wave components are anti-parallel. This is illustrated in Fig. 7.4(b), which shows ten electric field snapshots spaced $\Delta t = 0.166$ fs (10% of a cycle at E_0) apart. The wave packet is seen to move away from the source while the individual phase fronts travel towards the source, since for transverse modes a positive group velocity, required for causality, occurs at negative wavevectors $\boldsymbol{k}$, as seen in the dispersion relation of Fig. 7.2. Nanoparticle chain plasmon waveguides could thus serve as a relatively simple model system for the investigation of negative phase velocity structures.[15–16]

Local excitation and detection of guided light at optical frequencies with a lateral mode profile below the diffraction limit of light provides the most direct evidence for near-field coupling between particles in ordered arrays. Indeed, it has been shown both theoretically and experimentally that such array, depicted schematically in Fig. 7.5 and experimentally in Fig. 7.6 can guide electromagnetic energy over distances of several hundred nanometers via near-field particle interactions.[17] Highly ordered plasmon waveguides such as illustrated in Fig 7.6(a) can be fabricated by electron beam lithography and liftoff fabrication and can be imaged using atomic force microscopy, as shown in Fig 7.6(b), and near-field optical microscopy. Such structures could potentially be used in nanoscale all-optical networks, contributing to a new class of functional optical devices operating below the diffraction limit of light.

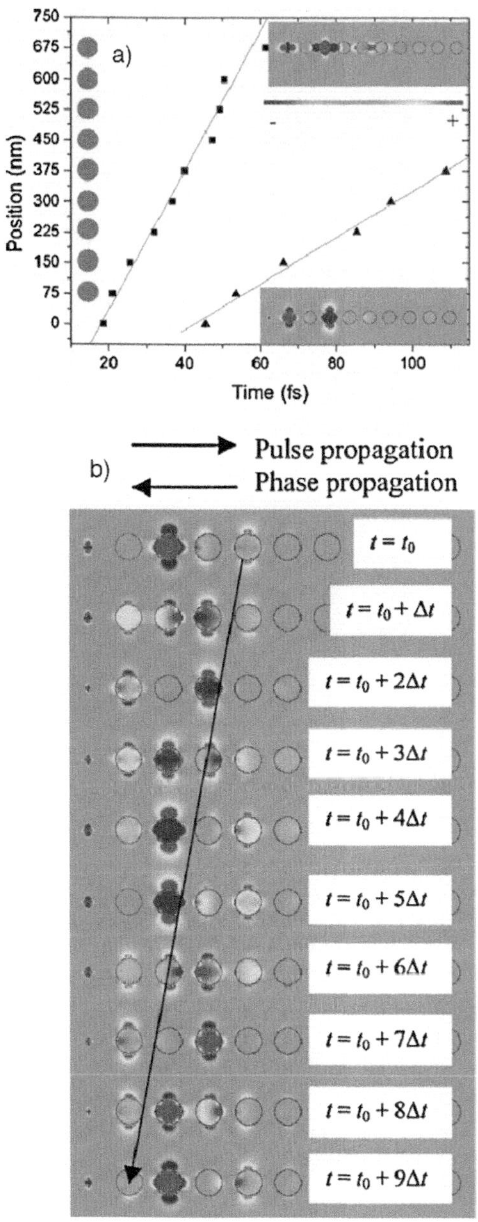

Figure 7.4. (a) Pulse peak positions over time in a plasmon waveguide consisting of spherical particles for both longitudinal (squares) and transverse (triangles) polarization. The spheres along the ordinate indicate the position of the Au nanoparticles. Snapshots of the $x(y)$ component of the electric field in the xy plane for longitudinal (transverse) polarization are shown in the upper (lower) inset. (b) Time snapshots of the electric field for transverse pulse propagation show a negative phase velocity with an antiparallel orientation of the phase and group velocities.

Figure 7.5. Schematic of local excitation and detection of energy transport in a subwavelength-scale plasmon waveguides by near-field optical microscopy. Light emanating from the tip of an illumination-mode near-field scanning optical microscope (NSOM) locally excites a plasmon waveguide. The waveguide transports the electromagnetic energy to a fluorescent polymeric nanosphere, and the fluorescence intensity for varying tip positions is collected in the far-field.

7.3. STRONGLY COUPLED PLASMON MODES IN NANOSTRUCTURES

Nanoparticles provide highly enhanced local fields which are promising for molecular sensors[18–21] or miniature nonlinear optical elements.[22–26] When arrays of particles are very closely spaced there are dramatic quantitative and qualitative changes in the collective eigenmodes and dispersion properties of the chains. Although precise, lithographically-based fabrication of such structures is presently difficult, closely spaced and contacting linear chain nanoparticle arrays in glass can be formed non-lithographically by use of high energy ion irradiation technique,[27] as illustrated in Fig. 7.7(a) and (b). Finite difference time domain simulations performed for arrays of touching particles (i.e., spacing 0 nm) illustrate this most dramatically, as seen in Fig. 7.7(c) and (d). In this case, two distinct modes were found in the spectrum, at 0.35 eV (free space wavelength 3500 nm) and 1.65 eV (750 nm), respectively. The physical significance of each of these modes becomes apparent by studying the spatial distribution of the electric field for the two cases. This is demonstrated in Fig. 7.7(c) and (d), which shows the longitudinal component E_x of the electric field in a system which consists of a linear array of 12 touching Ag spheres excited at the two resonance

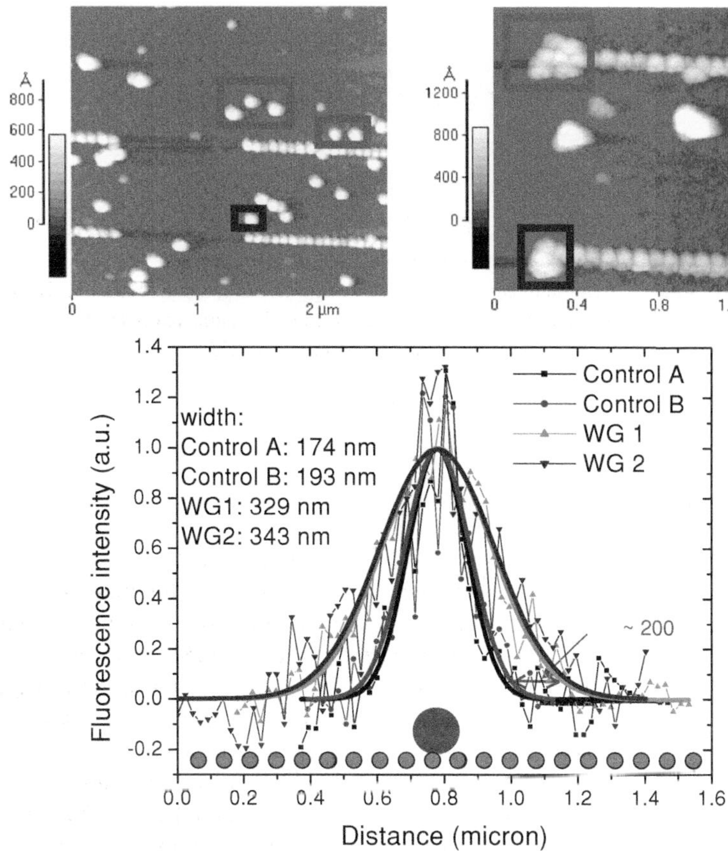

Figure 7.6. Above, noncontact atomic force micrographs of Au nanoparticle chain plasmon waveguides with fluorescent polymeric nanospheres which were positioned at the left end of each plasmon waveguide by scanning force manipulation prior to imaging. Left image illustrates polymeric nanospheres prior to manipulation; right images is nanospheres after manipulation. Below, Evidence for energy transport in plasmon waveguides by the width of the intensity of fluorescent nanospheres. Individual data sets represent averages of five parallel cuts along the plasmon waveguide direction through the fluorescent spots for isolated nanospheres (control A and B, square and diamond data points) and nanospheres located on top of plasmon waveguides (WG 1 and WG2, up-triangles and down-triangles data points) as depicted in the inset. Gaussian line-shape fits to the data show an increased width for nanospheres located on plasmon waveguides, indicating plasmon propagation.

frequencies. In a snapshot of the chain driven by a longitudinally polarized plane wave at 0.35 eV (panel (c)), regions of positive E_x are observed at either end, with negative E_x throughout the body of the array. This electric field pattern indicates that positive surface charge is concentrated on the rightmost particle, and negative charge on the leftmost particle. The mode is typical of a single wire antenna, and requires surface charge to flow from particle to particle along the entire length of the array. Alternatively, when the same structure is driven at 1.65 eV (panel (d)) the coupled-dipole

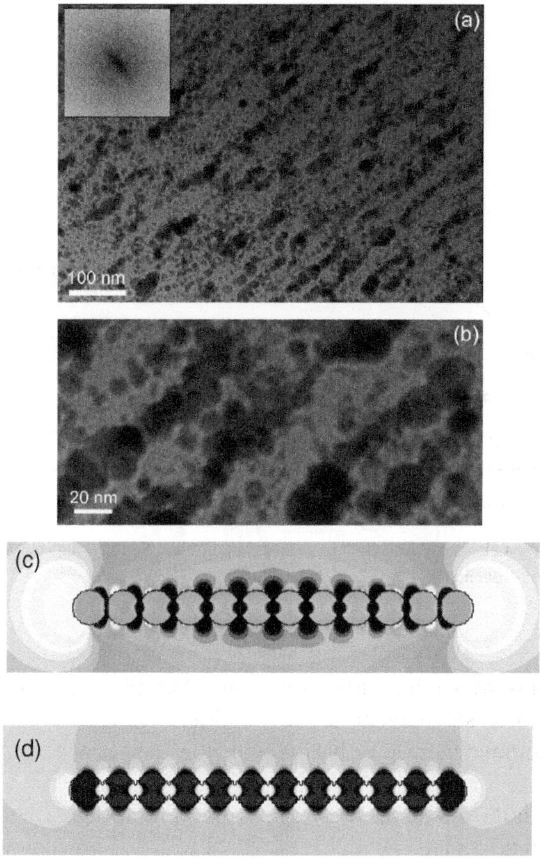

Figure 7.7. (a) Plan-view TEM image of Ag nanoparticles in Ag-doped soda lime glass after 30 MeV Si ion irradiation. Alignment of nanoparticles along the ion beam direction is observed, and verified by spatial fast Fourier Transform image, inset. (b) TEM indicates typical particle diameter of 10 nm, albeit with significant size polydispersity, and that the particles are arranged into quasi-linear arrays of up to ~10 particles. (c) and (d): Distribution of the x component of the electric field in the vicinity of an array of twelve Ag particles with 10 nm diameter, illustrating two distinct modes. In panel (c), an antenna-like mode resembling that of a single elongated wire is excited resonantly at 0.35 eV; in panel (d) a coupled particle-like mode resembling that of a chain of independent particles is excited resonantly at 1.65 eV. The slight axial asymmetry of the field distribution is caused by superposition of the resonant mode with the exciting plane wave.

resonance is selectively excited. The field diagram alternates from positive in each dielectric gap to negative inside each particle. This indicates an alternating surface charge distribution in which each individual particle is polarized but electrically neutral. Thus, in the touching-particle configuration, the system can support two kinds of longitudinal resonance: the particles can still act as individual coupled dipoles, or instead, as a single continuous wire antenna.[28] From Fig. 7.7(c) it is also clear that the wire-like resonance has a lesser degree of energy confinement than the coupled dipole mode.

7.4. METAL/INSULATOR/METAL NANOSCALE SLOT WAVEGUIDES

In planar metallodielectric structures, surface plasmon polaritons are the appropriate solutions to Maxwell's equations, with the complex wavevector determining both field symmetry and damping. For bound modes, field amplitudes decay exponentially away from the metal/dielectric interface with field maxima occurring at the surface. While the dispersion properties of long-ranging surface plasmons (SPs) in insulator/metal/insulator structures mimic those of a photon, multicentimeter propagation[29] is often accompanied by significant field penetration into the surrounding dielectric. For thin Ag films ($\sim$10 nm) excited at telecommunications frequencies, electric field skin depths can exceed 5 μm.[30,31] In order to realize highly integrated photonic and plasmonic structures with densely nested waveguide structures, a more favorable balance between localization and loss is required.

Surface plasmon polariton modes of single metal/dielectric interface exhibit strongly wavelength-dependent electric field penetration depths, increasing rapidly in the dielectric as the wavelength is varied away from resonance but the field penetration depths remain approximately constant ($\sim$25 nm) in the metal over a wide range of visible and near-infrared excitation frequencies. This observation has inspired a new class of plasmon waveguides that consist of an insulating core and conducting cladding. Not unlike conventional waveguides, including dielectric slab waveguides at optical frequencies, metallic slot waveguides at microwave frequencies, and the recently-proposed semiconductor slot waveguides,[32] these metal-insulator-metal (MIM) structures guide light via the strong refractive index difference between the core and cladding.[32] However, unlike dielectric slot waveguides, both plasmonic and conventional waveguiding modes can be accessed, depending on transverse core dimensions. MIM waveguides may thus allow optical mode volumes of planar surface plasmon waveguide to be reduced to subwavelength scales—with minimal field decay out of the waveguide physical cross-section—even for frequencies far from the plasmon resonance. Several theoretical studies have already investigated surface plasmon propagation and confinement in MIM structures.[33,34] Here identify the wavelength dependent surface plasmon and conventional waveguiding modes of MIM structure.[35]

When a plasmon is excited at a metallodielectric interface, electrons in the metal create a surface polarization that gives rise to a localized electric field. In insulator/metal/insulator structures, electrons of the metallic core screen the charge configuration at each interface and maintain a near-zero (or minimal) field within the waveguide. As a result, the surface polarizations on either side of the metal film remain in phase and a cutoff frequency is not observed for any transverse waveguide dimension. In contrast, screening does not occur within the dielectric core of MIM waveguides. At each metal/dielectric interface, surface polarizations arise and evolve independently of the other interface, and plasma oscillations need not be energy- or wavevector-matched to each other.

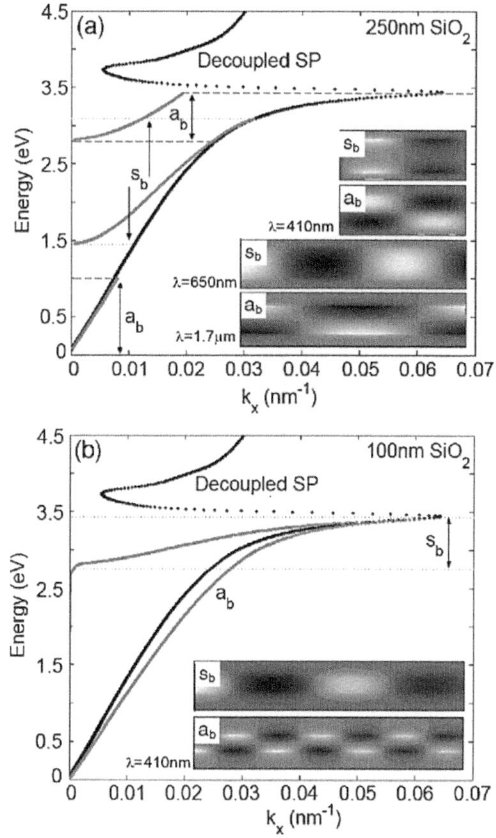

Figure 7.8. Dispersion relations for MIM planar waveguides with a SiO_2 core and an Ag cladding. Dispersion of an infinitely thick core is plotted in black and is in exact agreement with results for a single Ag/SiO_2 interface plasmon. (a) For oxide thicknesses of 250 nm, the structure supports conventional waveguiding modes with cutoff wavevectors observed for both the symmetric (s_b, dark gray) and antisymmetric (a_b, light gray) field configurations. (b) As oxide thickness is reduced to 100 nm, both conventional and plasmon waveguiding modes are supported. Accordingly, tangential electric fields are localized within the core for conventional modes but propagate along the metal/dielectric interface for plasmon modes, (Field plots for the allowed modes are shown in the insets to (a) at free-space wavelengths of $\lambda = 410$ nm ($\sim$3 eV) (top two panels), $\lambda = 650$ nm ($\sim$1.2 eV), and $\lambda = 1.7$ μm ($\sim$0.73 eV), and in (b) at $\lambda = 410$ nm ($\sim$3 eV)).

Therefore, for certain MIM dielectric core thicknesses, interface SPs may not remain in phase but will exhibit a beating frequency; as transverse core dimensions are increased, "bands" of allowed energies/wavevectors and "gaps" of forbidden energies will be observed.

This behavior is illustrated in Fig. 7.8, which plots the dispersion relations for a MIM waveguide with core thicknesses of 250 nm (Fig. 7.8(a)) and 100 nm (Fig. 7.8(b)). The waveguide consists of a three-layer metallodielectric stack with an SiO_2 core and an Ag cladding. The metal is defined by the empirical optical constants of Johnson and

Christy[36] and the dielectric constant for the oxide is adopted from Palik's Handbook.[37] Solution of the dispersion relations was achieved via application of a Nelder-Mead minimization routine in complex wavevector space; details of implementation and convergence properties are described elsewhere.[31] For reference, the figures include the waveguide dispersion curve in the limit of infinite core thickness, plotted in black. Allowed wavevectors are seen to exist for all free space wavelengths (energies) and exhibit exact agreement with the dispersion relation for a single Ag/SiO_2 interface SP.

Figure 7.8(a) plots the bound modes (i.e, modes occurring at frequencies below the SP resonance) of an $Ag/SiO_2/Ag$ waveguide with core thickness d = 250 nm. The asymmetric bound (a_b) modes correspond to solution of L^+ and are plotted in light gray; the symmetric bound (s_b) modes correspond to solution of L^- and are plotted in dark gray. As seen, multiple bands of allowed and forbidden frequencies are observed. The allowed a_b modes follow the light line for energies below ~1 eV and resemble conventional dielectric core/*conducting* cladding waveguide modes for energies above ~2.8 eV. Tangential electric fields in each a_b regime (considering free-space wavelengths of $\lambda = 410$ nm (~3 eV) and $\lambda = 1.7$ μm (~0.73 eV)) are plotted in the inset and illustrate the photonic and transverse electric nature, respectively, of the modes. In contrast, the s_b modes are only observed for energies between 1.5 and 3.2 eV. Dispersion for this mode is reminiscent of conventional dielectric core/dielectric cladding waveguides, with endpoint asymptotes corresponding to tangent line slopes (effective indices) of $n = 8.33$ at 1.5 eV and $n = 4.29$ at 3.2 eV. For energies exceeding ~2.8 eV, wavevectors of the s_b mode are matched with those of the SP, and the tangential electric field transits from a core mode to an interface mode (see the top 1st and 3rd panels of the inset, comparing $\lambda = 410$ nm (~3 eV) and $\lambda = 650$ nm (~1.9 eV)). As the core layer thickness is increased through 1 μm (data not shown), the number of a_b and s_b bands increases with the a_b modes generally lying at higher energies. In analogy with conventional waveguides, larger (but bounded) core dimensions increase the number of modes supported by the structure.

Figure 7.8 (b) plots the bound mode dispersion curves for an MIM waveguide with SiO_2 core thickness d = 100 nm. Again, the allowed a_b modes are plotted in light gray while the allowed s_b modes are dark gray. Although the s_b mode resembles conventional waveguide dispersion, the a_b mode is seen to exhibit plasmon-like behavior. Accordingly, the conventional waveguiding modes are found only at higher energies (over a range of ~1 eV), where photon wavelengths are small enough to be guided by the structure. The inset shows snapshots of the tangential electric field for both modes at a free-space wavelength $\lambda = 410$ nm (~3 eV). As seen, the s_b field is concentrated in the waveguide core with minimal penetration into the conducting cladding. In contrast, the a_b field is highly localized at the surface, with field penetration approximately symmetric on each side of the metal/dielectric interface. The presence of both of conventional and SP waveguiding modes represents a transition to subwavelength-scale photonics. Provided momentum can be matched between the photon and the SP, energy will be guided in a polariton mode along the metal/dielectric interface. Otherwise, the structure will support a conventional waveguide mode, but propagation will only occur over a narrow frequency band.

Surface plasmon polariton dispersion and propagation relations in planar structures are governed by the real and imaginary components, respectively, of the in-plane wavevector. Generally, propagation is high in regimes of near-linear dispersion where high signal velocities overcome internal loss mechanisms. In insulator/metal/insulator structures, long-range propagation is achieved at the expense of confinement: transverse field penetration typically exceeds microns in the surrounding dielectric. In MIM structures, SP penetration into the cladding will be limited by the skin depth of optical fields in the metal. This restriction motivates the question of how skin depth affects propagation, particularly for thin films.

Figure 7.9 illustrates this interdependence of skin depth and propagation in MIM structures for film thickness from 12 to 250 nm. The top panels plot propagation for the structure as a function of free space wavelength; the bottom panels plot the corresponding skin depth. Figure 7.9(a) plots propagation and skin depth for a 250 nm oxide layer. In accordance with the dispersion relations, wave propagation exhibits allowed and forbidden bands for the symmetric and anti-symmetric modes. The symmetric bound mode is seen to propagate for wavelengths between 400 and 850 nm, with maximum propagation distances of $\sim$15 μm. The skin depth for this mode is approximately constant over all wavelengths, never exceeding 22 nm in the metal. In contrast, the anti-symmetric bound mode is seen to propagate distances of 80 μm for wavelengths greater than 1250 nm. For wavelengths below 450 nm, a smaller band of propagation is also observed, though distances do not exceed 2 μm. In regions of high propagation (i.e., above 1250 nm), skin depth remains approximately constant at 20 nm; below 1250 nm, however, skin depths approach 30 nm. Interestingly, the figure indicates only a slight correlation between propagation and skin depth for both the a_b and s_b modes. This relation suggests that the metal (i.e., absorption) is not the limiting loss mechanism for wave propagation in MIM structures.

Figure 7.9(b) illustrates the propagation distance and skin depth for the asymmetric bound mode for oxide thicknesses of 12–100 nm. The continuous plasmon-like dispersion relations of Fig. 7.8 are well correlated with the observed propagation: decay lengths are longest for larger wavelengths, where dispersion follows the light line. Plasmon propagation generally increases with increasing film thickness, approaching $\sim$10 μm for a 12 nm oxide layer and $\sim$40 μm for a 100 nm thick oxide. Nevertheless, field penetration remains approximately constant in the Ag cladding, never exceeding 20 nm. Thus, unlike conventional plasmon waveguides, MIM waveguides can achieve micron-scale propagation with nanometer-scale confinement.

Figure 7.9(c) plots propagation and skin depth for the symmetric bound modes of thin films. As with the a_b modes, larger oxide thicknesses support increased propagation distances. However, the wave remains evanescent for thicknesses up through 50 nm, with propagation not exceeding 10 nm for longer wavelengths. As SiO_2 thicknesses approach 100 nm, a band of allowed propagation is observed at higher frequencies, reflecting the dispersion of 7.8(b): at $\lambda = 400$ nm, propagation lengths are as high as 0.5 μm. In addition, thin films exhibit a local maximum in propagation for wavelengths corresponding to the transition between quasi-bound and radiative modes (see inset), analogous to insulator/metal/insulator guides.[28–30] For films with

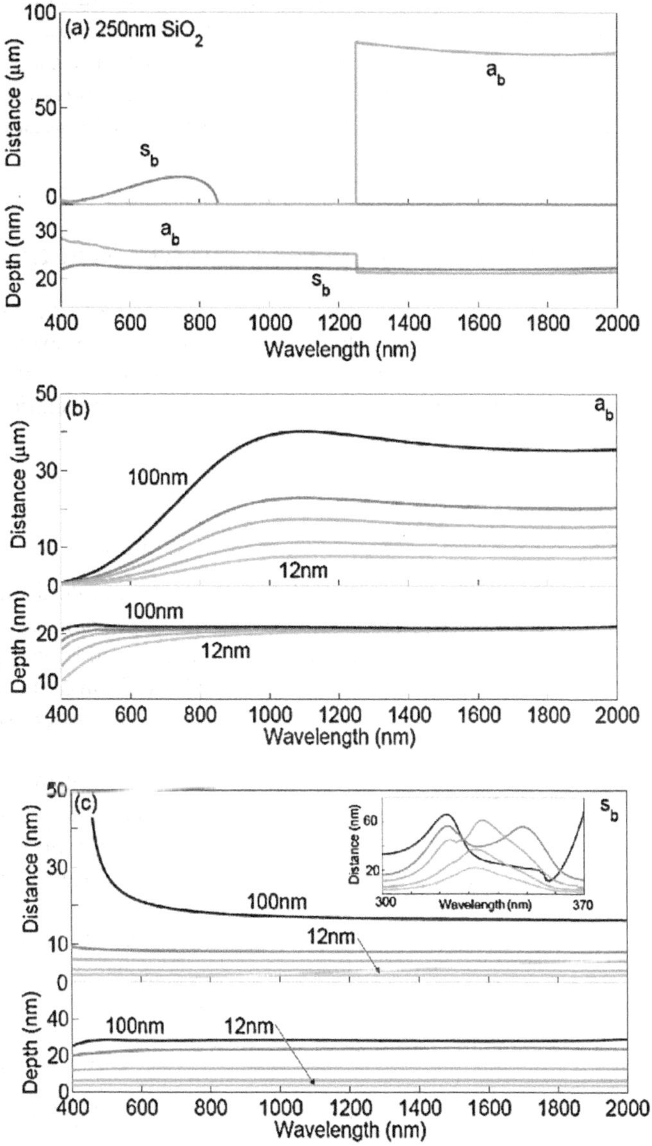

Figure 7.9. MIM (Ag/SiO2/Ag) propagation and skin depth plotted as a function of wavelength for core thicknesses of $d = 250$ nm (a) and $d = 12, 20, 35, 50$, and 100 nm (b, c). In panel (a), propagation lengths of conventional (as opposed to plasmonic) waveguiding modes are recovered and correlated with skin depth. In (b), the field anti-symmetric modes of MIM guides are seen to propagate over 10 microns with skin depth never exceeding 20 nm. In (c), the symmetric modes of thinner films ($d \leq 50$ nm) remain evanescent for all wavelengths. However, as d approaches 100 nm, conventional waveguiding modes can be accessed, and a region of enhanced propagation is observed for $\lambda \leq 400$ nm.Inset: Propagation distances of the symmetric mode for wavelengths characteristic of the quasi-bound regime. The dissociation of the thin-film single peak to the thick-film double peak indicates the onset of conventional waveguiding.

$d < 35$ nm, only a single peak is observed. However, as film thickness is increased, the peak begins to split with the lower energy peak forming the first band of allowed propagation. The transition indicates a dissociation of the quasi-bound modes and marks the onset of conventional waveguiding. While this regime is characterized by a slight increase in skin depth, field penetration for a given d remains generally constant over the full wavelength range. Thus, unlike IMI structures, extinction is determined not by ohmic losses but by field interference upon phase shifts induced by the metal. Whether MIM structures support propagating modes or purely evanescent fields, skin-depth is limited by absorption and will not exceed 30 nm.

7.5. SUMMARY

Today, photonic networks currently are comprised of largely dielectric materials and utilized the low index contrast between low loss dielectric media for signal storage and transmission. Conventional waveguides fabricated using this approach are now well understood and widely used in photonic networks for communications applications. In such networks, component size and in turn system compactness for complex systems is limited by diffraction usually to the micron or greater scale. However as this chapter has shown, another approach based on high index and high index contrast structures comprising metals and metallo-dielectric media may enable truly nanophotonic networks. Such structures are exemplified by the plasmon waveguides formed in nanoparticle chain networks and planar metal/insulator/metal structures discussed here, but future networks may employ a wide variety of topologically complex metallo-dielectric structures to optimize nanophotonic confinement as well as insertion and propagation loss characteristics of plasmon modes. Rapid advances in computer power for electromagnetic simulations coupled with rapid progress in nanofabrication technique development may ultimately enable a new class of nanophotonic networks whose critical and minimum dimension features are based on subwavelength plasmonic waveguides.

REFERENCES

1. H. Raether: *Surface Plasmons on Smooth and Rough Surfaces and on Gratings* (Springer-Verlag, 1988).
2. M. Quinten, A. Leitner, J.R. Krenn, F.R. Aussenegg: Electromagnetic energy transport via linear chains of silver nanoparticles, Opt. Lett. **23**, 1331 (1998)
3. M.L. Brongersma, J.W. Hartman, H.A. Atwater: Electromagnetic energy transfer and switching in nanoparticle chain arrays below the diffraction limit, Phys. Rev. B **62**, R16356 (2000).
4. B. Lamprecht, G. Schider, R.T. Lechner, H. Ditlbacher, J.R. Krenn, A. Leitner, F.R. Aussenegg: Metal nanoparticle gratings: influence of dipolar particle interaction on the plasmon resonance, Phys. Rev. Lett. **84**, 4721(2000).
5. S.A. Maier, M.L. Brongersma, P.G. Kik, S. Meltzer, A.A.G. Requicha, H.A. Atwater: Plasmonics—A route to nanoscale optical devices, Adv. Mater. **13**, 1501 (2001).
6. G. Mie: Articles on the optical characteristics of turbid tubes, especially colloidal metal solutions, Ann. Phys. **25**, 377 (1908).
7. U. Kreibig, M. Vollmer: *Optical Properties of Metal Clusters* (Springer-Verlag, Berlin, 1994).

8. C. Bohren, D. Huffman: *Absorption and Scattering of Light by Small Particles* (Wiley, New York, 1983).
9. S. Linden, J. Kuhl, H. Giessen: Controlling the interaction between light and gold nanoparticles: Selective suppression of extinction, Phys. Rev. Lett. **86**, 4688 (2001)
10. J.R. Krenn, A. Dereux, J.C. Weeber, E. Bourillot, Y. Lacroute, J.P. Goudonnet, G. Schider, W. Gotschy, A. Leitner, F.R. Aussenegg, C. Girard: Squeezing the optical near-field zone by plasmon coupling of metallic nanoparticles, Phys. Rev. Lett. **82**, 2590 (1999).
11. S.A. Maier, M.L. Brongersma, P.G. Kik, H.A. Atwater: Observation of near-field coupling in metal nanoparticle chains using far-field polarization spectroscopy, Phys. Rev. B **65**, 193408 (2002).
12. S.A. Maier, P.G. Kik, H.A. Atwater: Observation of coupled plasmon-polariton modes in Au nanoparticle chain waveguides of different lengths: Estimation of waveguide loss, Appl. Phys. Lett. **81**, 1714 (2002)
13. M.L. Brongersma, J.W. Hartman, and H.H. Atwater. Plasmonics: electromagnetic energy transfer and switching in nanoparticle chain-arrays below the diffraction limit. in *Molecular Electronics. Symposium*, 29 Nov.–2 Dec. 1999, Boston, MA, USA. 1999: Warrendale, PA, USA : Mater. Res. Soc, 2001, This reference contains the first occurrence of the word "Plasmonics" in the title, subject, or abstract in the Inspec® database.
14. S.A. Maier, P.G. Kik, H.A. Atwater: Optical pulse propagation in metal nanoparticle chain waveguides, Phys. Rev. B **67**, 205402 (2003)
15. D.R. Smith, N. Kroll: Negative refractive index in left-handed materials, Phys. Rev. Lett. **85**, 2933 (2000)
16. J.B. Pendry: Negative refraction makes a perfect lens, Phys. Rev. Lett. **85**, 3966 (2000)
17. S.A. Maier, P.G. Kik, H.A. Atwater, S. Meltzer, E. Harel, B.E. Koel, A.A.G. Requicha: Local detection of electromagnetic energy transport below the diffraction limit in metal nanoparticle plasmon waveguides, Nat. Mater. **2**, 229 (2003)
18. F.J. García-Vidal, J.B. Pendry: Collective theory for surface enhanced Raman scattering, Phys. Rev. Lett. **77**, 1163 (1996)
19. H. Xu, J. Aizpurua, M. Käll, P. Apell: Electromagnetic contributions to single-molecule sensitivity in surface-enhanced Raman scattering, Phys. Rev. E **62**, 4318 (2000).
20. A.D. McFarland, R.P. Van Duyne: Single silver nanoparticles as real-time optical sensors with zeptomole sensitivity, Nano Lett. **3**, 1057 (2003)
21. D.A. Genov, A.K. Sarychev, V.M. Shalaev, A. Wei: Resonant field enhancements from metal nanoparticle arrays, Nano Lett. **4**, 153 (2004).
22. Hache, D Ricard, C. Flytzanis: Optical nonlinearities of small metal particles: surface-mediated resonance and quantum size effects, J. Opt. Soc. Am. B **3**, 1647 (1986)
23. Y. Hamanaka, K. Fukata, A. Nakamura, L.M. Liz-Marzán, P. Mulvaney: Enhancement of third-order nonlinear optical susceptibilities in silica-capped Au nanoparticle films with very high concentrations, Appl. Phys. Lett **84**, 4938 (2004)
24. R.J. Gehr, R.W. Boyd: Optical properties of nanostructured optical materials, Chem. Mater. **8**, 1807 (1996)
25. Y. Shen, P.N. Prasad: Nanophotonics: a new multidisciplinary frontier, Appl. Phys. B **74**, 641 (2002)
26. D. Prot, D.B. Stout, J. Lafait, N. Pinçon, B. Palpant, S. Debrus: Local electric field enhancements and large third-order optical nonlinearity in nanocomposite materials, J. Opt. A **4**, S99 (2002).
27. J.J. Penninkhof, A. Polman, L.A. Sweatlock, S.A. Maier, H.A. Atwater, A.M. Vredenberg, B.J. Kooi: Mega-electron-volt ion beam induced anisotropic plasmon resonance of silver nanocrystals in glass, Appl. Phys. Lett. **83**, 4137 (2003)
28. L.A. Sweatlock, S.A. Maier, H.A. Atwater, J.J. Penninkhof, A. Polman: Highly confined electromagnetic fields in arrays of strongly coupled Ag nanoparticles, Phys. Rev. B **71**, 235408 (2005).
29. D. Sarid: Long-range surface-plasma waves on very thin metal films, Phys. Rev. Lett. **47**, 1927 (1981); A.E. Craig, G.A. Oldon, D. Sarid: Experimental observation of the long-range surface-plasmon polariton, Opt. Lett. **8**, 380 (1983)

30. J.J. Burque, G.I. Stegeman, T. Tamir: Excitation of surface-plasmon modes along thin metal-films, Phys. Rev. B **33**, 5186 (1985); P. Berini: Plasmon-polariton modes guided by a metal film of finite width, Opt. Lett. **24**, 15 (1999); P. Berini: Plasmon-polariton waves guided by thin lossy metal films of finite width: bound modes of asymmetric structures, Phys. Rev. B **61**, 15 (2000); P. Berini: Plasmon-polariton modes guided by a metal film of finite width bounded by different dielectrics, Opt. Express **7**, 10 (2000); P. Berini: Plasmon-polariton waves guided by thin lossy metal films of finite width: Bound modes of symmetric structures, Phys. Rev. B **63**, 12 (2001)
31. J.A. Dionne, L.A. Sweatlock, A. Polman, H.A. Atwater: Planar metal plasmon waveguides: frequency-dependent dispersion, propagation, localization, and loss beyond the free electron model Dionne, Phys. Rev. B **72**, 7 (2005); p. 075405
32. V. Almeida, Q. Xu, C. Barrios, M. Lipson: Guiding and confining light in void nanostructure, Opt. Lett. **29**, 1209 (2004).
33. E.N. Economou: Surface plasmons in thin films, Phys. Rev. **182**, 539 (1969).
34. R. Zia, M.D. Selker, P.B. Catrysse, M.L. Brongersma: Geometries and materials for subwavelength surface plasmon modes, J. Opt. Soc. Am. A **21**, 2442 (2004).
35. P.B. Johnson, R.W. Christy: Optical-constants of noble-metals, Phys. Rev. B **6**, 4370 (1972).
36. E. Palik, G. Ghosh: *Handbook of Optical Constants of Solids II* (Academic Press, Inc., New York, 1991).

CHAPTER EIGHT

OPTICAL SUPERLENS

X. ZHANG, M. AMBATI, N. FANG, H. LEE, Z. LIU, C. SUN AND Y. XIONG
Nanoscale Science and Engineering Center, 5130 Etcheverry Hall, University of California, Berkeley, CA 94720–1740, USA.

8.1. INTRODUCTION

The diffraction limit has long been a fundamental barrier for optical imaging. The ability to improve the resolving power of optical systems has attracted considerable interest. This ever-growing interest is due to the enormous potential benefit it offers in diverse fields such as bio-imaging, data storage, and lithography. Significant efforts have been made to enhance optical resolution. As an earliest effort to improve the resolution, contact mask imaging was proposed and demonstrated.[1,2] Immersion microscopy improves the resolution by increasing the refractive index of the surrounding medium; this method is limited by the availability of high index materials. Although scanning near-field optical microscopy (NSOM) provides subwavelength resolution, it does not project a whole image like a regular lens does. The optical information is collected by scanning a sharp tip in a point-by-point fashion near the surface which suffers from slow speed of serial scanning. It is often an "invasive" measurement that requires complicated post procedures for imaging reconstruction to remove the artifacts due to the tip-structure interaction. Recently, Pendry proposed an interesting "perfect lens theory"[3] in which a left-handed material (LHM) is used to obtain super-resolution well below the diffraction limit.

LHMs are classified as materials whose permittivity and permeability are simultaneously negative.[4] The phase velocity and the group velocity are in opposite directions in these LHMs, and the sign of the refractive index must be negative. Such negative-index materials are not readily available in nature; however metamaterials[5,6]—artificial materials—can be constructed in the form of periodic structures with size of the unit cell far less than the wavelength. These metamaterials are characterized by effective material properties—permittivity and permeability. These properties are highly dispersive, and they can turn negative simultaneously in certain frequency bands.[7] Lately, there has been a tremendous growth of interest in

M.L. Brongersma and P.G. Kik (eds.), Surface Plasmon Nanophotonics, 105–123.

the investigation of negative-index materials owing to their intriguing properties—reversed Doppler effect and negative Cerenkov radiation—and the exciting perfect lens theory of Pendry.[3]

Pendry proposed the theory of the perfect lens, in which a slab of negative-index material, under ideal conditions—loss free and perfect impedance matching to the surrounding media—acts as a perfect lens. This lens provides a perfect copy of an object by fetching both evanescent waves that carry subwavelength details of the object and propagating waves to the image plane. Evanescent waves that decay exponentially away from the object are significantly enhanced at the surfaces of the negative-index materials. This amplification of evanescent fields is attributed to the surface resonance modes that are present because of negative electromagnetic (EM) properties. The restrictions—loss free and negative index—for a perfect lens are rather stringent. In addition, there is no magnetic response in the natural materials at optical frequencies. Only very recently, metamaterials are fabricated that exhibit magnetic response at THz and infrared frequencies.[8,9] Significant efforts are underway to create artificial materials—metamaterials—that have negative responses to electric and magnetic fields at optical wavelengths. However, in the quasistatic limit—in the length scale much less than the wavelength—electric and magnetic fields decouple, and only one of the material properties has to be negative[3] to achieve the similar effect of a perfect lens. Negative electric permittivity is readily attainable in metals at optical frequencies; thus a thin slab of metal whose permittivity is equal and opposite to that of the surrounding media can be used as a "poor man's" perfect lens in the quasistatic limit. However, due to the losses in the metal film only a band of evanescent fields in the wave vector spectrum are enhanced. These enhanced electrostatic fields are used to construct an image with subwavelength resolution in the near field. Such a thin slab of metal that provides the resolution well below the diffraction limit is termed as a superlens.

The subject of this chapter is formulated in the following way. Section 8.2 discusses superlens theory and the imaging properties of the superlens. In section 8.3, experimental confirmation of thickness dependence on the enhanced transmission of evanescent waves through the silver superlens is presented, which is a key proposition in superlens theory. Finally, sect. 8.4 is devoted to the experimental study on optical superlens imaging beyond the diffraction limit. Detailed experimental design and fabrication procedures are presented. This chapter will extensively review work that has been done by many authors and their collaborators although many additional contributions to the development of the metamaterials and superlensing have been accomplished by various researchers.[10–13]

8.2. SUPERLENS THEORY AND IMAGING PROPERTIES

The performance of a conventional optical lens is limited by the diffraction limit. The rationale behind this limitation is that a conventional lens is only capable of transmitting the propagating components. The evanescent waves that carry subwavelength information about the object decay exponentially and cannot be collected in far field

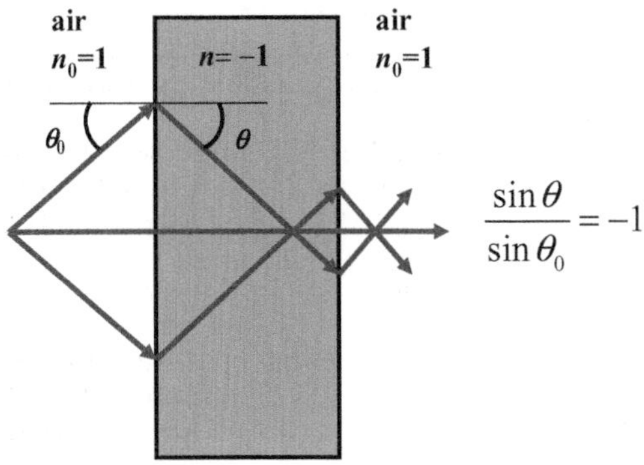

Figure 8.1. A slab of a negative refractive index ($n = -1$) material focuses the light diverging from a point source.

to reconstruct the image. These features can be explained by considering an infinitesimal dipole of frequency ω in front of a conventional lens.[3] The electric component of the field is given by Fourier expansion.

$$E(r,t) = \sum_{k_x,k_y} E\left(k_x, k_y\right) \exp\left(ik_z z + ik_x x + ik_y y - i\omega t\right) \tag{8.1}$$

$$k_z = \sqrt{\omega^2 c^{-2} - k_x^2 - k_y^2} \tag{8.2}$$

The axis of the lens is taken to be z-axis. For larger values of transverse wave vector that carries the finer details of the source,

$$k_z = +i\sqrt{k_x^2 + k_y^2 - \omega^2 c^{-2}}, \quad \omega^2 c^{-2} < k_x^2 + k_y^2 \tag{8.3}$$

These waves decay exponentially away from the source, and the information of these large transverse wave vectors is completely lost at the image plane. Only propagating wave information is restored at the image plane, and therefore the resolution is limited.

An unconventional alternative to the lens is proposed by Pendry.[3] Parallel sided slabs of negative index materials focus the light as shown in Fig. 8.1. In case of the negative refractive index medium, phase velocity is in the direction opposite to that of group velocity. An important characteristic of the negative-index medium is the double focusing effect shown in Fig. 8.1, which is obtained by satisfying Snell's law of refraction ($n = -1$, $\varepsilon = -1$ and $\mu = -1$). For the negative-index medium considered:

$$k_z' = -\sqrt{\omega^2 c^{-2} - k_x^2 - k_y^2} \tag{8.4}$$

A remarkable feature of the negative-index medium is that the evanescent waves which emerge from the far side of the negative-index medium are enhanced in amplitude by the transmission process as shown in Fig. 8.2. The transmission enhancement of evanescent waves from the object plane to image plane can be calculated by

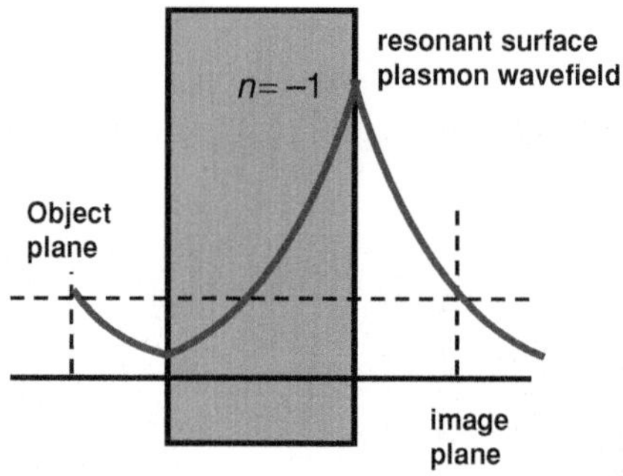

Figure 8.2. The restoration of the amplitude of an evanescent component is depicted schematically.

Fresnel equations,[3] accounting for multiple reflections. The degree of enhancement as a function of wave vector is termed as the transfer function. The characteristics of the transfer function determine the performance of the negative-index material. With negative-index materials, both propagating and evanescent waves contribute to the resolution of the image. Perfect reconstruction of an image is possible with a loss-free negative-index material, whose impedance exactly matches that of the surroundings.[3]

Negative-index metamaterials at optical frequencies have not been designed to date. So, an alternative to negative-index materials is a metal in the quasistatic limit. In such a case, a thin slab of metal whose electric permittivity is negative at optical frequencies can be used to enhance the transmission of the evanescent fields. This enhancement is assisted by the excitation of surface plasmons on the metal film. These surface plasmons are collective excitations of electron plasma at a metal surface coupled with photons.[14] The losses in the metal film and its thickness determine the band of evanescent fields in the wave vector spectrum that are enhanced. The detailed calculations are presented here. The transfer function is calculated using Fresnel equations, and this transfer function is used in calculating the imaging properties of a superlens. The system considered for these calculations is shown in Fig. 8.3, and the permeability is taken as '1' in all the media.

For a given transverse wave vector

$$k_{//} = \sqrt{k_x^2 + k_y^2} \tag{8.5}$$

one has $k_{zj} = \sqrt{\varepsilon_j \left(\frac{\omega}{c}\right)^2 - k_{//}^2}$ $(j = 1, 2)$ and $k_{zM} = i\sqrt{k_{//}^2 - \varepsilon_M \left(\frac{\omega}{c}\right)^2}$

Where ε_j is the dielectric constant in the media 1, $(j = 1)$, and media 2, $(j = 2)$ adjacent to the metal slabs.

The overall transmission coefficient across a superlens of thickness 'd' using Fresnel equations is:

$$T_p\left(k_{//}, d\right) = \frac{t_{1M} t_{M2} \exp\left(i k_{zM} d\right)}{1 + r_{1M} r_{M2} \exp\left(i 2 k_{zM} d\right)}, \tag{8.6}$$

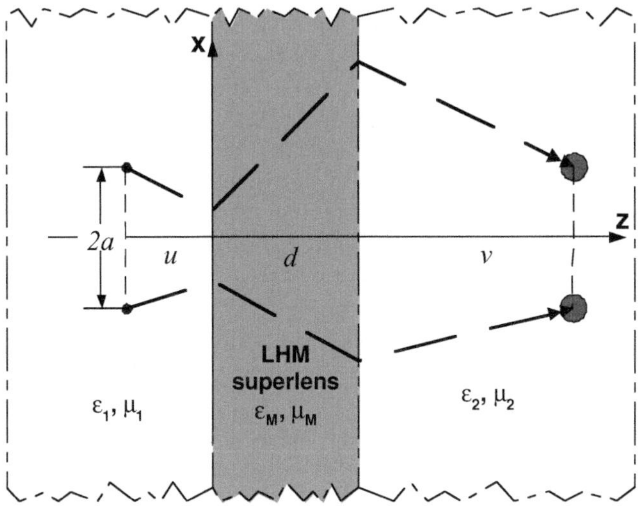

Figure 8.3. The model of a superlens imaging system under the radiation of two-line current sources.

where r_{1M} and r_{M2} are the reflection coefficients at the two interfaces,

$$r_{1M} = \frac{\frac{k_{z1}}{\varepsilon_1} - \frac{k_{zM}}{\varepsilon_M}}{\frac{k_{z1}}{\varepsilon_1} + \frac{k_{zM}}{\varepsilon_M}}, \quad r_{M2} = \frac{\frac{k_{zM}}{\varepsilon_M} - \frac{k_{z2}}{\varepsilon_2}}{\frac{k_{zM}}{\varepsilon_M} + \frac{k_{z2}}{\varepsilon_2}}, \tag{8.7}$$

and t_{1M} and t_{M2} are the corresponding transmission coefficients

$$t_{1M} = 1 + r_{1M}, \quad t_{M2} = 1 + r_{M2}. \tag{8.8}$$

The approximation of an exponential growth of the overall transmission T_p is valid when the following condition $|r_{1M} r_{M2}| \gg 1$ is satisfied, and the overall transmission in this case is:

$$T_p\left(k_{//}, d\right) = \frac{t_{1M} t_{M2}}{r_{1M} r_{M2}} \exp\left(-ik_{zM} d\right). \tag{8.9}$$

The condition for exponential growth of evanescent waves, $|r_{12} r_{23}| \gg 1$, which gives the condition that

$$\left(\frac{k_{z1}}{\varepsilon_1} + \frac{k_{zM}}{\varepsilon_M}\right)\left(\frac{k_{zM}}{\varepsilon_M} + \frac{k_{z2}}{\varepsilon_2}\right) \to 0. \tag{8.10}$$

This form is exactly the condition for surface plasmon resonance on either side of the slab.[15] Therefore surface plasmon conditions should be met to amplify the evanescent fields.

Superlenses are described theoretically by matching the electric permittivity of the slab with that of the surroundings in magnitude. An example of this configuration[3] shows considerable focusing. Enhancement of evanescent fields provides high contrast in superlens imaging; however, propagating waves are not focused at the image plane. The effect of loss, permittivity mismatch on the image quality and the depth of focus in the superlens are very important to investigate. Full-wave numerical calculations were performed to study these effects.[16] Figure 8.3 depicts the two-dimensional imaging system. The imaging quality of two monochromatic line current sources is considered. The sources are embedded in medium 1 with uniform and isotropic permittivity ε_1 and permeability μ_1, separated by distance $2a$; the separation from the current source to the slab is defined as the object distance, u. The EM field due to transverse magnetic (TM) sources $J(r) = \hat{z} I \delta(r - r')$, located at $r' = (x = \pm a, z = -u)$, travels through the superlens of thickness d with designed properties ε_M and μ_M, and reaches medium 2 where the images are formed at a distance $v$to the right-hand side of the lens.

As discussed before, the imaging quality of the superlens can be quantified using the transfer function. This optical transfer function (OTF) is defined as the ratio of image field to object field, $H_{\mathrm{img}}/H_{\mathrm{obj}}$, for different lateral wave vectors. The transfer function for a superlens is calculated by the Fresnel coefficients as in Eq. (8.6). The imaging properties are obtained by decomposing the incident field H_{obj} at $(-u < z < 0)$ into a superposition of transverse components with the help of the Weyl integral

$$H_{\mathrm{obj}}(x, -u < z < 0) = \frac{\nabla \times \hat{z}}{4\pi} \int_{-\infty}^{\infty} dk_x \frac{\exp(ik_x x + ik_{Z1}|z + u|)}{ik_{z1}} I(k_x, k_{z1}). \tag{8.11}$$

where $I(k_x, k_{z1})$ represents the Fourier transform of the line current source $I\delta(r - r')$, and $\beta_M = \sqrt{\varepsilon_M \mu_M \left(\frac{\omega}{c}\right)^2 - k_x^2}$.

The image field at the focal plane $z = d + v$ is the convolution of the source field and the OTF

$$H_{\mathrm{img}}(x, z = d + v) = \frac{\nabla \times \hat{z}}{4\pi} \int_{-\infty}^{\infty} dk_x \frac{\exp(ik_x x)}{ik_{z1}} I(k_x) \mathrm{OTF}(k_x). \tag{8.12}$$

The OTF for the system is given as:

$$\mathrm{OTF}(k_x) = T_p(k_x, d) \exp(ik_{z1}u) \exp(ik_{z2}v). \tag{8.13}$$

The sensitivity of image resolution depends on the material properties mismatch. The image collected at the paraxial focal plane with the original sources separated by $\lambda/6$ for a different set of parameters is shown in Fig. 8.4. In the numerical calculations, medium 1 is considered to be glass, $\varepsilon_1 = 2.368$, and medium 2 to be photoresist, $\varepsilon_2 = 2.79$. This is a slight asymmetric configuration with imag $(\varepsilon_M) = 0.4$ and $\mu_M = 1$. The simulation result of the average Poynting vector Real($\mathbf{S}(x, z)$) at the paraxial focal plane, $z = d + v$, is shown in Fig. 8.4. A resolution of $\lambda/6$ is obtained at *real*

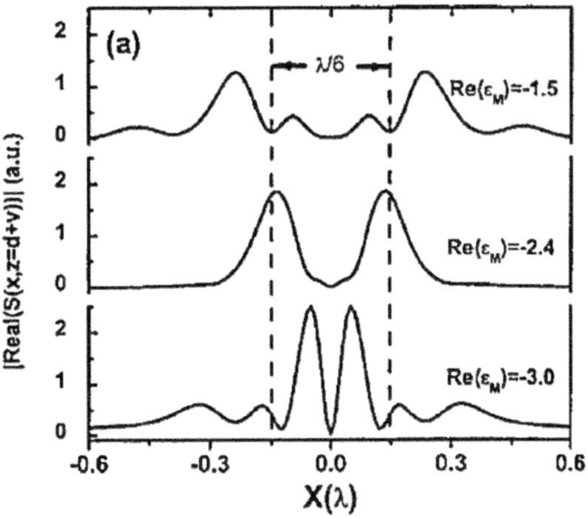

Figure 8.4. The image collected at the paraxial focal plane with the original sources separated by $\lambda/6$.

$(\varepsilon_M) = -2.4$ which corresponds to the 364 nm wavelength for silver. However, a compressed image is observed for *real* $(\varepsilon_M) = -3.0$. In contrast, an expanded image is obtained for *real* $(\varepsilon_M) = -1.5$. These phenomena can be attributed to the contribution of surface resonances, which detuned the lateral peak width and position.

The power density distribution for the case *real* $(\varepsilon_M) = -2.4$ is shown in Fig. 8.5. In contrast to the conventional imaging, the highest power flux is not observed at focal points. This effect is because of the decaying nature of evanescent waves; therefore, the field strength is much higher at the surface of the superlens. At the exit of the superlens, the enhanced evanescent field strength outweighs the contribution of propagating waves.

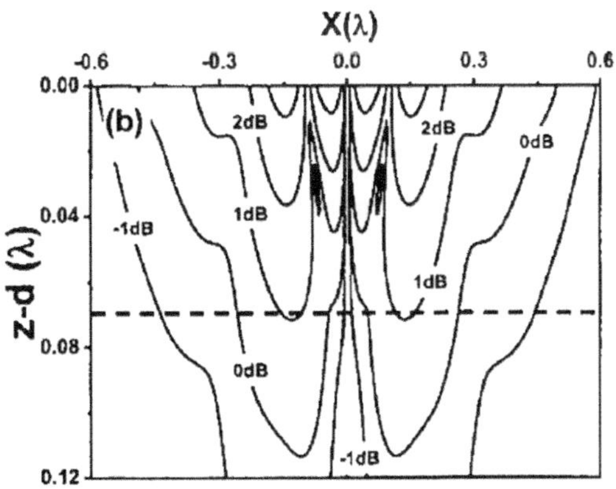

Figure 8.5. The logarithmic contour of power density in medium 2 for the case $2a = \lambda/6$. The dashed line corresponds to the paraxial focal plane.

8.3. TRANSMISSION ENHANCEMENT OF EVANESCENT WAVES

This section is a short review of the methods used for experimental investigation of the transmission enhancement of evanescent waves across the silver film—a key precursor of superlensing.[15,17,18] This enhancement is assisted by the excitation of surface plasmons on the silver film. These surface modes exist for the entire band of evanescent fields in the wave vector spectrum in the quasistatic limit, when the permittivity of metal and dielectric are of the same magnitude but opposite in sign. The loss in the metal film (silver has the lowest loss among the metals) limits the surface bound modes to a finite band of evanescent fields in the wave vector spectrum. The evanescent fields that carry subwavelength information of the object are resonantly enhanced at the surface of metal films, and the finite film thickness affects surface plasmon coupling and the enhancement of evanescent waves.

Experiments were performed to measure the transfer function of a thin silver slab superlens. The subwavelength surface roughness at the silver–air interface scatters the normal incident beam and produces EM waves of all possible transverse wave vectors, and these scattered waves provide the necessary momentum to excite surface plasmons at the same interface. It is important to note that the surface roughness acts as a broad band random grating to excite surface plasmons instead of evanescent fields from the near-field objects. Surface plasmons are characterized by the wave vector k_{sp} exceeding the wave vector of propagating photons of the same frequency in vacuum ($k = \omega/c$) and, therefore, cannot be excited directly by light on a perfectly smooth surface. An asymmetric configuration, where a thin slab of silver is sandwiched between air and glass, is considered. The evanescent waves scattered by the surface roughness as shown in Fig. 8.6(b) permeate the silver film. And when $\boldsymbol{k}_{//} < 2\pi n_{\mathrm{p}}/\lambda$, ($n_p$ is the refractive index of the prism), the evanescent waves are converted back to propagating waves. Therefore, the transmissivity of each evanescent mode across the silver slab can be measured in far field. The experimental scheme is presented in Fig. 8.6(a); Fig. 8.7(c) is image captured by placing a screen at far field, displaying a double crescent ring, with the center of the direct transmitted beam blocked by a circular disk.

In the experiments, a spectrum of surface roughness $/s(k_x)/$ is obtained with all k_x of interest by an accurate characterization with an atomic force microscope. This procedure helps in utilizing the random surface roughness of metal films as a natural grating with precisely determined Fourier components used for coupling light waves into the metal films. Silver films with thickness ranging from 30 to 90 nm were deposited onto BK7 glass hemispheres using an e-beam evaporator. The film thickness is chosen to be larger than 30 nm, ensuring that the bulk optical properties of silver can be applied in the calculations.[19] Cross-sectioning transmission electron microscopy (TEM) on a 50 nm thick silver film shows voids and volumetric cracks, which confirms that the contribution from bulk inhomogeneities in silver to scattered intensity are insignificant. In order to measure the relative transmission enhancement of the evanescent waves, a reversed attenuated total reflection (RATR) setup is used.[20] This setup includes a collimated Ar^{+} ion laser beam

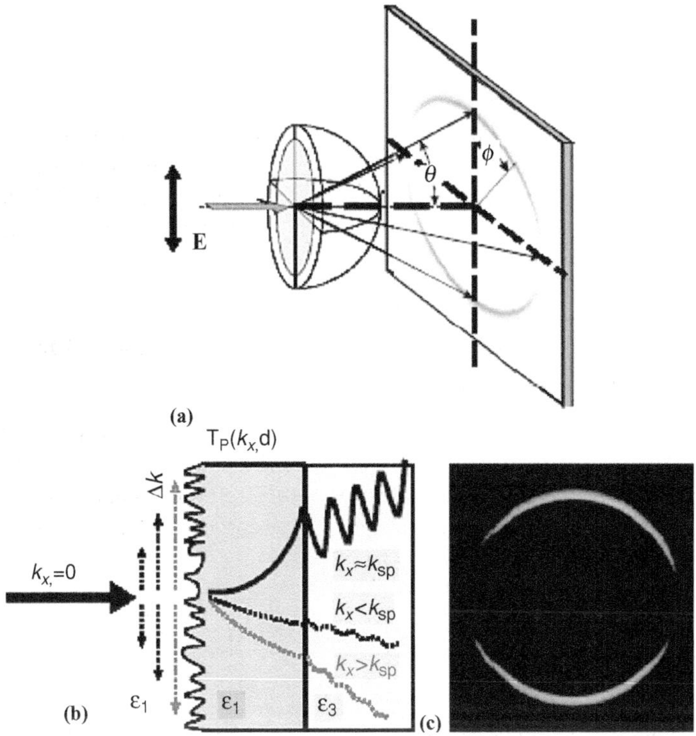

Figure 8.6. (a) Reversed attenuated total reflection setup for measuring the relative transmittivity of evanescent waves. (b) The assumed evanescent coupling by surface roughness scattering, and the schematic of the enhancement of evanescent fields (c) The scattered ring pattern observed at far field.

with wavelength λ and diameter < 1.5 mm, which illuminates the sample surface from the air side with normal incidence. A charge-coupled device (CCD) camera is placed about 5 cm away from the center of a BK7 glass prism to measure the relative light intensity along the azimuthal angle θ, while φ is centered at zero. The input laser power, I_0, is adjusted so that the power of transmitted light cone falls in the linear dynamic range of the CCD camera. The intensity profiles captured by CCD are averaged in $-3^\circ < \varphi < 3^\circ$ direction to further enhance intensity resolution. The angular resolution and repeatability in the setup is calibrated to within 0.1°.

The far-field light intensity dI per solid angle element $d\Omega$ is normalized by incident intensity I_0 and can be written as:[21,22]

$$\frac{dI}{I_0 d\Omega} = 4\left(\frac{\pi}{\lambda}\right)^4 \left|T_p\left(k_x, d\right)\right|^2 \left|s\left(k_x\right)\right|^2 \left|W\left(\theta, \varphi\right)\right|^2, \tag{8.14}$$

with T_p being the p-polarization transmission coefficient through the silver film, λ the incident wavelength, $/s(kx)/$ the roughness Fourier spectrum of the silver/air interface,

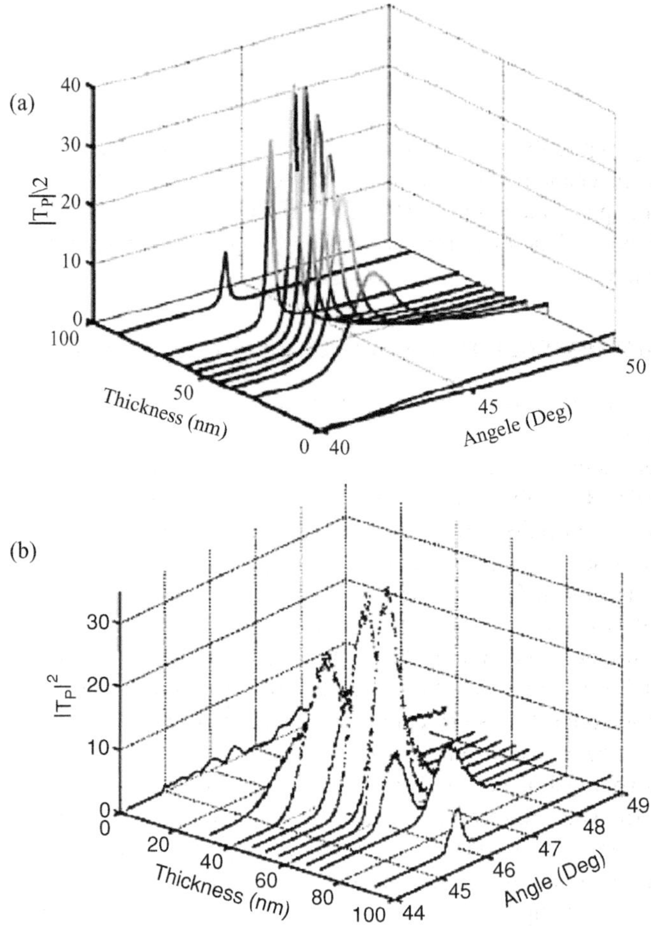

Figure 8.7. The computed (a) and measured (b) relative power transmissivity, as a function of azimuthal angle and silver film thickness. Incident wavelength = 514.5 nm.

and $W(\theta, \varphi)$ the dipole function. The above equation assumes the scattering occurs only once at the air/metal side; the roughness at the metal/prism interface and the scattering inside metal films do not contribute to the collection of evanescent waves. It is necessary to obtain the appropriate dipole function in addition to the surface roughness spectrum. The dipole function is calculated using theoretical values,[23] with the dielectric properties of silver and BK7 glass.[24,25] The measured relative transmittivity, $|T_p(k_x, d)|^2$ can be extracted from Eq. (8.15), and they are plotted in Fig. 8.6(b). In comparison, in Fig. 8.7(a), the theoretical transmissivity calculated from Eq. 8.6 as a function of silver thickness are plotted with the dielectric properties of silver and BK7 glass. In Figs. 8.7(a) and (b) one can see the good agreement of the peak shapes and strengths.

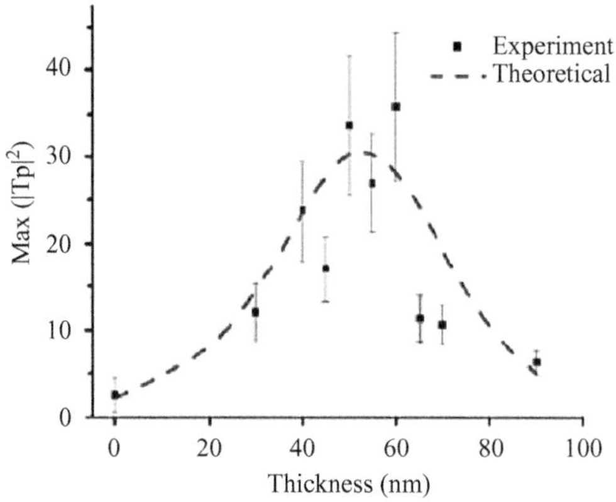

Figure 8.8. Measured peak field enhancement vs. the thickness of silver films at $\lambda = 514.5$ nm. The dashed line is calculated by the Fresnel equations.

The enhanced transmission of the evanescent components occurs only in a finite band centered at a surface plasmon wave vector $k_x = k_{sp}\sqrt{\varepsilon(\omega)/(1+\varepsilon(\omega))}$ known in metal optics and surface sciences. At this wave vector and frequency, the photon impinging on the metal surface can excite a collective oscillation of conduction electrons, that is, surface plasmon with maximum efficiency.

The peak enhancement is plotted as a function of sample thickness in Fig. 8.8. The enhancement factor $|T_p|^2$ rapidly grows with increasing silver thickness up to 50 nm. Above 50 nm, the enhancement is sharply suppressed due to the intrinsic loss inside the silver film,[17] so that the transmissivity decays as film thickness further increases. Experimental results provide direct evidence of the thickness dependence of enhanced evanescent field across the silver film, a key proposition in Pendry's superlens theory.

The transmittivity bandwidth broadens when the surface plasmon excitation frequency approaches resonance frequency. This broadening opens the avenue to access the subwavelength features of a near-field object. In order to obtain the sub-diffraction limited resolution, the object distance and its image distance from the interface of the superlens, and the thickness of the superlens must all be subwavelength. Otherwise, the evanescent waves from the object decay to the extent that their recovery becomes impractical from material losses and other material imperfections of the lens.

8.4. EXPERIMENTAL DEMONSTRATION OF THE SUPERLENS

After the confirmation of the key proposition of superlens theory in the last section, we now focus on the experimental demonstration of optical superlensing through a thin silver slab with $\lambda/6$ resolution. In addition to the detailed experimental design and fabrication procedures, simulation studies of silver superlens characteristics are also included in this section.[26,27]

In this study, the superlens is in the form of a thin slab of metal and subwavelength features imaged through the superlens are recorded on the opposite side in photo resist. The key to the experimental design is to maximize the enhancement of evanescent waves. Silver is chosen as the superlens material for its low absorption loss in the optical region, the selection of illumination wavelength is critical to the superlens design. The permittivity of the surrounding dielectric material should be equal and opposite in sign to that of silver for a wide band surface plasmon excitation. The wavelength of 365 nm is used, the permittivity of silver (ε_m) at this wavelength is $\varepsilon_m = -2.4012+i0.2488$. Polymethyl-metacrylate (PMMA) ($\varepsilon_1 = 2.30$) was used as a spacing layer between the superlens and the object. A commercial I-line negative Photoresist (PR) [NFR 105G, Japan Synthetic Rubber Microelectronics (JSR Micro), $\varepsilon_2 = 2.886 + 0.059i$] was directly spun on the other side for image recording. It was shown in the previous section that a slight asymmetric configuration still supports efficient coupling of broad band surface plasmon resonance of the two surfaces of the silver film.[16]

The enhancement of the evanescent field depends strongly on the thickness of the superlens, silver slab, and PMMA layer, which determines object distance from the superlens. Thicker silver films do not ensure an efficient plasmon coupling due to damping and result in a diffraction limited image.[28] Thinner PMMA shows larger enhancements, however, the use of a PMMA thickness of 40 nm in the superlensing experiment came from current fabrication limitations, as discussed later in the section. OTFs are obtained for various thicknesses of silver as shown in Fig. 8.8; such calculations are made by considering infinitely thick PR. The silver slab of 35 nm thick gives the optimum transfer function with resolution limit up to $4k_0$. Thinner sliver slabs, 15 nm and 25 nm, show higher but narrower enhancement bands. Thicker slabs with 45 nm and 55 nm show smaller enhancements compared to zero order transmission, which result in low image contrast of sub-diffraction limited scale features.

Transmissivity dependence for one particular wave number, $2\pi/120$ nm, with varying silver thickness is shown in Fig. 8.9(c). The optimum silver thickness of 35 nm exhibits maximum enhancement. Finally, Fig. 8.9(d) shows polarization dependence of the evanescent field enhancement through the designed structure, PMMA (40 nm)/silver (35 nm)/PR (thick). The solid curve shows enhanced transmission through the superlens with TM incident waves and the dashed curve is that of transverse electric (TE) waves where there is no enhancement. Also shown is the transfer function of a control sample, where silver is replaced with a dielectric PMMA of same thickness. As expected, the evanescent wave is strongly attenuated without a silver layer even with TM incident light (dashed line). The incident light scattered by an object consists of all possible polarizations; only TM component fields can excite surface plasmons on the silver surface, which can lead to sub-diffraction limited imaging.

The sample structure to demonstrate optical superlensing is shown in Fig. 8.10. A subwavelength object—a Cr nanowire grating with a subwavelength period (120 nm)—is located on one side of the silver slab. Upon normal incidence, the subwavelength features of the object generate evanescent fields. These evanescent fields of TM-polarized nature excite surface plasmons at the silver/PMMA interface. The evanescent field is then enhanced through the silver layer and the field of the object

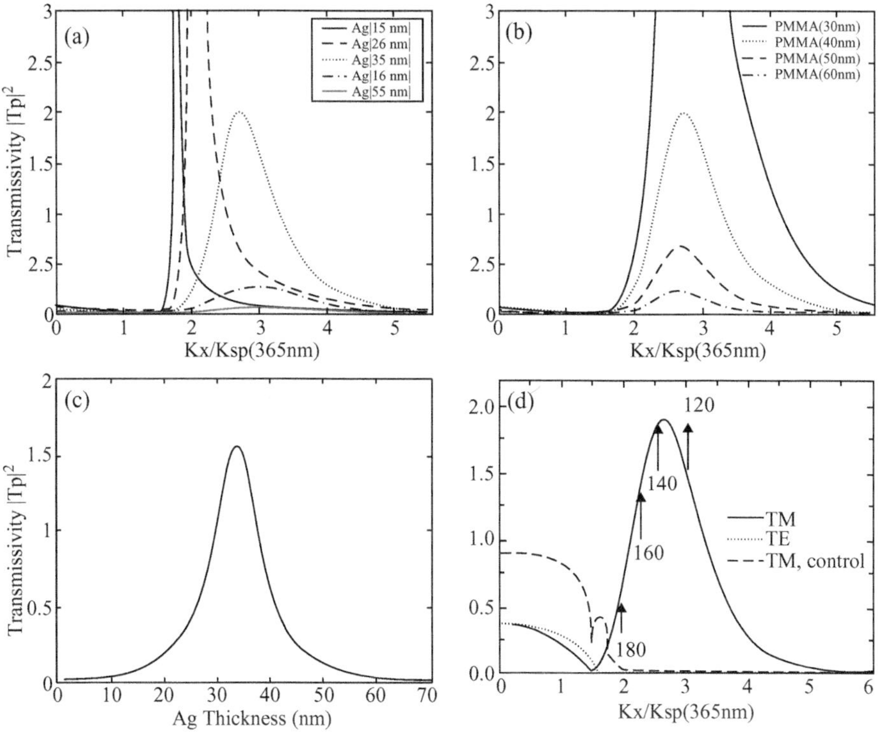

Figure 8.9. Transmissivity curves for superlens structures with (a) fixed PMMA (40 nm) and varying Ag (15 nm–55 nm), (b) fixed Ag (35 nm) and varying PMMA (30–60 nm). (c) Transmissivity of wave number $2\pi/120$ nm through fixed PMMA (40 nm) and varying Ag (0–70 nm). (d) Polarization-dependant transmissivity.

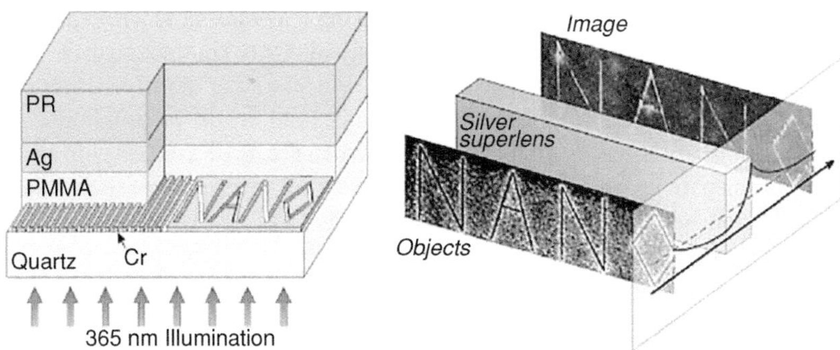

Figure 8.10. Optical superlensing experiment. The embedded objects are inscribed onto the 50 nm thick chrome (Cr) layer. (a) An array of 60 nm wide slots of 120 nm pitch, separated from the 35 nm thick silver film by a 40 nm PMMA spacer layer. (b) A schematic of the enhancement of evanescent waves in imaging the word "NANO".

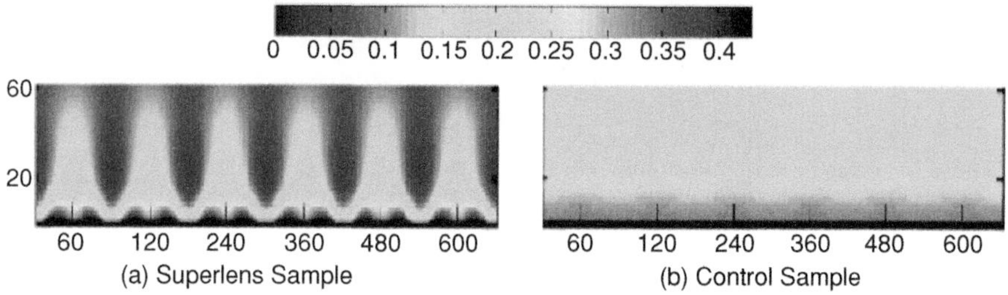

Figure 8.11. E-field intensity profile in photoresist of (a) superlens and (b) control sample. Six periods are simulated with periodic boundary conditions. The superlens sample restores the field of the object while the control sample shows no intensity contrast inside the photoresist.

is restored on the other side of the silver as an image, which is recorded by photoresist. While it is possible to image a grating object by restoring a narrow band of Fourier components, it is necessary to excite a wide band of large wave numbers in order to image an arbitrary object with sub-diffraction-limited resolution. An arbitrary line object, "NANO", with line width of 40 nm, is also to be imaged in this experiment which consists of broad band Fourier components. From Fig. 8.9(a), it can be determined that wave numbers in the range, $2k_0$–$4k_0$, can be transmitted through the superlens.[26]

Although numerical calculations using Fresnel equations provide valuable information to examine the characteristic behavior of the field transferring through the superlens structure, multiple reflections that might occur in each interface and the material property of the object (Cr) have not yet been considered. Therefore, a full vector simulation of the designed superlens structure is performed. A strong lateral field confinement in PR within ~60 nm (Fig. 8.11(a)) is observed, indicating the restoration of the evanescent field of the object. This simulation proves that the effect of the nature of the object on the performance of superlens is negligible. In the control structure, a 35 nm silver layer was replaced by a 35 nm PMMA layer, totaling the PMMA thickness to 75 nm. It is designed to prove that the enhancement of evanescent waves and subsequent optical superlensing is not possible without the silver. As shown in Fig. 8.11(b), there exists hardly any intensity contrast near the PR/PMMA interface. These two results provide strong evidences that the silver layer can indeed act as a superlens that is capable of imaging beyond the diffraction limit.

The fabrication process of the sample shown in Fig. 8.10(a) starts with an ultraviolet (UV) transparent quartz wafer. A 50 nm thick Cr layer is deposited using e-beam evaporation, and the objects are fabricated on the Cr film by Focused Ion Beam (Strata, FEI Company). Line gratings as well as a two-dimensional line pattern, "NANO" were written with subwavelength line width of ~40 nm. The periods of grating objects range from 120 nm to 180 nm in 20 nm steps with a 60 nm line width. Then a 40 nm thick planarized spacing layer is applied on the object using PMMA followed by the deposition of a 35 nm silver layer by e-beam evaporation. The silver surface is coated with negative PR, NFR105G, and is exposed from the substrate side to record the image reconstructed by the superlens.

The planarization of the spacing layer is a crucial step in the sample fabrication because imperfect surface conditions of the silver layer will modify its dispersion relation, which limits the final resolution.[29] The PMMA spacing layer needs to eliminate the surface modulation from the Cr and its surface needs to be smooth enough to minimize the surface roughness of silver down to 1 nm root mean square (RMS). According the superlens structure design discussed earlier, it is desirable to have the spacing layer as thin as possible. However, it is found that the film quality degrades when thinner than 40 nm. Such a thin layer with desirable flatness over 50 nm deep grating grooves is not achievable with a single spin coating process because the polymer layer tends to follow the profile of the surface topology.[30] In order to achieve a 40 nm thick planarized and smooth PMMA layer, multiple spin coatings of PMMA are used toproduce a thick layer ($\sim$0.7 μm) to eliminate surface irregularities. The peak-to-valley depth is gradually reduced after each coat, and after several coatings, a flat surface is achieved. A sufficiently long ($\sim$10 min, 180°C hot plate) bake is essential between each coating to remove the solvent from the polymer and harden the layer. The desired thickness is achieved through oxygen plasma blanket etch. A commercial photoresist asher (Tegal) is used to etch PMMA down to 40 nm. Because of the non-linear etch rate in the plasma, etch-and-measure step is repeated until 40 nm a thickness is reached. Finally, a reflow bake smoothes out the surface which was roughened by plasma bombardment during the blanket etch. Reflow effectiveness depends on the PMMA molecular weight and flow temperature. In this experiment, 495 PMMA is used and reflowed for 30 min at 180–200°C on a hotplate. After the planarization, less than 1 nm modulation and angstrom-scale ($\sim$0.5 nm) RMS roughness was achieved. Next, a 35 nm silver film was deposited on the spacing layer using electron beam evaporation. The surface quality of the film has to be controlled in order to maximize the image quality.[27] The average roughness analysis showed improvements from 3–4 nm RMS at a lower rate to 1–2 nm RMS at higher deposition rates. This improved surface quality avoids additional surface scattering, and thus improves the performance of superlens.

A negative I-line photoresist NFR 105G was used to record the image reconstructed by the superlens. The diluted photoresist was spin-coated to approximately 120–150 nm thickness and after 1 min bake at 100°C, the sample was exposed from the substrate side with 365 nm UV light on a commercial Karl Suss MA6 aligner. After a post-exposure bake of 1 min at 100°C the image was developed. Sub-diffraction limited imaging was expected because of enhancement of the evanescent fields. A control experiment was also carried out by replacing the silver layer with another layer of 35 nm thick PMMA. Significant decay of the evanescent field was expected without the silver superlens. After the development and hard bake of the resist ($\sim$1 min at 100°C), topography of the resist was imaged by Atomic Force Microscopy (AFM).

The superlens imaging results show that the nano-wire object with 120 nm period is clearly resolved (Fig. 8.12 (a)). The height modulation of the recorded image is observed in a cross-sion plot (Fig. 8.12(b)) and the Fourier analysis, showing sharp peaks at 120 nm (Fig. 8.12(c)) further confirms that the imaged period is indeed that of the object. These results indicate that sub-diffraction-limited imaging with half-pitch resolution down to 60 nm ($\lambda/6$) can be obtained using the silver superlens.

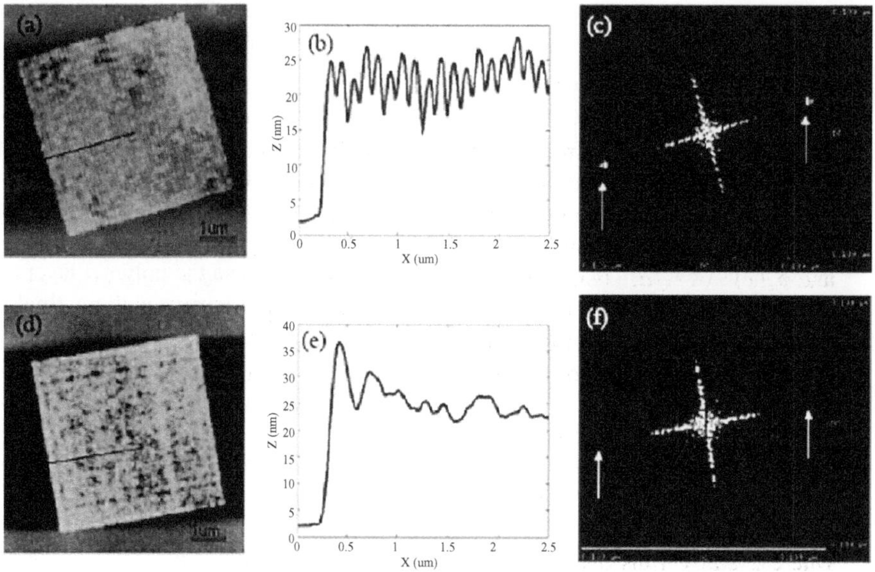

Figure 8.12. (a) AFM image of 120 nm period grating object recorded by PR in a superlensing experiment; (b) its average cross section perpendicular to the black line; and (c) 2-D Fourier analysis plot. (d) AFM image recorded by PR in a control experiment; (e) its average cross section and (f) 2-D Fourier plot.

The cross-section profile of the image shows the average peak-to-valley depth is 7 nm (Fig. 8.12(b)), while the entire height of the developed feature is ~30 nm. This difference has been expected from the simulation result shown in Fig. 8.11(a). A relatively stronger background signal is shown close to the silver surface and the image is recorded on top of the residual layer with smaller modulations than the entire exposed height. The reconstructed image on the silver surface is not perfect; the non-uniformity is due to lack of restoration of higher Fourier components, and also the non-ideal chemical process of photolithography. The line widening of the resist is another possible cause as negative resist tends to swell during the baking and development process. Surface scattering on silver surfaces is also a non-negligible source of noise.

The control experiment result supports the role of silver as a superlens. There is no evidence of imaging contrast without the silver as shown in Fig. 8.12(d). The cross section (Fig. 8.12(e)) and Fourier spectrum analysis (Fig. 8.12(f)) also show no evidence of periodic structure imaged by the control sample. As predicted in the numerical simulations, the evanescent wave generated by the object decayed significantly in a 75 nm thick PMMA layer. The decay length Z can be estimated by: $Z^{-1} = 4\pi\sqrt{a^{-2} - \varepsilon\lambda^{-2}}$ where, a is the grating period, ε is the permittivity of surrounding media, and λ is the incident wavelength. For ε ~2.4 for PMMA, a 120 nm period evanescent wave will decay to approximately 1/3 of the original field amplitude within 11 nm above the object. Because of this fast decay rate and with no means to enhance the field,

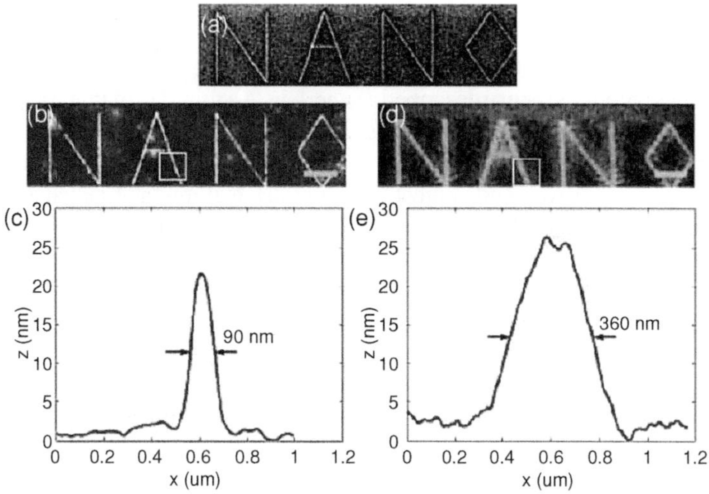

Figure 8.13. A 2-D arbitrary object imaging result. (a) FIB (Focused Ion Beam) image of the Cr line object "NANO" after fabrication. (b) Superlensing image recorded by PR (scale bar—2 μm) and (c) its cross-sectional profile. (d) Control imaging result of the same object (scale bar—2 μm) and (e) its average line cross section.

a significant decay occurs at 75 nm away from the object. This experiment confirms that the silver layer indeed acts as a superlens that enhances the evanescent field and images with sub-diffraction-limited resolution.

In order to verify that the result in Figs. 8.12(a)–(c) indeed resulted from the predicted transfer function of the superlens (Fig. 8.9(d)), several more objects with different periods have been imaged under the same exposure conditions. The wave numbers represented by different periodic objects are indicated by arrows in the transfer function curve in Fig. 8.9(d), which shows that for larger periods, the transmission intensity decreases. The imaging experiment results show good agreement. A 140 nm period image was recorded also with good fidelity, while, for 160 nm and 180 nm period objects, the image contrasts become lower. This is additional evidence of direct imaging by silver superlens with its own OTF which is unique.

In addition to the periodic gratings, an arbitrary, subwavelength structure, "NANO" was also imaged. The grating objects represent a very narrow band of wave numbers, but an object such as the word "NANO" has broad band Fourier components. The image of the object (Fig. 8.13(a)) captured by the superlens (Fig. 8.13 (b)) clearly shows far better resolution with a line width of ~90 nm (Fig. 8.13(c)) while that of the control experiment (Fig. 8.13(d)) still resulted in a diffraction-limited image with a line width of 360 nm (Fig. 8.13(e))—even with extended development times—which is comparable to the exposure wavelength (365 nm). Imaging of such arbitrary patterns proves that large wave vectors are transmitted through the superlens, enabling sub-diffraction-limited imaging.

8.5. CONCLUSIONS AND REMARKS

This chapter reviewed both the theoretical and experimental investigations of the superlens. Pendry's theory described a superlens—a thin slab of metal at the quasistatic limit—which is capable of enhancing a broad band of evanescent waves in the wave vector spectrum. Such enhanced evanescent waves can be used to reconstruct the image of a subwavelength object. The unique features associated with superlens imaging—compressed and expanded images—can be attributed to the characteristics of a superlens transfer function at different surface excitation conditions. Experiments were designed to obtain the transfer function of a silver superlens that confirmed the enhancement of evanescent waves, which is the key proposition of the superlens theory. This evanescent wave enhancement is largely due to the surface excitations or more specifically, plasmons, on the lens, and the degree of enhancement depends on the thickness of silver superlens and the loss associated with it. Finally an experimental demonstration of a silver superlens was presented with imaging resolution of 60 nm or $\lambda/6$, breaking the optical diffraction limit. Recent breakthroughs in this field are very encouraging and are just in the beginning stages. Truly, the far-field superlens remains a grand challenge. The concept of the superlens will have a significant impact in science and technology. The optical superlens has tremendous potential in ultra-high resolution imaging, high-density memory storage devices, and nanolithography.

ACKNOWLEDGEMENT

Support in preparation of this manuscript is provided by Multidisciplinary University Research Initiative (MURI) on Plasmonics (contract # FA9550-04-1-0434) funded by Air Force Office of Scientific Research (AFOSR).

REFERENCES

1. U.C. Fischer, H.P. Zingsheim: Sub-microscopic pattern replication with visible-light, J. Vac. Sci. Technol. **19** (4), 881–885 (1981).
2. H.I. Smith: Fabrication techniques for surface-acoustic-wave and thin-film optical devices, Proc. IEEE**62** (10), 1361–1387 (1974).
3. J.B. Pendry: Negative refraction makes a perfect lens, Phys. Rev. Lett. **85** (18), 3966–3969 (2000).
4. V.G. Veselago: Electrodynamics of substances with simultaneously negative values of sigma and mu, Soviet Phys. Uspekhi-USSR **10** (4), 509 (1968).
5. J.B. Pendry, A.J. Holden, W.J. Stewart, I. Youngs: Extremely low frequency plasmons in metallic mesostructures, Phys. Rev. Lett. **76** (25), 4773–4776 (1996).
6. J.B. Pendry, A.J. Holden, D.J. Robbins, W.J. Stewart: Magnetism from conductors and enhanced non-linear phenomena, IEEE Trans. Microwave Theory Tech. **47** (11), 2075–2084 (1999).
7. R.A. Shelby, D.R. Smith, S. Schultz: Experimental verification of a negative index of refraction, Science **292** (5514), 77–79 (2001).
8. T.J. Yen, W.J. Padilla, N. Fang, D.C. Vier, D.R. Smith, J.B. Pendry, D.N. Basov, X. Zhang: Terahertz magnetic response from artificial materials, Science **303** (5663), 1494–1496 (2004).
9. S. Linden, C. Enkrich, M. Wegener, J.F. Zhou, T. Koschny, C.M. Soukoulis: Magnetic response of metamaterials at 100 terahertz, Science **306** (5700), 1351–1353 (2004).

10. D.R. Smith, J.B. Pendry, M.C.K. Wiltshire: Metamaterials and negative refractive index, Science **305** (5685), 788–792 (2004).
11. S. Anantha Ramakrishna: Physics of negative refractive index materials, Rep. Prog. Phys. **68** (2), 449–521 (2005).
12. A. Grbic, G.V. Eleftheriades: Overcoming the diffraction limit with a planar left-handed transmission-line lens, Phys. Rev. Lett. **92** (11), 117403 (2004).
13. P.V. Parimi, W.T. Lu, P. Vodo, S. Sridhar: Photonic crystals—Imaging by flat lens using negative refraction, Nature **426** (4965), 404 (2003).
14. H. Raether: *Surface Plasmons* (Springer, Berlin, 1988).
15. N. Fang, Z.W. Liu, T.J. Yen, X. Zhang, Regenerating evanescent waves from a silver superlens, Opt. Express **11** (7), 682–687 (2003).
16. N. Fang, X. Zhang: Imaging properties of a metamaterial superlens, Appl. Phys. Lett. **82** (2), 161–163 (2003).
17. Z. Liu, N. Fang, T.J. Yen, X. Zhang: Rapid growth of evanescent wave by a silver superlens, Appl. Phys. Lett. **83** (25) 5184–5186 (2003).
18. N. Fang, Z. Liu, T.J. Yen, X. Zhang: Experimental study of transmission enhancement of evanescent waves through silver films assisted by surface plasmon excitation, Appl. Phys. A **80**, 1315–1325 (2005).
19. S. Heavens: *Optical Properties of Thin Solid Films* (Dover, Mineola, New York, 1991).
20. S. Hayashi, T. Kume, T. Amano, K. Yamamoto: A new method of surface plasmon excitation mediated by metallic nanoparticles, Jpn. J. Appl. Phys. **35** L331–L334 (1996).
21. E. Kretschmann: Determination of surface-roughness of thin-films using measurement of angular-dependence of scattered light from surface plasma-waves, Opt. Commun. **10** (4) 353–356 (1974).
22. H.J. Simon, J.K. Guha: Directional surface-plasmon scattering from silver films, Opt. Commun. **18** (3), 391–394 (1976).
23. R.W. Alexander, G.S. Kovener, R.J: Bell: Dispersion curves for surface electromagnetic-waves with damping, Phys. Rev. Lett. **32** (4), 154–157 (1974).
24. P.B. Johnson, R.W. Christy: Optical-constants of noble-metals, Phys. Rev. B **6** (12), 4370–4379 (1972)
25. Oriel Instruments: *The Book of Photon Tools*, Chapt. 15, (2002).
26. N. Fang, H. Lee, C. Sun, X. Zhang: Sub-diffraction-limited optical imaging with a silver superlens, Science **308** (5721), 534–537 (2005).
27. H. Lee, Y. Xiong, N. Fang, W. Srituravanich, M. Ambati, C. Sun, X. Zhang: Realization of optical superlens imaging below the diffraction limit, New J. Phys. **7**, 1–16 (2005)
28. D.O.S. Melville, R.J, Blaikie, C.R. Wolf: Submicron imaging with a planar silver lens. Appl. Phys. Lett. **84** (22), 4403–4405 (2004).
29. D.R. Smith, D. Schurig, M. Rosenbluth, S. Schultz, S.A. Ramakrishna, J.B. Pendry: Limitations on subdiffraction imaging with a negative refractive index slab, Appl. Phys. Lett. **82** (10), 1506–1508 (2003).
30. L.E. Stillwagon, R.G. Larson: Leveling of thin-films over uneven substrates during spin coating, Phys. Fluids A-Fluid Dynam. **2** (11), 1937–1944 (1990).

CHAPTER NINE

OPTICAL FIELD ENHANCEMENT WITH PLASMON RESONANT BOWTIE NANOANTENNAS

G.S. KINO, ARVIND SUNDARAMURTHY, P.J. SCHUCK, D.P. FROMM AND W.E. MOERNER
Stanford University, Stanford, California, USA

9.1. INTRODUCTION

When light of wavelength λ is focused into a medium of refractive index n, the minimum spot size due to diffraction is on the order of $\lambda/(2n)$. For example, at a wavelength of 405 nm, at the edge of the visible band, diffraction limits the minimum spot size to be greater than 100 nm. Near-field techniques based on the idea that light can be passed through a tapered metal-covered optical fiber, which acts as a cut-off waveguide when the guide diameter is less than $\lambda/(2n)$, have made it possible to obtain spot sizes on the order of 50 nm with power transmission on the order of 10^{-3} to 10^{-6} of the incident power.[1] Similarly, passing light through a small pinhole on the order of 50 nm diameter leads to a field intensity at the end of the guide which is greatly reduced from that of the incident field.

An alternative technique, "Apertureless Imaging," illuminates nanoparticles or sharp tips (like atomic force microscope probes) with an incident beam.[2,3] The field intensity in nanosized subwavelength regions near the sharp edges of metal nanoparticles or sharp tips can be much larger than the intensity of the incident wave due to the so-called lightning rod effect. Antenna-like resonances or plasmon resonances can enhance the fields still further. These effects can be used for a variety of applications like near-field imaging,[4] Raman spectroscopy,[5] and guiding electromagnetic energy in subwavelength-sized optoelectronic devices.[6]

Crozier *et al.* studied metallic antennas at mid-infrared wavelengths with shapes and sizes determined by electron-beam lithography, and explored the reasons for electric field enhancement by computing the current distribution in the antennas.[7] They found good agreement between the experimental extinction efficiency and resonant wavelength values with those obtained from finite difference time domain (FDTD) computations. Typically, at infrared wavelengths the resonant length of these antennas,

M.L. Brongersma and P.G. Kik (eds.), Surface Plasmon Nanophotonics, 125–137.

along the direction of incident polarization, is approximately $\lambda/(2n_s)$ where n_s is the refractive index of the substrate on which the metal is deposited. However, in the optical range, as we shall show, the resonance for antennas made from Au and Ag is highly dependent on the plasmon resonant frequency, and the required size for resonance is much smaller than $\lambda/(2n_s)$.

Many groups have studied nanoparticles of various shapes to enhance the field intensity and have used various theoretical techniques to predict the behavior. Genov *et al.*[8] studied gold nanodisks separated by small gaps and predicted near-field intensity enhancements greater than 10^3. Hao *et al.*[9] studied nanostructures of different shapes and sizes, including two triangular nanoprisms facing each other tip-to-tip and separated by 2 nm, using the discrete dipole approximation method. They predicted intensity enhancements as high as $\sim 5 \times 10^4$. Rechberger *et al.*[10] considered coupling between gold nano disks experimentally and showed that the resonant wavelength depended on the spacing between the particles.

9.2. BOWTIE ANTENNAS

Grober *et al.* suggested that a receiving antenna shaped like a bowtie could be used to receive an incident wave with its E-field in the y-direction, along the line between the two opposing triangles (see Fig. 9.1 for an example of bowties fabricated to resonate at optical wavelengths).[11] This produces a large field between the facing tips of the two triangles comprising the bowtie. They demonstrated this principle at microwave frequencies and suggested extending its use to visible wavelengths. Following this suggestion, we have fabricated and experimentally characterized metallic Au bowtie structures with sizes small enough to resonate at visible/near-IR wavelengths.[12]

Our experiments were aimed at making reproducible sub-100 nm sized bowties which could be located at specified positions. To achieve this, Au bowties were

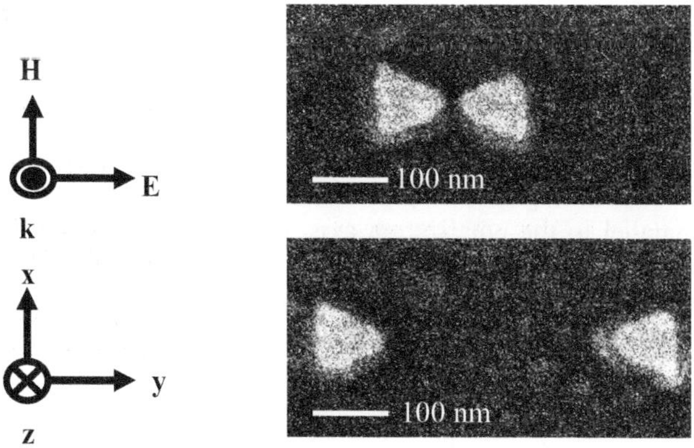

Figure 9.1. SEM images of two representative gold bowties on a fused silica–ITO substrate. The polarization of the incident beam and the coordinate system are shown.

fabricated with electron beam lithography (EBL) on a transparent substrate using a commercial EBL tool (Raith 150), at the Stanford Nanofabrication Facility. Substrates consisted of 160 μm thick fused silica coverslips (refractive index, $n = 1.47$), overcoated with 50 nm of indium tin oxide (ITO) to reduce charging effects during EBL. Goodberlet *et al.* demonstrated the ability to write line widths of 19 nm with this tool using single pass lines.[13] The bowties were laid out using single pass lines to reduce proximity effects, thus increasing EBL resolution. Prior to EBL writing, the sample was cleaned in acetone and baked in an oven at 150°C for 2 h to remove moisture. A 50-nm thick layer of poly-methyl-methacrylate (PMMA) (1% in chlorobenzene, 950,000 m.w.) was spun on the ITO layer and the sample was baked in an oven at 150°C for 2 h to harden the resist and remove residual solvent. During EBL, the resist was patterned at an acceleration voltage of 10 kV and a dose of 275 pC/cm. After exposure, the resist was developed at 22°C using a 25% methyl-iso-butyl-ketone in isopropyl alcohol solution for 28 s. A 5 nm Cr adhesion layer and 20 nm Au were then deposited using an electron-beam evaporator, and the patterns were transferred to the substrate via a lift-off process.

Scanning electron microscope (SEM) images of representative bowties are shown in Fig. 9.1. For this study, the shape of each constituent triangle of a bowtie was held constant at 75 nm in length (the distance from the midpoint of the triangle base to its apex), gap lengths were varied from 16 nm to nearly 500 nm, and the triangular tips were observed to have a radius of curvature of 16 nm. Single bowties were separated by a 10 μm pitch to eliminate long-range coupling effects and to ensure that only scattering from a single bowtie was collected.[12]

9.2.1. Single-Photon Effects

Single bowtie scattering spectra were measured with far-field total internal reflection (TIR) microscopy. This method has the advantage that the excitation beam is trapped in the evanescent field until scattered toward the detector by the bowtie. Broadband light from a 30-W halogen lamp was passed through a 455 nm long-pass filter and coupled into a multimode fiber (65 μm core), which served as a spatial filter. The fiber output was collimated and focused into a fused silica prism. Index-matching fluid (refractive index $n = 1.47$) coupled the prism to the bottom of the fused silica substrate, placing the Au bowties at the ITO–air interface which was also the location of the evanescent field. The excitation light was *s*-polarized and the axis of each bowtie was carefully oriented parallel to the polarization axis. (The perpendicular polarization lead to unremarkable results, similar to the scattering from isolated triangles.[12]) Scattered light was collected with an air objective (100×, 0.8 numerical aperture, NA) and a 50 μm pinhole was placed at the microscope image plane to spatially filter the collected light to an area ~0.5 μm in diameter on the sample plane. The antennas were located by scanning the sample stage and collecting broadband scattered light with a single photon counting avalanche photodiode. Spectra were acquired during 60 s exposures by dispersing the scattered light with a 150 lines/mm grating spectrograph with a liquid nitrogen cooled charged coupled device Si detector at the image plane. All spectra were corrected for the wavelength and polarization dependences of the lamp emission, optical system throughput, and detector quantum efficiency. The measured results were compared to finite-difference time-domain FDTD calculations of the electromagnetic field distributions.

The electric field variation in the vicinity of a bowtie antenna was determined theoretically through three-dimensional FDTD simulations[14] (and experimentally by the use of two-photon scattering effects, as described below). The FDTD simulations did not make any approximations to compute the final field intensity in the vicinity of the bowties. The size of each node in the simulation grid was 4 nm along the x-, y- and z-directions. In the simulations, the bowtie antennas were illuminated from the substrate side by a plane wave polarized along the y-direction and the antennas radiated into free space. The wavelength-dependent dielectric constants used for the bowtie antennas and the substrate were taken from Palik *et al.*[15] Perfectly matched layers (PML) were used at the top and bottom of the simulation domain to completely absorb waves leaving the simulation domain in the direction of propagation. The FDTD simulations assumed periodic boundary conditions in the x–y directions, thereby simulating an array of antennas while the experimental results were obtained for individual bowtie antennas. The inter-element spacing between the antennas in the FDTD simulation were chosen to be sufficiently large so that coupling between the antennas had a negligible effect on the near-field distribution in each antenna. This enabled a direct comparison between theory and experiment.

All the bowtie antennas simulated have constituent triangles that are equilateral, where each side is 88 nm in length with tip to base length 75 nm, and the radius of curvature at the triangle apex equal to 12 nm. The gap between the triangles in the bowtie is varied from 16 to 500 nm. The thickness of each bowtie antenna simulated is 24 nm (20 nm gold layer + 4 nm chrome sticking layer). The substrate is fused silica ($n = 1.47$) with a 52 nm layer of indium tin oxide (ITO, $n = 1.95$) evaporated on it. These values are very close to the specifications of the fabricated bowties.[12,16]

The figure in the overall introduction shows a surface plot of the calculated intensity enhancement 4 nm above a bowtie antenna with a 16 nm gap. Figure 9.2 shows the peak near-field intensity ($|E|^2$) enhancement values for bowties with 16 and 160 nm gaps as a function of excitation wavelength. The bowtie antenna with a 16 nm gap has a peak near-field intensity enhancement of 1645 relative to the incident beam at a wavelength of 850 nm and is spatially confined to a 20 nm region in the x–y direction, measured as a full width half maximum (FWHM) value. The bowtie with 160 nm gap has a far smaller enhancement peaking at 730 nm. The maximum intensity value, 4 nm above the bowtie surface, is found to be close to each triangle apex with a slight dip in the middle of the gap. The direction of the maximum field for gaps smaller than 60 nm is along the plane of the bowties in the y-direction, and for gaps greater than 60 nm it is in the direction of propagation (z-direction), perpendicular to the surface of the metal.[16] It should be noted that with smaller gap spacings than 16 nm and smaller tip radii of curvature than 12 nm, it should in principle be possible to obtain stronger fields. However, smaller and well controlled gap spacings and tip radii may be difficult to reliably fabricate in practice.

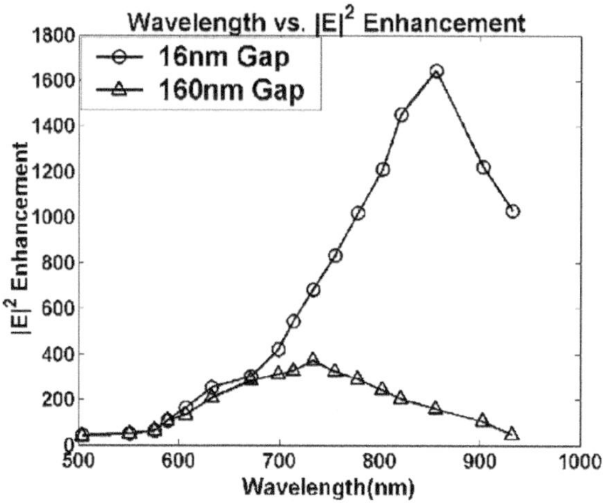

Figure 9.2. Peak near-field intensity enhancement for bowtie antennas with gap widths equal to 16 and 160 nm as a function of excitation wavelength. The maximum enhancement in the case of a 16 nm gap bowtie is 1645.

In the experimental measurements, using TIR excitation with s-polarized light, there is no transmitted wave when the bowtie antenna is not present, while a scattered wave is produced in the presence of the antenna. In the FDTD calculation for triangles much smaller in extent than a wavelength, we calculate the scattering efficiency, Q_{scat}, as the ratio of scattering cross section, C_{scat}, to the metal antenna area, A. We compute the current density distribution in the antennas and use this to compute far-field radiated power, W_{ext} (in Watts), thereby obtaining the scattering cross section, C_{scat} which is defined by the relation:

$$C_{\text{scat}} = W_{\text{ext}}/I, \tag{9.1}$$

where $I = E^2/2\eta$ is the calculated incident power density (Watts/μm^2) on an antenna with a known current distribution, E is the incident electric field on the antenna and η is the impedance of the medium containing the incident wave.

To calculate C_{scat}, we use the fact that antenna theory provides a systematic method to compute average radiated power (in Watts) from an antenna, which has a known current distribution.[17] The total current in the y-direction in the metal of the antenna is obtained by integrating the complex polarization current density (J_{ym}) in the metal over x and z. The current density in the antenna, the scattering efficiency, and the maximum intensity in the gap are all calculated to be largest at the same wavelength.

Figures 9.3(a) and (b) show the calculated and experimentally determined variation in scattering efficiency with wavelength for bowtie antennas with two different gap widths. To make a fair comparison with the experimental data, the radiated power obtained from the FDTD simulations is multiplied by a constant factor to model

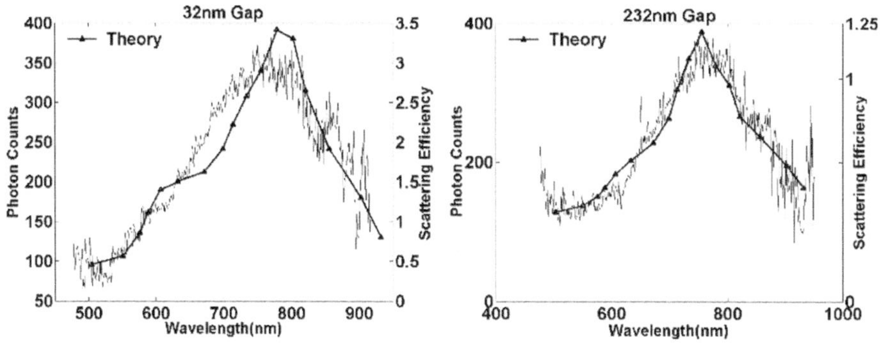

Figure 9.3. Plots (a) and (b) compare the spectra obtained from FDTD calculations (full curve) and experiments (curve with experimental noise) for 32 and 232 nm gap lengths respectively. The theoretical and experimental peak resonant wavelengths are λ_{rT} and λ_{rE} are $\lambda_{rT} = 779$ nm and $\lambda_{rE} \sim 776$ nm for the 32 nm gap and $\lambda_{rT} = 756$ nm and $\lambda_{rE} \sim 748$ nm for the 232 nm gap.

the finite acceptance angle of the collection lens (0.8 NA) used in the experiment. The theoretical scattering efficiency at resonance is highest for the smallest gaps ($Q_{\text{ext}} = 4.1$ for a 16 nm gap) indicating that antennas with smaller gaps radiate more power relative to their area. All the experimental photon counts were scaled to match the peak theoretical FDTD calculated scattered power for a 16 nm gap in Fig. 9.3(a) and (b). The absolute experimental scattering cross section could not be accurately computed because it was not possible to determine the equivalent incident power on the bowtie with TIR excitation. The experimental scattering spectra shown here rest on the assumption that as a function of wavelength, the incident power is proportional to the measured illumination power from the source scaled by the detection efficiency of our microscope.

The displacement current density, $J_{\text{Dn}} = j\omega\varepsilon_0 E_n$, in the x–y plane in the gap normal to the edge of the metal is continuous with the current density J_{mn} normal to the edge inside the metal, where

$$J_{\text{mn}} = j\omega\varepsilon_0(\varepsilon_{\text{r}} - 1)E_{\text{mn}}, \tag{9.2}$$

$$(n + jk)^2 = \varepsilon_{\text{r}} \tag{9.3}$$

and where ω is the angular frequency, ε_0 is the dielectric constant of free space, ε_{r} is the relative complex permittivity of the metal, the subscript n stands for the normal component of field, and $(n + jk)$ is the complex refractive index of Au taken from Palik *et al.*[15] and E_{mn} is the E field in the metal normal to the edge. The relation between the field components normal to the surface is $\varepsilon_{\text{r}} E_{\text{mn}} = E_{\text{n}}$. Consequently, the enhanced field in the gap is primarily due to the enhanced current density near the tips of the constituent triangles of the bowtie.

Figure 9.4 shows the resonant wavelength as a function of gap length for both theory and experiment, and the agreement is excellent. At resonance the bowtie with the smallest gap has the largest peak current density, and the calculated peak current density decreases as the gap width increases. For a bowtie with a 16 nm gap width,

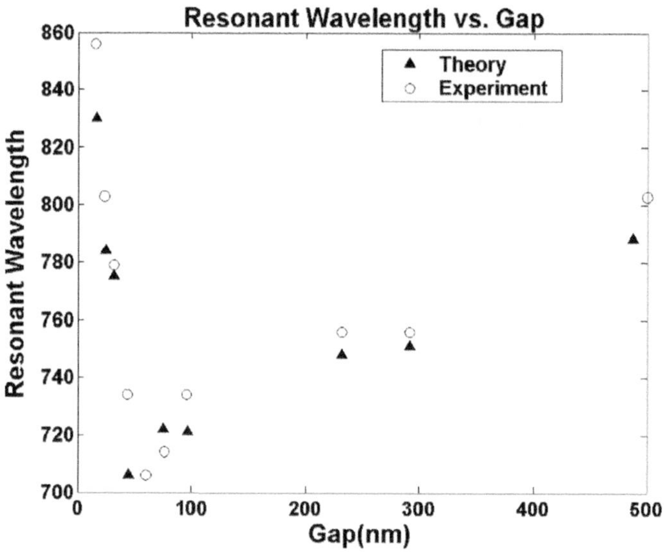

Figure 9.4. The variation of resonant wavelength with gap length for light polarized along the line between the two triangles, with experiment shown as open circles and FDTD simulations as filled triangles.[12]

the peak current density lies at the apex of the triangle and the current density is fairly uniform with x across the tip region. In this situation, the resonance wavelength is 856 nm, where the real part of the dielectric constant of the metal is negative and hence the impedance of the metal is inductive. Consequently, we might expect to observe a coupled plasmon resonance controlled by the capacity of the gap and the inductance of the tip regions rather than by the dimensions of the relatively large triangles. For gaps smaller than 60 nm the antenna resonance is dominated by the capacitance of the air gap with its associated field in the y-direction. In this regime the resonant wavelength decreases with increasing gap (decreasing capacity).

At the other extreme of a 500 nm gap, since the capacity of the air gap is small, the total current flowing into the gap is small. In this regime, the E-field outside the metal is at a maximum in the z-direction normal to the plane of the triangles, the coupling between the triangles is decreased and the current distribution approaches that of a single triangle. For a gap width greater than 60 nm, the field lines near the tip tends to return either to infinity or to the metal triangle from which they left and the resonant wavelength increases with gap width.

9.2.2. Two-Photon Effects

It is difficult to directly measure the electromagnetic fields in the gap because the incident pumping field extends over a much larger (diffraction-limited) area and tends to leak into the detector. Instead, we have found it useful to use two-photon effects which only occur in the region where the optical fields are very strong. Using this approach, we have experimentally determined the optical intensity enhancement values for the fields *in the metal* of these structures, which closely approximate fields outside the metal near the surface.[22]

Strongly enhanced local fields due to the excitation of surface plasmons in rough films, sharp tips, and nanoparticles give rise to detectable two-photon absorption in Au.[18–20] The resulting excitation of electrons from the d valence band to the *sp* conduction band leads to a broadband emission continuum, termed two-photon-excited photoluminescence (TPPL) in Au. Due to its nonlinear (intensity squared) dependence on excitation intensity, TPPL is a sensitive probe of excitation field strength and distribution. Variations between the TPPL spectra from differently shaped Au nanoparticles provide evidence for the localized origin of the absorption and subsequent emission.[21] We have used TPPL to directly determine absolute values for optical field enhancements of single Au bowties by comparing the strength of TPPL from bowties with TPPL from a smooth Au film.[22]

The TPPL from individual bowties was measured with a sample-scanning microscope. A mode-locked Ti:sapphire laser producing 120 fs pulses at $\lambda = 830$ nm with a repetition rate of 75 MHz was used for excitation through the epifluorescence port of an inverted optical microscope (Nikon TE300). The value $\lambda = 830$ nm was chosen since the smallest gap bowties were measured to be resonant close to this wavelength (see Fig. 9.4). The laser is focused through the coverslip to a diffraction-limited spot on the bowtie–air interface using a 1.4 NA, $100\times$ oil objective. Au TPPL is collected by the same objective and passed through three spectral filters, which effectively transmit emission with wavelengths between 460 and 700 nm while attenuating 830 nm laser light with an optical density (OD) > 18. The luminescence is focused onto a single-photon counting avalanche photodiode (APD) for broadband collection. After optical experiments, the sample is studied with an atomic force microscope (AFM), then coated with a thin Cr layer ($\sim$4 nm) for particle analysis in the SEM to carefully measure gap sizes.

To directly determine optical field enhancements from bowtie nano-antennas, the TPPL emission from smooth Au films was used for calibration. The optical spectra of the TPPL were identical for the smooth film and the bowties in the collected wavelength range. Figure 9.5(a) and (b) show that the TPPL is strongly polarization dependent. Figure 9.5(c) shows TPPL from an array of bowties with 20 nm nominal gap, where the inhomogeneity in brightness results from variations in the actual gap width, and where bowties which appear dark were later found to be shorted by SEM. Fig. 9.5(d) shows an array of 400 nm gap bowties, and both the weaker signal and the ability to optically resolve the two constituent triangles are evident (note the higher pumping power required). Figure 9.5(e) is a TPPL image from an Au film square, which required even higher pump power. As expected, TPPL is brightest at the edges due to the field enhancement caused by localized plasmon resonances at the rough edge caused by the lift-off procedure. Less intense and remarkably uniform TPPL is also observed from the interior of the square. In addition, five hot spots of localized field enhancement are visible with less than one hot spot per 10 μm^2, providing optical evidence of the film's smoothness.

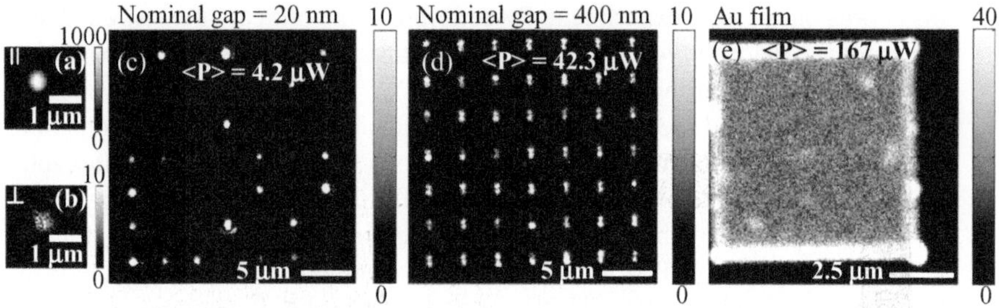

Figure 9.5. Visible luminescence images from bowties. Incident polarization dependence is shown in (a) and (b) for a single 20 nm gap bowtie (note different scales). Images of arrays with nominal gap sizes of 20 nm (c) and 400 nm (d), as well from a smooth Au film (e), with 10 ms per pixel dwell time, 830 nm pump, and average incident powers (c) 4.2 μW, (d) 42.3 μW, (e) 167 μW.[22]

The intensity variation of the TPPL with input power is shown in Fig. 9.6 for a smooth Au film, a single 400 nm gap bowtie, and a single bowtie from the array with nominal gap size of 20 nm. The APD count rates from the Au film have been normalized for the relative surface area of Au illuminated by the focused laser. For the Au film and 400 nm gap bowtie, the collected emission shows a quadratic dependence, as expected for TPPL. For the bowtie with a 20 nm gap, the collected emission follows a quadratic dependence at low average powers. However, as incident power increases, TPPL intensity deviates from this dependence and begins to decrease until, at still higher powers, it again shows a quadratic dependence but along an I^2 curve shifted to the right of the original curve. This behavior occurs for almost all small gap bowties (gaps <40 nm). As is confirmed by SEM observations, this is because the field intensity becomes large enough to physically damage the tips of the bowtie, thus reducing its field-enhancing capabilities. Thereafter, the emission follows a new quadratic curve defined by the lower enhancement factor. For subsequent

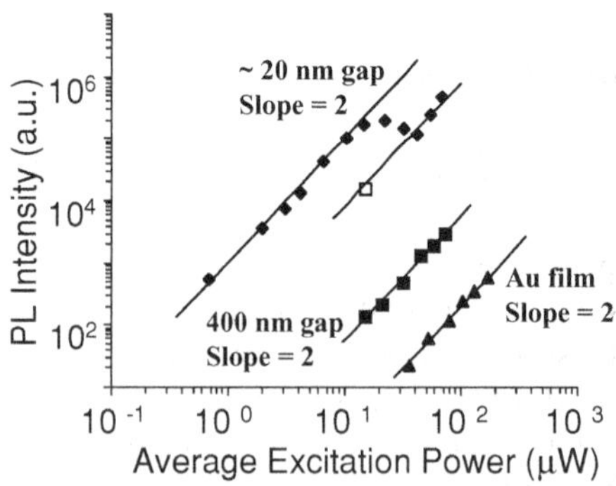

Figure 9.6. TPPL intensity dependences for a smooth Au film, a 400 nm gap bowtie, and a bowtie from an array with nominal gap size of 20 nm. Open square: measurement at lower power after the intensity scan.[22]

measurements at low powers, the measured TPPL lies on the shifted curve (open square in Fig. 9.6), indicating an irreversible change. Note that the original gap size for this particle is not known since it was damaged before SEM analysis, thus we use the phrase "nominal gap size of 20 nm."

The intensity enhancement in the metal of the bowtie i, α^{i}_{bt}, can be calculated from the ratio of TPPL intensities from the bowtie and the film:

$$\frac{\langle TPPL^{i}_{\mathrm{bt}}\rangle}{\langle TPPL_{\mathrm{film}}\rangle} = \frac{A_{\mathrm{bt}}}{A_{\mathrm{film}}} \times \frac{\left(\alpha^{i}_{\mathrm{bt}}\right)^2 \left\langle P^{i}_{\mathrm{bt}}\right\rangle^2}{\langle P_{\mathrm{film}}\rangle^2} \tag{9.4}$$

where $\langle TPPL^{i}_{\mathrm{bt}}\rangle$ is the (time-averaged) TPPL signal when the bowtie i is centered in the focus of the laser excitation spot, $\langle TPPL_{\mathrm{film}}\rangle$ is the TPPL signal when the excitation spot is anywhere within the uniform emission region from the Au film, $\langle P^{i}_{\mathrm{bt}}\rangle$ is the average incident power at bowtie i that yields $\langle TPPL^{i}_{\mathrm{bt}}\rangle$, $\langle P_{\mathrm{film}}\rangle$ is the average incident power at the film that yields $\langle TPPL_{\mathrm{film}}\rangle$, A_{bt} is the surface area of the bowtie from which the $\langle TPPL^{i}_{\mathrm{bt}}\rangle$ originates and A_{film} is total surface area of the Au film from which the $\langle TPPL_{\mathrm{film}}\rangle$ originates.

We assume the area A_{film} excited by two-photon absorption of the focused laser is equal to a circular region with a diameter equal to the full width at half maximum (FWHM) of the square of the intensity point spread function (PSF) of our microscope objective. For $\lambda = 830$ nm and NA $= 1.4$, the measured FWHM from a diffraction-limited TPPL spot is equal to 214 nm, so $A_{\mathrm{film}} = 35{,}600\ \mathrm{nm}^2$. FDTD calculations show that plasmonic current densities and optical near-field intensities are concentrated in a confined region of each bowtie, e.g. within the Au nearest the gap, particularly for bowties with small gap sizes, thus A_{bt} is also gap-dependent and must be calculated.

As mentioned above, the strongest optical fields are found in the gap between the constituent triangles. However, the experimentally observed TPPL is certainly arising from electromagnetic fields *in the metal*. Therefore, the intensity of the TPPL should be proportional to the fourth power of electric field in the metal, E_{m}, and the total TPPL power will be correlated to the integral of $|E_{\mathrm{m}}|^4$.

For comparison with these experimental results, plane wave excitation normally incident on the bowties is assumed in FDTD calculations. For an incident electric field amplitude E_0, the tangential electric field at the metal surface is $E_{\mathrm{inc}} = 2E_0/[1 + (n + jk)]$. In the metal film, the electric field falls off as $\exp(-kz)$. FDTD simulations for the bowties show that, to a good approximation, the E-field in the bowties also falls off as $\exp(-kz)$. Since all fields have the same z-dependence in the metal, the integral of $|E_{\mathrm{m}}|^4$ need only be performed along the x- and y-directions just below the surface of the metal in the FDTD computations.

We used FDTD calculations to determine the square of the intensity enhancement in the bowtie i, $\left|\alpha^{i,\mathrm{FDTD}}_{\mathrm{bt}}\right|^2$ using

$$\left|\alpha^{i,\mathrm{FDTD}}_{\mathrm{bt}}\right|^2 = \frac{\iint \left|E_{\mathrm{bt,m}}\right|^4 \mathrm{d}x\,\mathrm{d}y}{\iint E^4_{\mathrm{incident}} \mathrm{d}x\,\mathrm{d}y} \tag{9.5}$$

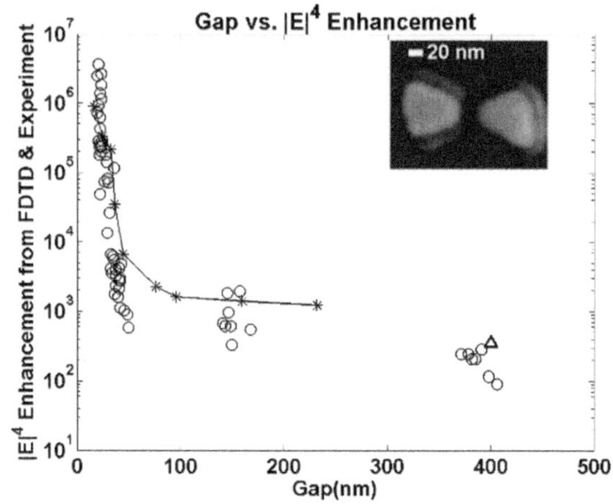

Figure 9.7. Comparison of experimental (circles) and theoretical (asterisks and triangle) values of the square of intensity enhancement, $(\alpha^i_{\rm bt})^2$, for bowties with gaps of 16–406 nm. FDTD simulations for $(\alpha^i_{\rm bt})^2$ from an individual triangle are used for comparison to the bowties with nominal gap size of 400 nm. Excellent agreement is observed, especially for the largest and smallest gap sizes where experimental conditions most closely approximate the theoretical treatment.

where $|E_{\rm bt,m}|^4 = \left(E^2_{\rm bt,mx} + E^2_{\rm bt,my} + E^2_{\rm bt,mz}\right)^2$ and the integrals are over the area of the bowtie. The experimental and theoretical results for $|\alpha^i_{\rm bt}|^2$ are shown in Fig. 9.7 for various gap widths. The effective area $A_{\rm bt}$ of the TPPL source is determined from the following integral:

$$A_{\rm bt} = \int E^4_{\rm m} {\rm d}x\,{\rm d}y / E^4_{\rm max} \tag{9.6}$$

where $E_{\rm max}$ is the maximum field in the bowtie. $A_{\rm bt}$ is a function of gap size, increasing as gap width increases. The $A_{\rm bt}$ obtained from theory for the smallest gap width (16 nm) was 642 nm^2 indicating that the field was confined to approximately one-tenth of the ~6530 nm^2 area of the metal in the bowtie. Using these values for $A_{\rm bt}$ in Eq. (9.1) along with the experimentally determined TPPL intensities yields E^2 enhancement factors of greater than 10^3, or greater than 10^6 for α^2 for bowties with the smallest gaps (Fig. 9.7). These are the largest such factors reported to date for lithographically produced nanoantennas.

We observe good agreement between the E^4 values for theory and experiment for gap widths less than 30 nm, but slightly less so for gap widths of 40–60 nm. For nominal gap widths of 400 nm, we again get good agreement. This can be explained by the fact that the plane wave excitation used for FDTD closely approximates the experimental conditions for small gap bowties, but does not accurately model intermediate gaps, where the triangles are not uniformly pumped by the focused laser spot. To avoid this difficulty for gap widths of 400 nm, we excited and collected TPPL from each triangle of the bowtie separately and summed the two since the coupling is minimal. Furthermore, since a bowtie with a 400 nm gap was too large to simulate, the E^4 enhancement value was obtained by doubling the value from FDTD simulations for a single triangle (Fig. 9.7, triangle).

9.3. CONCLUSION

We have fabricated Au bowtie nanoantennas by e-beam lithography which have resonances in the visible/near-IR wavelength range. FDTD calculations have been used to predict resonance frequencies, field profiles and local field enhancements. Experimental measurements of peak scattering wavelength by TIR microscopy for bowties with a range of gap sizes from 20 to 500 nm are in good agreement with theory.

We have also experimentally measured optical intensity enhancements at Au bowtie nanoantennas of various gap sizes using TPPL, and find good agreement with FDTD simulations. For small gap bowties, the field intensity enhancement is $>10^3$ confined to a region $\sim$650 nm^2, which may be interpreted as a dramatic improvement in the mismatch between conventional optical excitations and nanoscale structures.

It is to be expected that with optimization of the fabrication, Au bowties will be reproducibly manufactured either individually on a scanning probe or in large arrays on a single substrate. The large enhanced fields in the metal will also lead to similarly enhanced, localized fields on the metallic surface and in between the two bowties. This will yield extremely intense near-field optical light sources with high local contrast that have applications ranging from the elucidation of SERS mechanisms, ultra-sensitive biological detection, single-molecule spectroscopy, and nanometer-scale lithography to high-resolution optical microscopy and spectroscopy.

ACKNOWLEDGMENTS

This work was supported in part by the U.S. Department of Energy Grant No. DE-FG03-00ER45815 and by the National Institutes of Health Grant No.GM65331-R21. The authors wish to thank Prof. Andy Neureuther, University of California (Berkeley) for use of TEMPEST 6.0 FDTD software developed by his group.

REFERENCES

1. E. Betzig, J.K. Trautman, T.D. Harris, J.S. Weiner, R.L. Kostelak: Breaking the diffraction barrier: optical microscopy on a nanometric scale, Science **251**, 146 (1991).
2. F. Zenhausern, Y. Martin, H.K. Wickramasinghe: Scanning interferometric apertureless microscopy: optical imaging at 10 angstrom resolution, Science **269,** 1083 (1995).
3. J.L. Bohn, D.J. Nesbitt, A. Gallagher: Field enhancement in apertureless near-field scanning optical microscopy, J. Opt. Soc. Am. A. **18**, 2998–3006 (2001).
4. L. Novotny, E.J. Sanchez, X.S. Xie: Near-field optical imaging using metal tips illuminated by higher-order Hermite-Gaussian beams, Ultramicroscopy **71**, 21 (1998).
5. A. Hartschuch, E.J. Sanchez, X.S. Xie, L.Novotny: High-resolution near-field Raman microscopy of single-walled carbon nanotubes, Phys. Rev. Lett. **90**, 095503 (2003).

6. S.A. Maier, P.G. Kik, H.A. Atwater: Observation of coupled plasmon-polariton modes in Au nanoparticle chain waveguides of different lengths: Estimation of waveguide loss, Appl. Phys. Lett. **81**, 1714 (2002).
7. K.B. Crozier, A. Sundaramurthy, G.S. Kino, C.F. Quate: Optical antennas: resonators for local field enhancement, J. Appl. Phys. **94**, 4632 (2003).
8. D.A. Genov, A.K. Sarychev, V.M. Shalaev, A. Wei: Resonant field enhancements from metal nanoparticle arrays, Nano Lett. **4**, 0343710 (2004).
9. E. Hao, G.C. Schatz: Electromagnetic fields around silver nanoparticles and dimers, J. Chem. Phys. **120**, 357 (2004).
10. W. Rechberger, A. Hohenau, A. Leitner, J.R. Krenn, B. Lamprecht, F.R. Ausenegg: Optical properties of two interacting gold nanoparticles, Opt. Commun. **220**,137 (2003).
11. R.D. Grober, R.J. Schoelkopf, D.E. Prober: Optical antenna: Towards a unity efficiency near-field optical probe, Appl. Phys. Lett. **70**, 1354 (1997).
12. D.P. Fromm, A. Sundaramurthy, P.J. Schuck, G. Kino, W.E. Moerner: Gap-dependent optical coupling of single "Bowtie" nanoantennas resonant in the visible, Nano Lett. **4**, 957 (2004).
13. J.G. Goodberlet, J.T. Hastings, H.I. Smith: Performance of the Raith 150 electron-beam lithography system, J. Vac. Sci. Technol. B **19**, 2499 (2001).
14. TEMPEST 6.0, Electronics Research Laboratory, University of California, Berkeley, California.
15. E.D. Palik: *Handbook of Optical Constants* (Academic Press, Orlando, Florida, 1985).
16. A. Sundaramurthy, P.J. Schuck, D.P. Fromm, W.E. Moerner, G. Kino: Field enhancement and gap-dependent resonance in a system of two opposing tip-to-tip Au nanotriangles, Phys. Rev. B. **72**, 165409 (2005).
17. S. Ramo, J.R. Whinnery, T. Van Duzer *Fields and Waves in Communication Electronics*, 2nd edn (Wiley, New York, 1984), pp. 586–589.
18. C. K. Chen, A.R.B. de Castro, Y.R. Shen: Surface-Enhanced Second-Harmonic Generation, Phys. Rev. Lett. **46**, 145 (1981).
19. G.T. Boyd, Z.H. Yu, Y.R. Shen: Photoinduced luminescence from the noble metals and its enhancement on roughened surfaces, Phys. Rev. B **33**, 7923 (1986).
20. M.R. Beversluis, A. Bouhelier, L. Novotny: Continuum generation from single gold nanostructures through near-field mediated intraband transitions, Phys. Rev. B **68**, 115433 (2003).
21. A. Bouhelier, M.R. Beversluis, L. Novotny: Characterization of nanoplasmonic structures by locally excited photoluminescence, Appl. Phys. Lett. **83**, 5041 (2003).
22. P.J. Schuck, D.P. Fromm, A. Sundaramurthy, G.S. Kino, W.E. Moerner: Improving the mismatch between light and nanoscale objects with gold bowtie nanoantennas, Phys. Rev. Lett. **94**, 017402 (2005).

CHAPTER TEN

NEAR-FIELD OPTICAL EXCITATION AND DETECTION OF SURFACE PLASMONS

ALEXANDRE BOUHELIER[1] AND LUKAS NOVOTNY[2]
[1]Laboratoire de Physique de l'Université de Bourgogne, 21000 Dijon, France
alexandre.bouhelier@u-bourgogne.fr
[2]The Institute of Optics, University of Rochester, Rochester, NY 14627, USA
novotny@optics.rochester.edu

10.1. INTRODUCTION

By definition, surface plasmons are the quanta of surface-charge-density oscillations, but the same terminology is commonly used for collective oscillations in the electron density at the surface of a metal. Because the surface charge oscillations are intimately coupled to electromagnetic fields, surface plasmons are polaritons. In the past, surface plasmons have attracted considerable attention due to their application in optical sensor devices.[1,2] Because of their localized nature, surface plasmons have recently also been explored in integrated optical circuits and optical waveguides.[3,4] However, one of the key properties of surface plasmons is the associated light localization, which can be explored for localized photon sources in optical spectroscopy and microscopy.[5,6] Surface enhanced Raman scattering (SERS) is a prominent example of the latter application. Recently, it was demonstrated that the SERS effect can be spatially controlled with a laser-irradiated metal tip. This combination of SERS and microscopy provides high spatial resolution and simultaneous chemical information in the form of vibrational spectra.[7]

Starting with experiments on planar metal surfaces, we will discuss in this chapter general properties of surface-plasmon polaritons (SPPs) and then concentrate on light localization associated with laser-irradiated metal tips. This localization is accomplished through an interplay of the quasi-static lightning-rod effect associated with the sharpness of the tip and surface plasmons.[8] The localized field can be used as a secondary light source for highly confined optical interactions with a sample surface. Vice versa, a localized field such as the field at a metal tip or the field emanating from a tiny aperture, possesses the large spatial frequencies that are necessary

M.L. Brongersma and P.G. Kik (eds.), Surface Plasmon Nanophotonics, 139–153.

to couple to surface plasmons on a sample surface. This approach of launching surface plasmons on a plane metal surface has been first experimentally demonstrated by Hecht *et al.*[9]

The use of a light-irradiated particle for achieving subwavelength light localization has already been suggested by Synge in a letter that he wrote to Einstein.[10] The idea was never published in its original form. Instead, Synge published an article in which he proposed to use a tiny aperture in a metal screen to achieve a subwavelength-scale light source.[11] In 1984, even before the invention of the atomic force microscope (AFM), John Wessel proposed a concept very similar to the original idea of Synge.[5] A laser-irradiated, elongated metal particle is used to establish an enhanced, localized light field. By raster-scanning the metal particle over the surface of a sample, and detecting a spectral response due to the interaction with the sample surface point-by-point, a spectroscopic map can be recorded of the sample surface. Wessel's proposal was soon forgotten and it resurfaced again in the context of near-field optical microcopy. In fact, within this field of study, several research groups reinvented Wessel's idea. Denk and Pohl suggested to use the enhanced field at a metal tip in combination with nonlinear optical spectroscopy[12] and first experimental results demonstrating the field enhancement effect associated with a single metal nanoparticle have been published by Fischer and Pohl.[13] In the following years, metal tips were introduced as scattering centers to convert the nonpropagating evanescent fields near the surface of an object into propagating radiation that can be recorded by a remote detector. This scattering-type near-field optical microscopy has recently found various important applications ranging from localized measurements of dopants to the characterization of surface phonon polaritons.[14] However, scattering-type near-field microscopy does not make explicit use of the light localization associated with the field enhancement effect. The use of the field enhancement effect near a metal tip for localized spectroscopic measurements was theoretically formulated[6] and subsequently experimentally demonstrated by Sanchez *et al.* using two-photon excited fluorescence.[15] Following these experiments, the same principle was extended to other spectroscopic interactions such as Raman scattering.[7] In essence, Synge's and Wessel's original ideas have made their way into real applications, and enabled chemically specific measurements with nanoscale spatial resolution.

10.2. LOCALIZED EXCITATION OF SURFACE PLASMONS

Before discussing the field enhancement effect at laser-irradiated metal tips, we first review some important experiments aimed at locally exciting SPPs propagating on planar metal surfaces. The optical excitation of surface plasmons on flat metal interfaces is challenged by the phase matching condition between the plasmons and the exciting radiation. The surface plasmon dispersion $\omega(k)$ is located outside the light cone $\omega = ck$ and hence, no SPPs can be excited with freely propagating radiation. The excitation of SPPs can only occur if the photon momentum—or the wave vector—can be artificially increased. Various experimental techniques have been developed

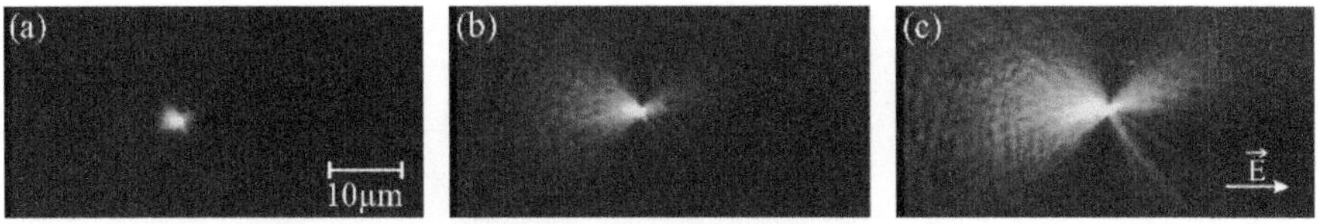

Figure 10.1. Spatial distribution of the leakage radiation of a propagating surface plasmon excited by a near-field probe. (a) the tip is far away from the film interface, (b) the tip gets closer and (c) optimum distance for plasmon excitation. The polarization of the excitation light is horizontal. Some inhomogeneities in the plasmon intensity distribution is seen in the right emission lobe.

to accomplish this task, such as (1) increasing the index of refraction of the incident medium, or (2) engineering the surface of the film (grating coupler). While these approaches provide very efficient coupling between the incident photons and the SPP waves, the interaction area is usually comparable or greater than the SPP propagation distances.

With the advance of near-field optics, the localized excitation of surface plasmons became feasible. In the first experiments, tiny apertures were used as SPP excitation sources.[9,16–19] To describe the excitation efficiency in these experiments, the Bethe-Bouwkamp model[20] is instrumental. It considers a small circular aperture in a perfectly thin, metallic, infinite, and flat screen, illuminated by a plane wave. The angular spectrum representation of the emitted field demonstrates that the emitted field is composed of both homogeneous plane waves ($k < \omega/c$) and of evanescent waves ($k > \omega/c$). Here, k denotes the projection of the k-vector on the aperture plane. Because the dispersion of evanescent fields lies outside the light-cone, it is possible to excite surface plasmons on a plane metal surface by the field emitted by the aperture, provided the separation between aperture and surface is much less than the wavelength of light. Thus, surface plasmons can be excited by the large in-plane component of the wave vector much like in the Otto configuration[21] with the difference that the excitation area is extremely local owing to the very small size of the aperture.

Figures 10.1 (a), (b) and (c) are snapshot images from a video sequence acquired, as a near-field probe approaches a thin silver film deposited on a glass substrate. These images were recorded by focusing a high numerical aperture (NA) oil immersion objective at a metal/glass interface. When the tip is far away from the sample [Fig. 10.1 (a)], the light emitted from the tip extremity can be seen through the finite thickness of the silver film ($\approx$ 60 nm). As the tip comes closer to the surface, a drastic change in the spatial distribution of the emission occurs: surface plasmons are resonantly excited by the large evanescent wavectors emitted from the tip and they propagate along the silver/air interface [Fig. 10.1 (b) and (c)].

The two-lobe patterns seen in the images originate from intrinsic lossy waves associated with the SPP. The wave amplitude of the bounded SPP mode is exponentially decaying along the metal surface. As the field penetrates to the opposite interface (silver/glass), it couples to radiative leaky waves that can be detected with the objective lens. The far-field observation of this leakage radiation (LR) gives a direct

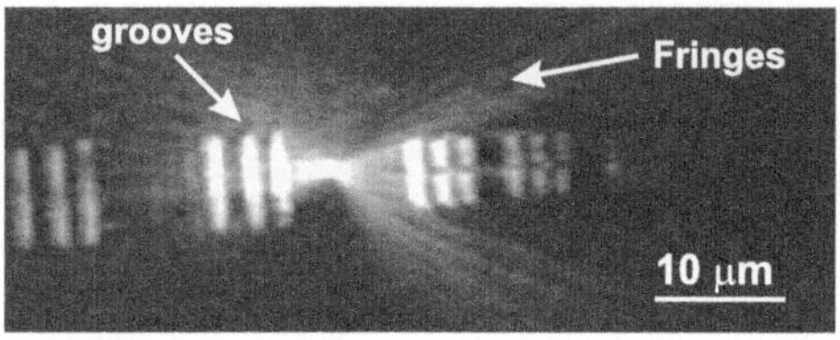

Figure 10.2. Surface plasmons in interaction with triplets of grooves. The distance between the two centered triplets is 10 μm. The hyperbola fringes are the signature of an interference between the SPP beam and its reflection on the left-hand side triplet.

measurement of the nonradiative surface plasmon propagation at the opposite interface. The intensity of the radiation, at a given lateral position in the film, is proportional to that of the SPP at the same position.[9,18,22,23]

The SPP emission pattern of Fig. 10.1 (c) has the characteristics of a two-dimensional dipole. The intensity distribution, $I_{r,l}$, can be heuristically compared to the following functional dependence[9,18]:

$$I_{r,l}(\rho, \psi) = \frac{\Gamma_{r,l}}{\rho} \, e^{-2\alpha\rho} \cos^2 \psi, \tag{10.1}$$

where $\Gamma_{r,l}$ is a measure for the right-hand side and the left-hand side lobe intensities, respectively. The coordinates ρ and ψ are the distance from the source and the azimuth with respect to the central axis of the right-hand side lobe, respectively. For the radial decay of the intensity, a $1/\rho$ dependence due to spreading in two dimensions is expected. In addition, damping resulting from *intrinsic* losses and potentially *extrinsic* ones, is to be accounted for by an exponential decay term (decay constant α). $\psi = 0$ is defined as the central direction of the right-hand side surface plasmon lobe. The angular representation of the two-lobe pattern is included in the term $\cos^2 \psi$.

The near-field probe acts as a localized SPP source that can be positioned precisely over a selected spot near a metal surface. This opens new perspectives for investigating plasmonic structures fabricated on the film. An example is shown in Fig. 10.2. The SPP intensity distribution shows evidence of SPP reflection and transmission through an array of grooves cut in the metal film. Here, the near-field tip was positioned between two triplets of grooves. Attenuated transmission and reflection of the SPP is quite obvious, the signature of the latter being extended interference fringes outside the range of the grooves. The triplets apparently act as a SPP multilayer mirror. Analysis of the visibility of the fringes yields a groove reflectivity of $\sim 10\%$.[18] In reflective optics, the reflectance of such a mirror is a function of wavelength and phase shift between the different beams. Higher reflectivity might be achieved by optimizing the groove separation for a given wavelength.

The excitation area of the surface plasmons is defined by the aperture size of the near-field probe. Because of the low throughput of aperture probes, it is not possible to arbitrarily reduce the aperture size. Better field localization can be achieved with laser-irradiated metal tips as discussed in later sections of this chapter.

10.3. CONICAL NEAR-FIELD PROBES SUPPORTING SURFACE PLASMONS

In its classical embodiment, the heart of a near-field optical microscope consists of a tapered optical fiber where the extremity is used as a local optical probe. In order to achieve good imaging properties and subdiffraction resolution, the surface of the fiber is usually coated with a metal layer, and a nanometer-size opening—or aperture—responsible for the optical confinement, is situated at the tip apex. While these types of optical probes have been successfully employed to produce subwavelength optical resolutions in a large number of studies, it is commonly accepted that they suffer from fundamental limitations. When the diameter of the opening becomes comparable to the skin-depth of the coating material, the field distribution at the aperture can be significantly larger than its physical size, resulting in an increased effective opening. Furthermore, as the diameter of the aperture decreases, the intensity of the light transmitted through the opening also rapidly decreases. Consequently, it becomes increasingly difficult to maintain a good signal-to-noise ratio for very small apertures.

While it is most certain that the current design of aperture probes will not be able to sustain routinely ultrahigh resolution imaging (<20 nm), the manufacturing techniques of fiber-based probes are fairly mature, and provide a cost-effective alternative to more involved top-down approaches. Therefore, it would be of practical interest to restore an electromagnetic confinement that is localized at the apex of metal-coated optical fibers.

There are a certain number of approaches making use of tapered optical fibers to achieve field localization, and in some cases an enhancement effect. However, the majority of these techniques require a postprocessing of the tip extremity such as the attachment of a resonant metallic nanoparticle,[24,25] or the controlled growth of a nanotip.[26]

Different research groups suggested that subwavelength field localization may be achieved by focusing surface plasmon polaritons (SPPs).[27–29] Usually, SPPs are bound to planar surfaces, but it was realized that cylindrical metallic waveguides can also sustain SPP waves.[29,30] A metal coating surrounding an optical fiber can therefore act as a SPP waveguide, and for proper phase-matching conditions, focusing of the SPPs at the extremity of the tip can occur.

Below the cut-off radius of an entirely metal-overcoated tapered optical fiber, optical modes are no longer guided and are decaying exponentially. The large wave vectors that are associated with the electric fields after the cut-off can match the phase conditions for surface plasmons propagating on the outer side of the metal layer.[29] As a result, the electromagnetic energy associated with the initial waveguided mode can be coupled to the surface plasmons. After being excited, the surface plasmons propagate along the metal coating toward the tip apex. Despite the slow (almost adiabatic) change of the waveguide radius, the propagation of surface plasmons is accompanied by radiation losses. To create a strong field enhancement at the very end of the tip, the surface plasmons propagating along the circumference must interfere constructively.

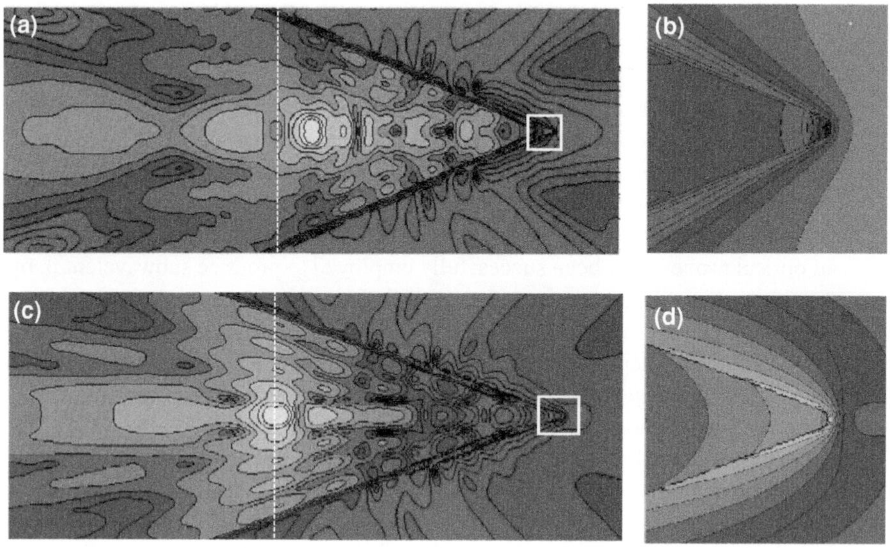

Figure 10.3. (a) Intensity distribution at the end of a gold-coated glass tip. An excitation dipole orientated perpendicular to the tip axis is placed at the cut-off radius as indicated by the dashed line and no field enhancement is observed near the tip apex. Intensity plot with a logarithmic scale (factor of 2.33 between successive lines). Picture size is 2 μm by 6 μm. (b) Magnified view (120 nm by 120 nm) of the very end of the fiber tip (factor of 1.62 between successive lines). (c) Intensity distribution for an excitation dipole oriented along the tip axis at a location indicated by the dashed line. The polarization state is such that an enhanced field is created at the tip apex. (d) Magnified view (120 nm by 120 nm) of the very end of the fiber tip.

The condition of constructive interference is met only for certain symmetries and polarization states of the initial guided mode.[31,32]

If a linearly polarized mode is injected into the fiber, the electric fields associated with the traveling surface plasmons cancel at the end of the tip and hence, do not provide any field enhancement. This situation is shown in Fig. 10.3(a) where the intensity distribution at the extremity of a gold coated tip was calculated for a linearly polarized fundamental HE_{11} waveguided mode. For simplicity, the mode was modeled by a dipole located on the tip axis at the cut-off radius (dashed line in the figure) and aligned perpendicular to it. Figure 10.3(a) demonstrates that surface modes are excited and that they converge toward the very end of the tip. Due to the phase mismatch introduced by the tapered geometry, surface plasmons decay weakly in the form of radiation, giving rise to a far-field background. The calculation also shows that there is no field enhancement produced at the end of the tip. Figure 10.3(b) shows a magnified view of the tip extremity and demonstrates that the polarization conditions give rise to a cancellation of the electric field at the end of the tip.

In order to produce constructive interference at the tip extremity, the electric fields of the surface plasmons must overlap in phase. For a radially polarized waveguide mode possesses the desired rotational symmetry. The different plasmon waves overlap in phase at the apex of the tip thereby producing a high surface charge density, and thus, a strong field enhancement. Figure 10.3(c) shows the calculated intensity distribution at the tip extremity for a radially polarized mode. To mimic the field distribution of a guided radial mode, a dipole oriented parallel to the tip axis was placed at the cut-off radius (dashed line). Similar to Fig. 10.3(a), the surface plasmons propagate toward the tip apex. While in the previous situation, the symmetry of the surface plasmons gave rise to a cancellation of the electric field at the end of the tip, the present case is characterized by a strongly localized field at the end of the tip as shown in the magnified view of Fig. 10.3(d). The localization originates from the fact that the SPP fields polarized along the tip axis add up constructively due to their phase relationship.

It is worth noticing that a recent theoretical work described the effect of SPP superfocusing in a conical geometry through a radius-dependent wave number.[33] It was found that the magnitude of the wave number increases when the SPP propagates toward smaller radius. As a consequence of the dispersion relation, the SPP wavelength decreases for smaller and smaller radius, reducing thus its group velocity to finally achieve localization at the extremity of the tip.[27]

The practical implementation of plasmon guiding probes is challenged by the high nanofabrication requirements. Slight imperfections will distort the axial symmetry of the probes and disturb the phase relationship between the plasmon waves. Thus, while one probe might work, the next one might not, which poses a challenge for reproducible experiments. In principle, the ideal localized excitation or detection source in near-field optical microscopy is a point dipole. As discussed in the following section, a point dipole source is experimentally reasonably realized by a laser-irradiated metal tip. The field confinement only depends on the tip size. Localized fields as small as 8 nm have been demonstrated.

10.4. FIELD DISTRIBUTIONS NEAR METAL TIPS

A metal is characterized by free electrons that interact with external electromagnetic radiation. Depending on the type of the metal and its geometry, the collective response of the free electrons can greatly enhance the electric field strength of the incoming radiation. The coupled excitation of electrons and electromagnetic field is generally referred to as *surface plasmons*. In this section, we discuss the role of surface plasmons associated with the field enhancement effect at laser-irradiated metal tips.

The field enhancement at a sharp tip arises from a combination of the quasi-static lightning-rod effect and surface plasmon excitations.[8] The former is a result of near singularity at the tip; since Maxwell's equations are second-order differential equations, the fields can become singular when the first or second derivative is not defined. This situation is encountered for a perfectly conducting tip. But real metals have finite conductivity and the radius of curvature at the tip is finite. Consequently, there is no real field singularity, but the field at the tip can be strongly enhanced. The contribution of surface-plasmon excitations arises from collective resonances on the tip's surface. Because of the open geometry, one does not expect a pronounced resonance for a tip geometry. A tip is a challenging geometry and one is forced to use numerical methods

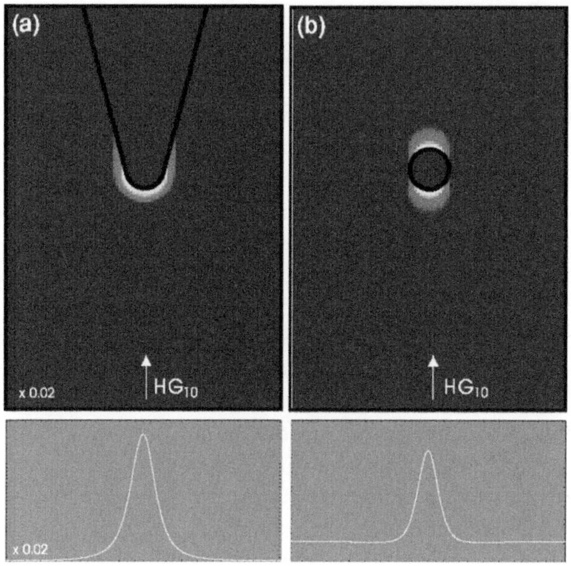

Figure 10.4. Comparison of field distributions ($\mathbf{E}^2$) near a gold tip (a) and a gold particle (b) irradiated by an on-axis focused Hermite-Gaussian (1,0) laser mode. At the apex of tip, the exciting laser field is polarized along the tip axis resulting in a *longitudinal polarization*. The resulting field is enhanced near the tip apex. The line scans at the bottom represent the field distribution on a transverse line 1 nm in front of tip and particle, respectively. However, the field distribution in case of the metal tip has been scaled by a factor of 0.02. The qualitative agreements between the field distributions indicate that in the case of longitudinal polarization, the tip can be modeled as a polarizable particle, but with a modified polarizability that depends on the field enhancement factor.

to analyze the field distributions near a tip. Figure 10.5(a) and Fig. 10.4(a) show calculations for a gold tip performed with the multiple multipole method (MMP).[34] In these examples, two different excitation polarizations are used, transverse to the tip axis and longitudinal to the tip axis, respectively. For transverse polarization, no field enhancement is observed whereas for longitudinal polarization, the enhancement depends strongly on tip geometry, tip material, and excitation wavelength.[8]

To gain a more physically intuitive understanding for the optical response of a metal tip it is favorable to find an approximative model that is analytically exactly solvable. The simplest model for the tip is a quasi-static sphere. Figure 10.5(b) and Fig. 10.4(b) show the optical response of a small gold sphere with the same radius of curvature as the tip geometry. For excitation with transverse polarization, the field distributions in Figs. 10.5(a), and (b) demonstrate that the replacement of the tip with a polarizable sphere provides a very good approximation. The curves in the figures represent the field strength ($|\mathbf{E}|^2$) evaluated 1 nm beneath the tip and sphere, respectively. It turns out that for transverse polarization even the phases of the secondary fields are equal. However, for longitudinal polarization the agreement is only qualitative as shown in Figs. 10.4(a,b). For the gold tip the field enhancement is much stronger than the field enhancement associated with a small gold sphere. Therefore, one needs to adopt an

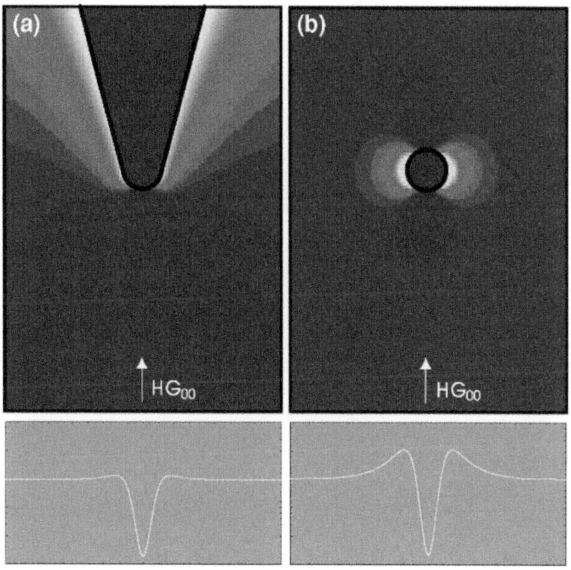

Figure 10.5. Comparison of field distributions ($\mathbf{E}^2$) near a gold tip (a) and a gold particle (b) irradiated by an on-axis focused Gaussian laser beam. At the apex of tip, the exciting laser field is polarized transverse to the tip axis resulting in a *transverse polarization*. The resulting field is depleted near the tip apex. The line scans at the bottom represent the field distribution on a transverse line 1 nm in front of tip and particle, respectively. The qualitative agreements between the field plots indicate that in the case of transverse polarization, the tip can be modeled as a polarizable particle.

empirical polarizability for the sphere that depends on the field enhancement strength at the gold tip. In short, we find that the fields near a metal tip can be described by a polarizable sphere with anisotropic polarizability that accounts for the field enhancement effect.

Figure 10.6(a) shows the calculated surface charge density on a gold tip, when the tip is excited by longitudinal polarization (polarization along z). The surface charge oscillates with the same frequency as the exciting field. The incident light drives the free electrons in the metal along the direction of polarization. While the charge density is zero inside the metal at any instant of time ($\nabla \cdot \mathbf{E} = 0$), charges accumulate on the surface of the metal. When the incident polarization is parallel to the tip axis (Fig. 10.6(a)), the induced surface charge density is rotationally symmetric and has the highest amplitude at the end of the tip. The surface charges form an oscillating standing wave (surface plasmon) with wavelengths shorter than the wavelength of the illuminating light.[8] On the other hand, there is no field enhancement when the polarization is perpendicular to the tip axis. In this case, the tip is simply polarized in the transverse plane and there is no surface charge accumulation at the tip. Let us now invoke the dipole model for the tip using the coordinates defined in Fig. 10.6(b). As discussed before, we find that no matter what the magnitude of the enhancement factor is, the field in the vicinity of the tip can be quite accurately described by the fields of an effective dipole $\mathbf{p}(\omega)$ oscillating at the angular frequency ω, and located

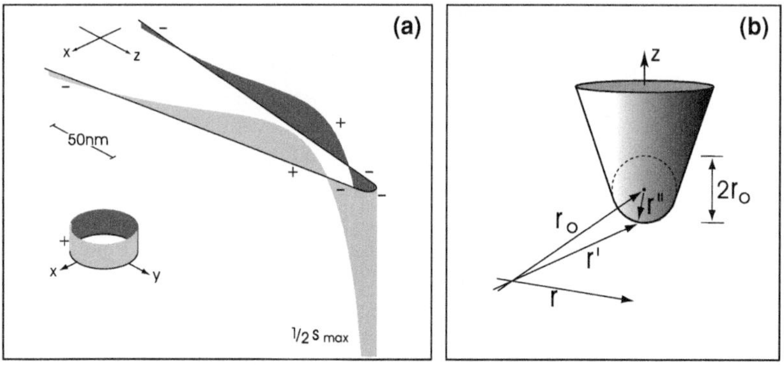

Figure 10.6. (a) Induced surface charge density on the surface of a laser-irradiated gold tip captured at a certain instant of time. Polarization along the tip axis gives rise to a large surface charge accumulation at the apex of the tip. The solid line indicates the outline of the tip whereas the shaded areas represent the surface charge. (b) Coordinates used for the dipole model.

at the center of the tip apex and with the magnitude[35]

$$\mathbf{p}(\omega) = \begin{bmatrix} \alpha_\perp & 0 & 0 \\ 0 & \alpha_\perp & 0 \\ 0 & 0 & \alpha_\parallel \end{bmatrix} \mathbf{E}_o(\omega) \tag{10.2}$$

where we chose the z-axis to coincide with the tip axis. $\mathbf{E}_o$ is the exciting electric field in the absence of the tip. The transverse polarizability $\alpha_\perp$ is identical to the quasi-static polarizability of a small sphere

$$\alpha_\perp(\omega) = 4\pi\varepsilon_o r_o^3 \frac{\varepsilon(\omega) - 1}{\varepsilon(\omega) + 2}. \tag{10.3}$$

where r_o is the tip radius and ε and ε_o are the dielectric constant of the tip and the surrounding medium, respectively. On the other hand, the longitudinal polarizability $\alpha_\parallel$ is given by

$$\alpha_\parallel(\omega) = 8\pi\varepsilon_o r_o^3 f_e(\omega). \tag{10.4}$$

with f_e being the complex field enhancement factor. For a wavelength of $\lambda = 830$ nm, a gold tip with $\varepsilon = -24.9 + 1.57i$, and a tip radius of $r_o = 10$ nm, our numerical calculations based on the multiple multipole method[34] lead to $f_e = -2.9 + 11.8i$. The expression for $\alpha_\parallel$ originates from the requirement that the magnitude of the field produced by $\mathbf{p}(\omega)$ at the surface of the tip is equal to the computationally determined field which we set equal to $f_e\mathbf{E}_o$. The electric field $\mathbf{E}$ at a given position of the tip relative to the exciting laser beam r is now approximated as

$$\mathbf{E}(\mathbf{r}, \omega) = \mathbf{E}_o(\mathbf{r}, \omega) + \frac{1}{\varepsilon_o}\frac{\omega^2}{c^2}\overleftrightarrow{\mathbf{G}}^o(\mathbf{r}, \mathbf{r}_o, \omega)\,\mathbf{p}(\omega), \tag{10.5}$$

where $\mathbf{r}_o$ specifies the origin of $\mathbf{p}$ and $\overleftrightarrow{\mathbf{G}}^o$ is the free space dyadic Green's function.

In summary, we find that the localized field near a laser-irradiated metal tip is well approximated by the field of an oscillating dipole located at the tip apex. The size of the tip defines how close the tip can be brought to a sample surface and how strong the field localization is. Because the external illumination not only excites the metal tip but also irradiates the sample surface, the field enhancement needs to be strong enough to suppress the signal associated with direct sample irradiation. As discussed in the next section, nonlinear signal generation at the apex of a bare metal tip provides ultrahigh field localization that is basically background free and suitable for ultrahigh resolution imaging.

10.5. LOCALLY EXCITED LUMINESCENCE FROM METAL NANOSTRUCTURES

The local field enhancement associated with structural resonances has been found to be a prerequisite for efficient surface-enhanced processes and nonlinear responses of a material. Near-field scattering studies of rough silver films showed that the field-enhancement predominantly originates from a few locations (hot spots) where local plasmon modes are present.[36] It was found that a visible broad-band photoluminescence originating from the surface itself was emitted from these regions.[37,38] Similar emission has been reported for gold nanoparticles[37], and more recently, for elliptically shaped particles[39,40] and nanometric tips.[37] The white-light photoluminescence emitted in the visible part of the spectrum originates from interband recombination of holes in the d-band with electrons in the conduction band near the Fermi surface. This signal can be excited directly with high energy photons, or through a nonlinear two-photon absorption process. The spectrum of the photoluminescence represents the convolution of the electronic joint density of states between the two bands, and the optical coupling efficiency due to the particular structured geometry and associated resonances. In particular, it was found that the photoluminescence yield is drastically increased in regions of strong electromagnetic enhancement thus providing a probe to investigate site-specific enhancement properties.[37]

Recent experiments have shown that nanoparticles excited by the field of a tightly focused femtosecond laser beam generate a photoluminescent continuum through a two-photon absorption mechanism.[37] Figure 10.7(a) shows the photoluminescence emitted from a gold particle scanned through the focal region. The photoluminescence pattern indicates that the particle responds to the total field of the focal region, and not to any particular polarization component present in the focus.[43] The photoluminescence spectrum is shown in Fig. 10.7(b). The spectrum is peaked at 644 nm in agreement with the predicted surface plasmon resonance indicating a strong relationship with the intrinsic response of the particle.[41,42]

In order to obtain spatial information about the local distribution of the photoluminescence, and consequently regions of large enhancement, a metal tip was used to locally scatter the photoluminescence. The tip was held at a fixed position within the focal area while the particle was laterally raster-scanned underneath it. Figure 10.8(a) depicts the topography of the particle revealing its ellipsoidal shape. Figure 10.8(b)

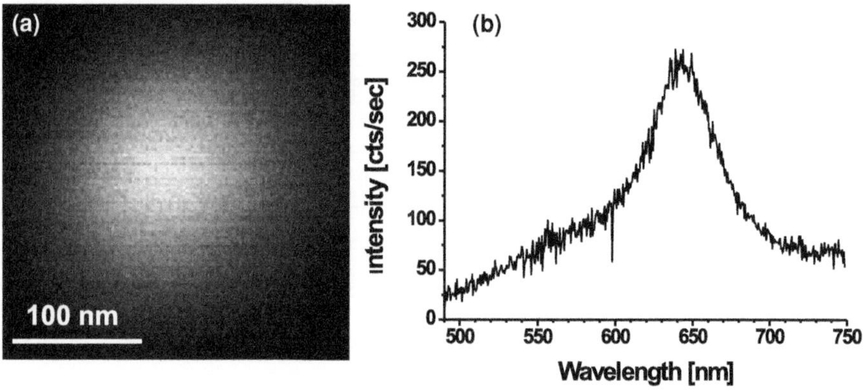

Figure 10.7. (a) Two-photon excited photoluminescence image of gold particles scanned through the focal region of a strongly focused, 120 fs pulsed Gaussian laser beam. (b) Spectrum of the photoluminescence.

shows the near-field spatial distribution of the photoluminescence emitted by the particle as a function of its position. The original far-field pattern seen in Fig. 10.7(a) is still visible as a faint background. Superimposed to this background are high resolution details originating from the tip-particle interaction. At the extremities of the particle, the photoluminescence response is strongly enhanced relative to the far-field background. On the other hand, the signal intensity is reduced at the two diametrically opposed points along the short axis of the ellipsoid. Interestingly, these two effects are not influenced by the polarization direction of the excitation and were observed on all investigated particles.

The image of Fig. 10.8(b) strongly resembles the calculated surface intensity distribution of an ellipsoid excited at resonance as shown in Fig. 10.8(c). The image demonstrates the dipolar character of the excited particle: charge accumulation at both ends along the long axis and charge depletion along the short axis are in a very nice agreement with the experimental image of Fig. 10.8(b). The calculation

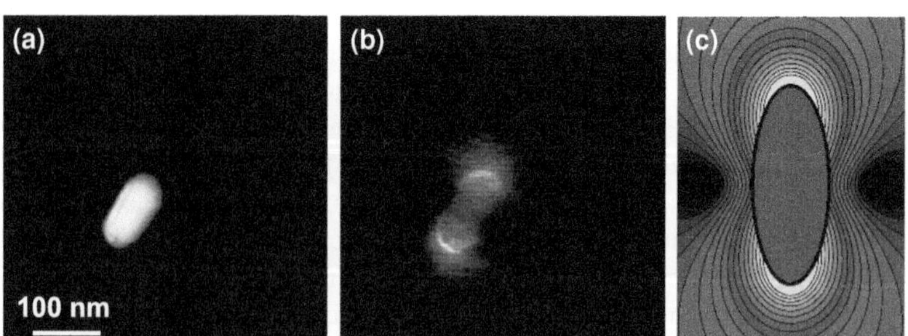

Figure 10.8. (a) Topography of the investigated gold particle (b) Two-photon excited photoluminescence distribution of the particle imaged with a stationary gold tip. (c) Calculated intensity distribution of ellipsoid with the same dimensions excited near resonance (650 nm).

was based on MMP for an excitation wavelength of 650 nm and a gold permittivity $\varepsilon_{Au} = -12.9+1.09i$. The width of the photoluminescence peaks at the particle extremities is approximatively 17 nm (full width at half maximum) indicating a high degree of confinement of the photoluminescence. As a consequence, and due to the sensitivity of the photoluminescence yield on the field strength, Fig. 10.8(b) also reveals the regions of large field enhancement localized on the nanoparticle. In summary, the detection of localized photoluminescence near metal nanostructures provides an efficient strategy for the localization and characterization of hot spots.

10.6. CONCLUSIONS AND OUTLOOK

The confined nature of surface plasmons and their relatively long propagation length make them suitable for integration in metallic planar circuitry. Elementary plasmonic functions are currently being developed and studied with techniques relying on extended, unlocalized excitation. With the increasing complexity of plasmonic structures, it is an advantage to understand the functional properties *locally* by confining the surface-plasmon excitations to regions comparable to the dimensions of the nanocircuitery. A near-field probe fulfills this requirement. Surface plasmons can be locally excited near arbitrary structures on metal surfaces. Furthermore, the detection of the intrinsic leakage radiation associated with plasmons propagating on metal films deposited on dielectric substrates renders a direct measurement of the propagation length of surface plasmons and the various damping mechanisms.

In the context of near-field optical microscopy, surface plasmons play an important role in creating both a highly confined source of photons and an enhanced electromagnetic field. Focusing of surface plasmons, for instance, might be instrumental to restore the electromagnetic confinement of metal-coated, fiber-based, near-field probes. Surface plasmons can be excited on the outer surface of near-field probes and propagated toward the probe apex creating a virtually background-free localized excitation. Alternatively, localized surface-plasmon resonances occurring at metal tips can produce significant field enhancement used for high-resolution optical imaging. It is found that the field in the vicinity of a laser-irradiated metal tip can be described by the field of a dipole with an anisotropic polarizability. Furthermore, for excitation with high-peak power laser pulses, the field enhancement effect gives rise to local nonlinear effects. It was shown that both second-harmonic generation and continuum generation (luminescence) at the tip apex create a highly confined light source. This property was utilized to produce a direct map of the distribution of local fields (hot spots) of plasmonic structures, and is a promising approach for the characterization of field-enhancing structures used in future devices based on the general properties of surface plasmons.

REFERENCES

1. J. Homola, S. S. Yee, G. Gauglitz: Surface plasmon resonance sensors: review, Sensors and Actuators B **54**, 3 (1999).
2. A. J. Haes, W. P. Hall, L. Chang, W. L. Klein, R. P. Van Duyne: A localized surface plasmon resonance biosensor: First steps toward an assay for Alzheimers disease, Nano Lett. **4**, 1029 (2004).

3. J. C. Weeber, A. Dereux, C. Girard, J. R. Krenn, J. P. Goudonnet: Plasmon polaritons of metallic nanowires for controlling submicron propagation of light, Phys. Rev. B **60**, 9061 (1999).
4. H. Ditlbacher, J. R. Krenn, G. Schider, A. Leitner, F. R. Aussenegg: Two-dimensional optics with surface plasmon polaritons, Appl. Phys. Lett. **81**, 1762 (2002).
5. J. Wessel: Surface-enhanced optical microscopy, J. Opt. Soc. Am. B **2**, 1538 (1985).
6. L. Novotny, E. J. Sanchez, X. S. Xie: Near-field optical imaging using metal tips illuminated by higher-order Hermite-Gaussian beams, Ultramicroscopy **71**, 21 (1998).
7. A. Hartschuh, E. J. Sanchez, X. S. Xie, L. Novotny: High-resolution near-field Raman microscopy of single-walled carbon nanotubes, Phys. Rev. Lett. **90**, 95503 (2003).
8. L. Novotny, R. X. Bian, X. S. Xie: Theory of nanometric optical tweezers, Phys. Rev. Lett., **79**, 645 (1997).
9. B. Hecht, H. Bielefeldt, L. Novotny, Y. Inouye, D. W. Pohl: Local excitation, scattering, and interference of surface plasmons, Phys. Rev. Lett. **77**, 1889 (1996).
10. D. McMullan: SPIE Milestone Series: Selected Papers on Near-field Optics, **172**, 31 (2002).
11. E. H. Synge: A suggested model for extending microscopic resolution into the ultra-microscopic region, Phil. Mag. **6**, 356 (1928).
12. W. Denk, D. W. Pohl: Near-field optics: microscopy with nanometer-size fields, J. Vac. Sci. Technol. B **9**, 510 (1991).
13. U. Ch. Fischer, D. W. Pohl: Observation of single-particle plasmons by near-field optical microscopy, Phys. Rev. Lett. **62**, 458 (1989).
14. F. Keilmann, R. Hillenbrand: Near-field microscopy by elastic light scattering from a tip, Phil. Trans. R. Soc. Lond. A **362**, 787 (2004).
15. E. J. Sanchez, L. Novotny, X. S. Xie: Near-field fluorescence microscopy based on two-photon excitation with metal tips, Phys. Rev. Lett. **82**, 4014 (1999).
16. L. Novotny, B. Hecht, D. W. Pohl: Interference of locally excited surface plasmons, J. Appl. Phys. **81**, 1798 (1997).
17. A. Bouhelier, Th. Huser, H. Tamaru, H. J. Güntherodt, D. W. Pohl: Plasmon transmissivity and reflectivity of narrow grooves in silver films, J. Microscopy **194**, 571 (1999).
18. A. Bouhelier, Th. Huser, H. Tamaru, H. J. Güntherodt, D. W. Pohl, F. Baida, D. Van Labeke: Plasmon optics of structured silver films, Phys. Rev. B **63**, 155404 (2001).
19. F. I. Baida, D. Van Labeke, A. Bouhelier, Th. Huser, D. W. Pohl: Propagation and diffraction of locally excited surface plasmons, J. Opt. Soc. Am. A **18**, 6 (2001).
20. C. J. Bouwkamp: On Bethe's theory of diffraction by small holes, Rep. Phys. **5**, 321 (1950).
21. A. Otto: Excitation of nonradiative surface plasma waves in silver by the method of frustrated total reflection, Z. Angew. Phys. **216**, 398 (1968).
22. A. Bouhelier, G. P. Wiederrecht: Surface plasmon rainbow jets, Opt. Lett. **30**, 884 (2005).
23. A. Bouhelier, G. P. Wiederrecht: Excitation of broadband surface plasmon polaritons: Plasmonic continuum spectroscopy, Phys. Rev. B, **71**, 195406 (2005).
24. O. Sqalli, M. P. Bernal, P. Hoffmann, F. Marquis-Weible: Gold elliptical nanoantennas as probes for near field optical microscopy, Appl. Phys. Lett. **76**, 2134 (2000).
25. Th. Kalkbrenner, M. Ramstein, J. Mlynek, V. Sandoghdar: A single gold particle as a probe for apertureless scanning near-field optical microscopy, J. Microscopy **202**, 72 (2001).
26. H. G. Frey, F. Keilmann, A. Kriele, R. Guckenberger: Enhancing the resolution of scanning near-field optical microscopy by a metal tip grown on an aperture probe, Appl. Phys. Lett. **81**, 5030 (2002).
27. M. I. Stockman: Nanofocusing of optical energy in tapered plasmonic waveguides, Phys. Rev. Lett. **93**, 137404 (2004).
28. F. Keilmann: Surface polaritons propagation for scanning near-field optical microscopy applications, J. Microscopy **194**, 567 (1999).
29. L. Novotny, C. Hafner: Light propagation in a cylindrical waveguide with a complex, metallic, dielectric function, Phys. Rev. E **50**, 4094, (1994).
30. G. Goubau: Surface waves and their application to transmission lines, J. Appl. Phys. **21**, 1119 (1950).

31. A. Bouhelier, J. Renger, M. R. Beversluis, L. Novotny: Plasmon-coupled tip-enhanced near-field optical microscopy, J. Microscopy **210**, 220–224 (2003).
32. L. Vaccaro, L. Aeschimann, U. Staufer, H. P. Herzig, R. Dändliker: Propagation of the electromagnetic field in fully coated near-field optical probes, Appl. Phys. Lett. **83**, 584 (2003).
33. A. J. Babadjanyan, N. L. Margaryan, Kh. V. Nerkararyan: Superfocusing of surface polaritons in the conical structure, J. Appl. Phys. **87**, 3785 (2000).
34. Ch. Hafner: *The Generalized Multiple Multipole Technique for Computational Electromagnetics* (Artech, Boston, 1990).
35. A. Bouhelier, M. Beversluis, A. Hartschuh, L. Novotny: Near-field second-harmonic generation induced by local field enhancement, Phys. Rev. Lett. **90**, 13903 (2003).
36. V. A. Markel, V. M. Shalaev, P. Zhang, W. Huynh, L. Tay, T. L. Haslett, M. Moskovits: Near-field optical spectroscopy of individual surface-plasmon modes in colloid clusters, Phys. Rev. B **59**, 10903 (1999).
37. M. R. Beversluis, A. Bouhelier, L. Novotny: Continuum generation from single gold nanostructures through near-field mediated intraband transitions, Phys. Rev. B. **68**, 115433 (2003).
38. G. T. Boyd, Z. H. Yu, Y. R. Shen: Photoinduced luminescence from the noble metals and its enhancement on roughened surfaces, Phys. Rev. B **33**, 7923 (1986).
39. M. B. Mohamed, V. Volkov, S. Link, M. A. El-Sayed: The ‘lightning’ gold nanorods: fluorescence enhancement of over a million compared to the gold metal, Chem. Phys. Lett. **317**, 517 (2000).
40. A. Bouhelier, M. R. Beversluis, L. Novotny: Characterization of nanoplasmonic structures by locally excited photoluminescence, Appl. Phys. Lett. **83**, 5041 (2003).
41. A. Bouhelier, R. Bachelot, G. Lerondel, S. Kostcheev, P. Royer, G. Wiederrecht: Surface Plasmon Characteristics of Tunable Photoluminescence in Single Gold Nanorods, Phys. Rev. Lett. **95**, 267405 (2005).
42. H. Wang, T. B. Huff, D. A. Zweifel, W. He, P. S. Low, A. Wei, J. X. Cheng: In vitro and in vivo two-photon luminescence imaging of single gold nanorods, Proc. Nat. Acad. Sci. **102**, 10552 (2005).
43. A. Bouhelier, M. R. Beversluis, L. Novotny: Near-field scattering of longitudinal fields, Appl. Phys. Lett. **82**, 4596 (2003).

CHAPTER ELEVEN

PRINCIPLES OF NEAR-FIELD OPTICAL MAPPING

ALAIN DEREUX
Laboratoire de Physique de l'Université de Bourgogne, BP 47870, F-21078 Dijon, France
adereux@u-bourgogne.fr

11.1. NEAR-FIELD OPTICAL MICROSCOPY

Before the emergence of the actual interest for the potential applications of *nanophotonics*, a generation of surface physicists worked on the development of a new class of instruments known today as near-field optical microscopes.[1] Many fundamental phenomena discovered in this context are directly relevant for the development of miniaturized optical devices. As time passed and as we will show in this chapter, it was recognized that one main practical feature of near-field optical microscopes is the ability to map the electromagnetic fields associated with optical waves. Today, this functionality provides an essential way of characterizing miniaturized optical devices, such as, but not limited to, plasmonic devices. In such devices, where the sizes of the objects are of the order of the wavelength λ or smaller, phenomena involving evanescent electromagnetic waves, i.e. decaying within a range given by λ, dominate. To fully exploit the potential of optical nanodevices, it is clear that signal detection or any signal conversion process should be controlled at the subwavelength scale. Therefore, the development of nanophotonics demands a clear understanding of the fundamental issue of the detection of optical fields at the subwavelength scale.

Since the 1980s, various configurations of near-field optical microscopes, operating under stationary illumination conditions, have been developed around two generic experimental configurations: the Scanning Near-Field Optical Microscope (SNOM) and the Photon Scanning Tunneling Microscope (PSTM).

The SNOM exploits the analogy to the electron scanning tunneling microscope: a nanometer size light source scans the sample surface. Depending on the nature of the sample, the outgoing light is detected in transmission or in reflection. Although reflection SNOM devices use the tip both as local emitter and as local probe, the discussion below will make clear that SNOMs are fundamentally *illuminating* probe devices.

M.L. Brongersma and P.G. Kik (eds.), Surface Plasmon Nanophotonics, 155–168.

The PSTM operates differently. The sample lies on a glass prism, which enables illumination via total internal reflection. The nanometer size tip, scanning the surface, then frustrates the total reflection. The PSTM probe tip is thus used as a detector of the optical field close to the surface. This is referred to as the *collection* mode. Note that it has been proven[2] that a third type of near-field optical microscope, called an "apertureless" near-field optical microscope, may be viewed as a collection mode microscope in the sense discussed here. Both configurations generally use tips obtained by pulling optical fibers which may be coated with a metal. The structure of the metallization at the tip apex is not trivial. In a simplified description of this structure, the tip apex is modeled as a subwavelength (sub-λ) aperture. During the 1990s, the development of near-field optical microscopy was hindered by the absence of a criterion that rigorously defines this sub-λ nature. The lack of any reliable criterion led to controversies about the interpretation of near-field optical images.[3]

11.2. INTERPRETATION OF NEAR-FIELD OPTICAL IMAGES

In order to interpret the images obtained by either of the general setups, we propose a practical point of view relying on the Heisenberg uncertainty principle, which deals with a measurement in a volume δl^3 such that δl is of subwavelength (sub-λ) size.

By going through a sub-λ structure, the incident wave faces the consequence of the Heisenberg uncertainty principle ($i = x, y, z$):

$$\Delta x_i \, \Delta p_i \geq \frac{\hbar}{2} \tag{11.1}$$

For an electromagnetic wave, this leads to an uncertainty principle which, through the cyclic permutation of the indexes ($i, j = x, y, z$), links the components of the electric E_i and magnetic H_j fields of the optical wave and the typical size δl (SI units):

$$\Delta E_i \, \Delta H_j \geq \frac{\hbar}{2} \frac{c^2}{(\delta l)^4} \tag{11.2}$$

If $\delta l \leq 0.1$ μm, the right hand side of this formula becomes large (Fig. 11.1). This uncertainty principle means that a simultaneous (in the sense of quantum theory, i.e. without any reciprocal influence) measurement of the electric field and of the magnetic field is not possible if the detection occurs in a volume δl^3 such that δl is sub-λ. Consequently, the energy of an electromagnetic wave, containing both electric and magnetic contributions, also becomes uncertain (in the sense of quantum theory) in a sub-λ volume. Therefore, using near-field optical microscopy one cannot achieve a sub-λ measurement that could be interpreted like the far-field measurement of the reflected or the transmitted energy as is done in a standard microscope. The practical interpretation that we propose, *defines* sub-λ resolution by the detection of the spatial distribution of the intensity of either the electric field or (exclusively) the magnetic

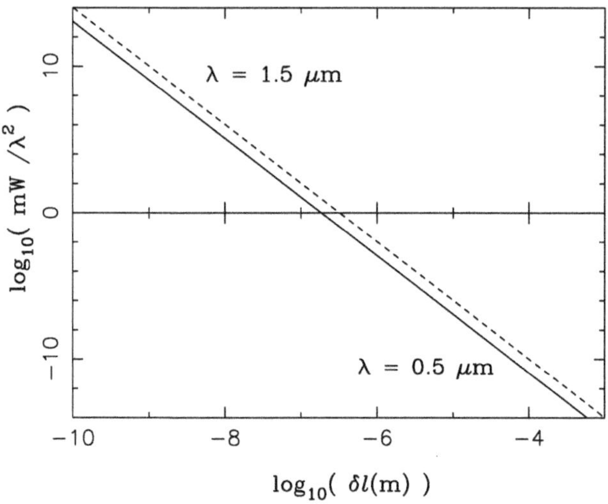

Figure 11.1. Evaluation of the right hand side of equation 11.2 for two typical wavelengths in vacuum.

field close to the sample surface. Practically, we suggest that sub-λ resolution is achieved if [4,5]:

- *in collection mode,* the experimental images agree with the theoretical distribution of the electric *or* the magnetic field scattered by the sample surface, as computed without including any tip.
- *in illumination mode,* the experimental images agree with the theoretical distribution of the electromagnetic Local Density of States (LDOS) at optical frequencies, as computed without including any tip.

This criterion allows for a rigorous determination of not only the quality of the probe tip, but also of the quality of the entire experimental setup. It requires the ability to compute the electromagnetic field and the electromagnetic LDOS in the near-field zone. The scattering theory explained in the next chapter enables the computation of both within a single framework.

11.3. SCATTERING THEORY OF ELECTROMAGNETIC WAVES

From a mathematical point of view, scattering theory (also known as the field susceptibility or Green's dyadic technique) is based on the Green's function theory applied to the wave equation when a source term is introduced. It simply describes the most general analytical solution of the inhomogeneous wave equation as an integral equation where the kernel is a Green's function.[6] Several formulations of electromagnetic scattering theory have been applied successfully to the modeling of near-field optical phenomena. Although the Green's function may be expanded in Fourier or multipole series, most formulations favor a discretization in real space, since near-field optical phenomena occur on a subwavelength scale.[7]

Until now, most applications of near-field optics involve stationary laser illumination. This experimental mode allows one to restrict the theoretical description to electromagnetic fields that depend harmonically on time. With this $\exp(-i\omega t)$ time dependence, the Maxwell's equations in the absence of any external source lead to the following vector wave equation for the electric field $\mathbf{E}$ (SI units):

$$-\nabla \times \nabla \times \mathbf{E}(\mathbf{r}) + \frac{\omega^2}{c^2}\,\epsilon(\mathbf{r})\,\mathbf{E}(\mathbf{r}) = 0 \tag{11.3}$$

which may be written as

$$-\nabla \times \nabla \times \mathbf{E}(\mathbf{r}) + q^2\,\mathbf{E}(\mathbf{r}) = \boldsymbol{V}(\mathbf{r})\,\mathbf{E}(\mathbf{r}) \tag{11.4}$$

with

$$q^2 = \frac{\omega^2}{c^2}\,\varepsilon_{\text{ref}}. \tag{11.5}$$

Any complicated behavior due to the anisotropy or due to the low-symmetry of the geometrical shape of the original dielectric tensor profile $\epsilon(\mathbf{r})$, is described as a difference relative to the reference system ε_{ref}:

$$\boldsymbol{V}(\mathbf{r}) = \frac{\omega^2}{c^2}\,(\boldsymbol{I}\,\varepsilon_{\text{ref}} - \epsilon(\mathbf{r}))\,. \tag{11.6}$$

The solution of (11.4) is obtained from the implicit Lippmann–Schwinger equation

$$\mathbf{E}(\mathbf{r}) = \mathbf{E}_o(\mathbf{r}) + \int_D \mathrm{d}\mathbf{r}'\,\boldsymbol{G}_o(\mathbf{r},\mathbf{r}')\,\boldsymbol{V}(\mathbf{r}')\,\mathbf{E}(\mathbf{r}'). \tag{11.7}$$

In scattering theory, the first term $\mathbf{E}_o(\mathbf{r})$ is referred to as the incident field, while the second term is called the scattered field, obtained from the integration over the domain D where $\boldsymbol{V}(\mathbf{r}')$ is non-zero. D defines the volume of the scatterer relative to the reference system.

To solve the Lippmann–Schwinger equation, we need to know the analytical solution $\mathbf{E}_o(\mathbf{r})$ satisfying

$$-\nabla \times \nabla \times \mathbf{E}_o(\mathbf{r}) + q^2\,\mathbf{E}_o(\mathbf{r}) = 0 \tag{11.8}$$

and the associated Green's dyadic defined by ($\boldsymbol{I}$ being the unit dyadic)

$$-\nabla \times \nabla \times \boldsymbol{G}_o(\mathbf{r},\mathbf{r}') + q^2\,\boldsymbol{G}_o(\mathbf{r},\mathbf{r}') = \boldsymbol{I}\,\delta(\mathbf{r}-\mathbf{r}'). \tag{11.9}$$

The reference structure ε_{ref} is usually a homogeneous background material or a semi-infinite surface system. For homogeneous media, the analytical form of $\boldsymbol{G}_o(\mathbf{r},\mathbf{r}')$ is known from ancient works[8,9] to be:

$$\boldsymbol{G}_o(\mathbf{r},\mathbf{r}') = -\left[\boldsymbol{I} - \frac{1}{q^2}\nabla\nabla\right]\frac{\exp(iq\,|\mathbf{r}-\mathbf{r}'|)}{4\pi\,|\mathbf{r}-\mathbf{r}'|}. \tag{11.10}$$

For a surface system, the expression of the propagator is somewhat more complicated.[10–12]

11.4. ELECTROMAGNETIC LDOS

The electromagnetic LDOS of vacuum $\rho_o(\mathbf{r},\omega)$ is well-known as the factor multiplying the Bose–Einstein distribution in Planck's law describing the black-body radiation.

$$U(\omega)\,d\omega = \rho_o(\mathbf{r},\omega)\,\frac{\hbar\omega}{e^{\frac{\hbar\omega}{k_B T}}-1}\,d\omega \tag{11.11}$$

It also forms the basis of the *Fermi Golden Rule*, which describes the decay rate Γ as the problem of coupling a discrete system with a continuum. Indeed, from

$$\Gamma = \frac{2\pi}{\hbar}\,|\langle f|\mathbf{p}\cdot\mathbf{E}(\mathbf{r})|i\rangle|^2\,\delta(\omega=\omega_f-\omega_i) \tag{11.12}$$

one can show that

$$\Gamma = \frac{2\pi}{\hbar}\,|\langle f|\mathbf{p}|i\rangle|^2\,\rho_o(\mathbf{r},\omega). \tag{11.13}$$

Application of standard calculus of distributions shows that the factor $\rho_o(\mathbf{r},\omega)$ follows from the electric Green's dyadic $\boldsymbol{G}_o$ of vacuum ($\Im$ stands for the imaginary part):

$$\rho_o(\mathbf{r},\omega) = -\frac{1}{\pi}\,\Im\,\mathrm{Trace}\,\boldsymbol{G}_o(\mathbf{r},\mathbf{r}',\omega) = \sum_{j=x,y,z}\rho_{o;j}(\mathbf{r},\omega) = \frac{1}{\pi^2}\,\frac{\omega^2}{c^3} \tag{11.14}$$

where, to account for the vector nature of electromagnetic fields, we have defined the "partial" LDOS by:

$$\rho_{o;j}(\mathbf{r},\omega) = -\frac{1}{\pi}\,\Im\,\boldsymbol{G}_{o;jj}(\mathbf{r},\mathbf{r},\omega) \tag{11.15}$$

Close to sub-λ structures deposited on surfaces, this LDOS may vary from point to point and may depend on the polarization of the exciting dipole. In the case of a system described by its dielectric function $\epsilon(\mathbf{r},\omega)$, the LDOS is related to (but not equal to) the dipolar point source corresponding to the Dirac δ function appearing in the wave equation that defines Green's dyadic of the system:

$$-\nabla\times\nabla\times\boldsymbol{G}(\mathbf{r},\mathbf{r}',\omega) + \frac{\omega^2}{c}\,\epsilon(\mathbf{r},\omega)\,\boldsymbol{G}(\mathbf{r},\mathbf{r}',\omega) = \boldsymbol{I}\,\delta(\mathbf{r}-\mathbf{r}') \tag{11.16}$$

The Green's dyadic of the actual system $\boldsymbol{G}(\mathbf{r},\mathbf{r}',\omega)$ may be deduced numerically from one of a reference system after applying a Dyson equation. In principle, this reference system can be chosen as a homogeneous medium such as vacuum, therefore:

$$\boldsymbol{G}(\mathbf{r},\mathbf{r}') = \boldsymbol{G}_o(\mathbf{r},\mathbf{r}') + \int_V \boldsymbol{G}_o(\mathbf{r},\mathbf{r}'')\,\boldsymbol{V}(\mathbf{r}'')\,\boldsymbol{G}(\mathbf{r}'',\mathbf{r}')\,d\mathbf{r}'' \tag{11.17}$$

The LDOS is deduced from the electric Green's dyadic $\boldsymbol{G}$:

$$\rho(\mathbf{r},\omega) = -\frac{1}{\pi}\,\Im\,\mathrm{Trace}\,\boldsymbol{G}(\mathbf{r},\mathbf{r}',\omega) = \sum_{j=x,y,z} \rho_j(\mathbf{r},\omega) \tag{11.18}$$

where, again, we have to define the "partial" LDOS by:

$$\rho_j(\mathbf{r},\omega) = -\frac{1}{\pi}\,\Im\,\boldsymbol{G}_{jj}(\mathbf{r},\mathbf{r},\omega) \tag{11.19}$$

Through the unit dyadic $\boldsymbol{I}$, each partial LDOS is related to a given orientation x, y, or z of the point-like source. The above summary points out that in order to detect a signal proportional to the LDOS, the adopted point of view requires finding the experimental conditions such that, *in practice*, one can consider the probe tip as a point-like dipole oscillating at the angular frequency ω.

11.5. MAPPING THE OPTICAL NEAR-FIELD

11.5.1. PSTM Detection of the Electric or Magnetic Components of Optical Waves

In the context of the interpretation of PSTM images, the tip design is of primary importance. It turns out that dielectric tips obtained by pulling optical fibers provide a signal proportional to the electric field associated with the optical wave. These same tips, when coated with a thin film of Au (10 nm to 50 nm), provide a signal proportional to the magnetic field associated with the optical wave. Experiments have reproduced the phenomenon of the detection of the magnetic component of the optical wave at several incident wavelengths. However, to observe this phenomenon at a given wavelength, the thickness of the gold coating surrounding the dielectric core of the tip has to be adjusted precisely in order to excite a circularly symmetric plasmon in the coating.

Figures 11.2, 11.3 and 11.4 illustrate the detection of the optical magnetic field close to nanostructures as observed by PSTM. Figure 11.2 shows an AFM image of the topography of the reference nanostructure. Figure 11.3 displays the theoretical distributions of the intensities of the electric (a) and magnetic (b) fields. For this specific sample, the distributions keep the same features if $\lambda = 543$ nm. The calculations assume that the nanostructures are deposited on a perfectly planar surface. This leads to strong interference patterns in the vicinity of the pads. In experiments, realistic surfaces degrade these interference fringes, resulting in a speckle pattern. For this reason, a comparison between theory and experiment must be limited to the contrast on top of each pad. Figure 11.4 combines PSTM images recorded above the sample of Fig. 11.2 using optical fiber tips that were coated with different thicknesses d of Au. In Fig. 11.4(a) and (c), the thicknesses have been selected in order to excite a circularly symmetric plasmon. Images (a) and (c) agree with the distribution of the optical magnetic field (Fig. 11.3(b)), while images (b) and (d) agree with the distribution

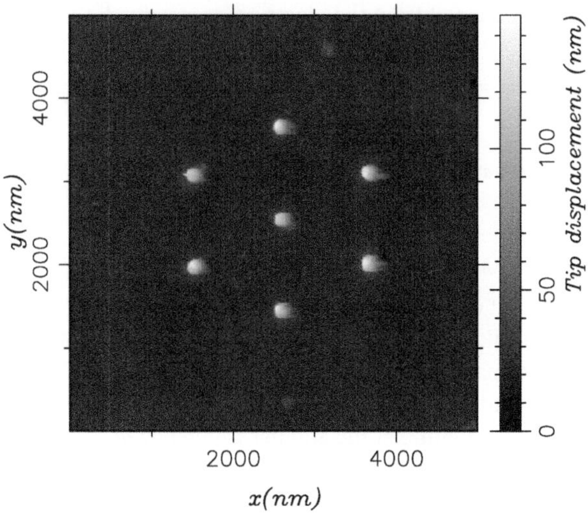

Figure 11.2. AFM image of the topography of the reference nanostructure: seven gold glass pads (130 nm × 130 nm × 100 nm) are deposited on a planar glass surface.

of the electric field (Fig. 11.3(a)). At both wavelengths $\lambda = 543$ nm et $\lambda = 633$ nm, dielectric (uncoated) tips provide images similar to (b) and (d).

Using another incident polarization or observing a different kind of sample such as gold pads supporting plasmon resonances leads to similar conclusions.[13]

The unprecedented agreement of the experimental PSTM results with the relevant theoretical distributions validates the practical point of view proposed in sect. 11.2 for collection mode near-field optical microscopes.[14–16] Moreover, the excitation of

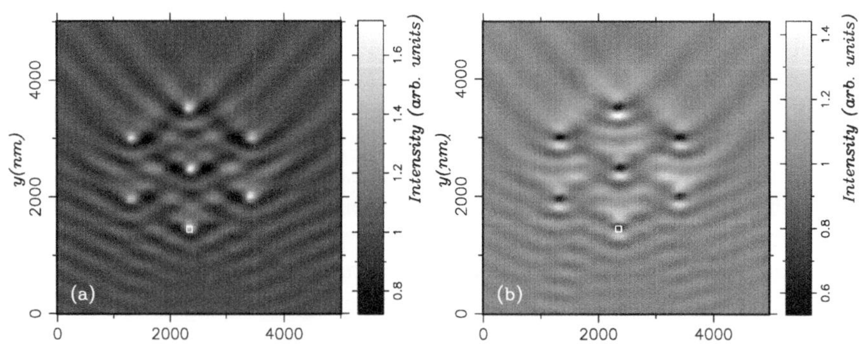

Figure 11.3. Theoretical distributions (angle of incidence = 60 degrees, TM polarization, $\lambda = 633$ nm) of the intensity of the electric (a) and magnetic (b) fields scattered in the near-field zone close to the surface of the sample shown in Fig. 11.2.

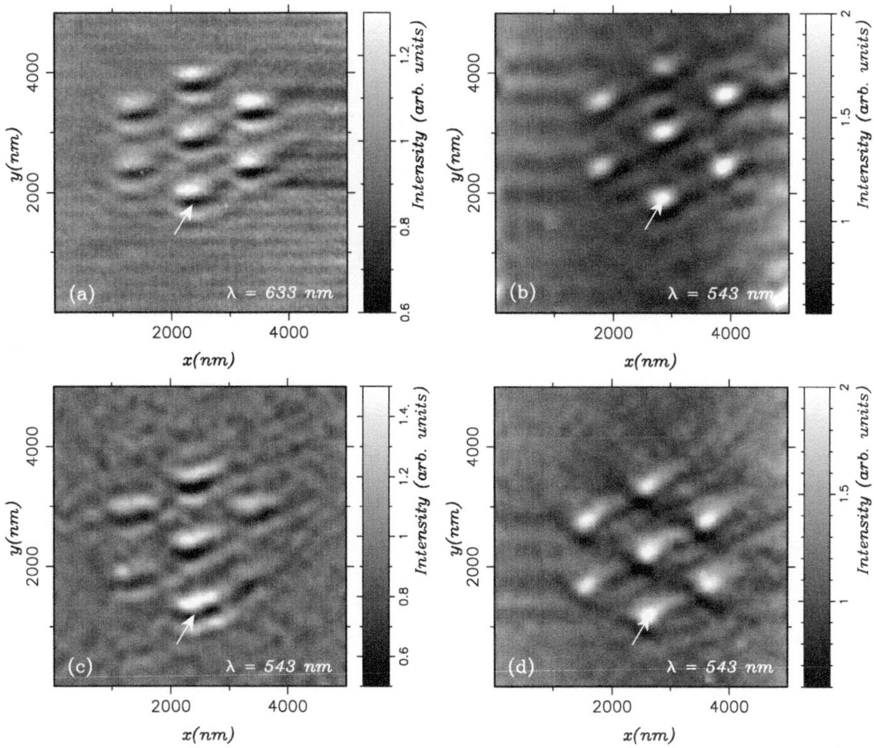

Figure 11.4. PSTM images of the sample shown in Fig. 11.2 (TM polarization, angle of incidence = 60 degrees). (a) $d = 20$ nm, $\lambda = 633$ nm; (b) $d = 20$ nm, $\lambda = 543$ nm; (c) $d = 30$ nm, $\lambda = 543$ nm; (d) $d = 35$ nm, $\lambda = 543$ nm.

circularly symmetric localized surface plasmons supported by the metal coating on the probe tip turns out to be essential to confirm the validity of the proposed point of view.

11.5.2. SNOM Detection of the Electromagnetic LDOS

Applying the practical point of view of Sect. 11.2 to SNOM configurations is somewhat less intuitive. Indeed, since the local probe is also the source of incident light, how can one identify the field when no tip is present, which is assumed in the practical point of view? To answer this question, one should remember that the field existing independently of any external excitation corresponds to the ground state of the electromagnetic field. This ground state is described by the electromagnetic LDOS at the frequency of the laser used in the experiment. The practical point of view assumes that SNOM configurations meet the criterion of sub-λ resolution if a signal proportional to the electromagnetic LDOS is detected.

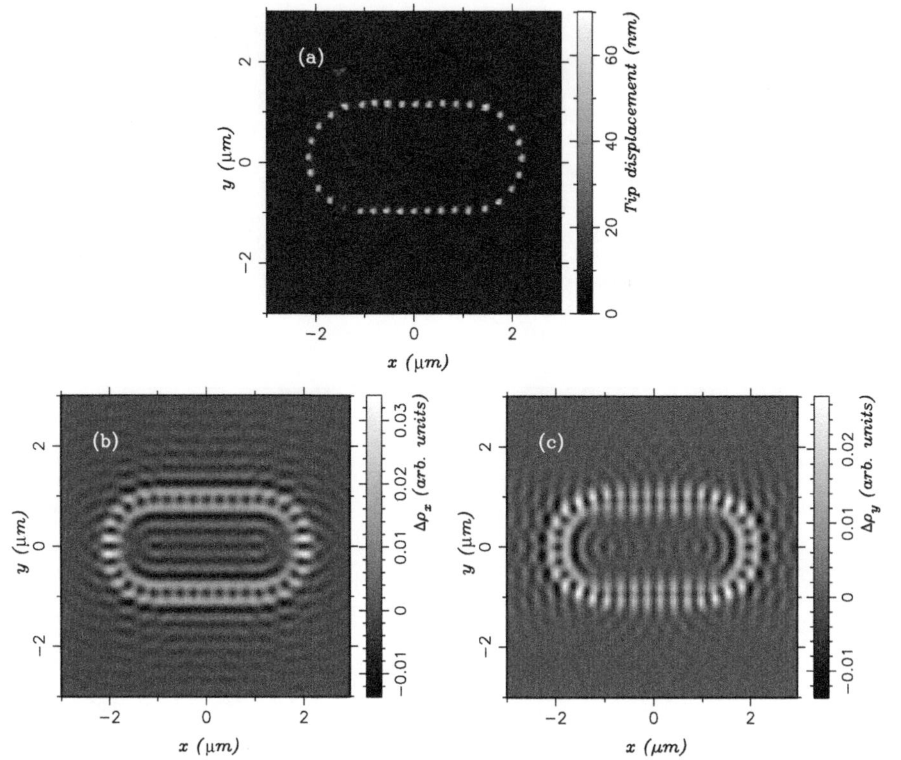

Figure 11.5. (a) AFM images of the reference sample: gold particles (100 nm × 100 nm × 50 nm) deposited on a flat glass surface. Theoretical distributions of partial LDOS change close to the sample relative to the constant value of vacuum (height of calculation plane: $z = 160$ nm above the substrate) (b) $\Delta\rho_x(\mathbf{r}, \omega)$, (c) $\Delta\rho_y(\mathbf{r}, \omega)$.

The experimental verification of the hypothesis of detection of the electromagnetic LDOS by SNOM configurations requires the realization of specific nanostructures. Numerical simulations of the spatial distribution of the variation of the LDOS $\Delta\rho_x(\mathbf{r}, \omega)$ et $\Delta\rho_y(\mathbf{r}, \omega)$ relative to the constant value in vacuum (see Fig. 11.5(b) and (c)), have established that the "stadium" geometry (Fig. 11.5(a)) provides an easy way to study the effects related both to the polarization and to the sub-λ tailoring of the LDOS. Inside the stadium, $\Delta\rho_x(\mathbf{r}, \omega)$ displays a pattern of concentric replica of the stadium shape, while $\Delta\rho_y(\mathbf{r}, \omega)$ features two "focal" zones.

The experiments (Fig. 11.6) lead to the following conclusions about the necessary conditions to detect the LDOS:

- Including a device to detect the scattered light at angles larger than the critical angle for total reflection in the substrate;

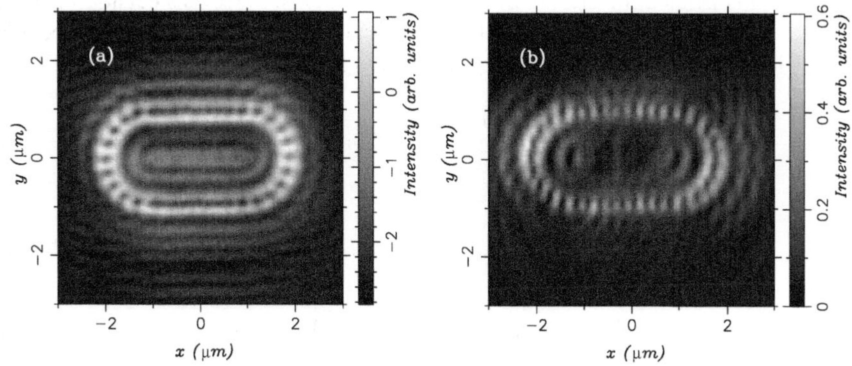

Figure 11.6. SNOM images recorded above the sample shown in Fig. 11.5(a) for two polarization states giving rise to an effective dipole oriented along x (a) and y (b) at the tip apex.

- Using a specific tip whose emission features can be considered as a point-like dipole. Bare (uncoated) optical fiber tips turn out to be inefficient for this purpose. The same optical fiber tip coated with Au provides a signal proportional to the LDOS. Contrary to an assumption commonly spread in the literature, it was not found necessary to form a small hole in the coating at the apex of the tip.

The agreement between the experimental results of Fig. 11.6 with the theoretical distributions (Figs. 11.5(b) and (c)) is unprecedented in the context of SNOM research and confirms the possibility of detecting a signal proportional to the electromagnetic LDOS. The practical point of view suggested in sect. 11.2 is thus also valid for illumination mode near-field optical microscopes.[17–21]

11.6. OBSERVATION OF LOCALIZED PLASMONS

We now discuss the phenomenon of optical confinement close to metal nanostructures deposited on surfaces. These phenomena were observed by operating a PSTM equipped with probe tips that detect the intensity of the electric field.

11.6.1. Squeezing of the Near-Field by Localized Plasmon Coupling

Metal nanostructures were fabricated on glass surfaces. Very sensitive measurements were realized with a PSTM operated at constant height above single gold particles (100 nm $\times$ 10 nm $\times$ 40 nm) (Fig. 11.7). In the calculation (Fig. 11.7(a)), such a particle is centered at the origin of the coordinate system while it is slightly displaced to the right in the experimental image (Fig. 11.7(b)). The agreement between the patterns of the calculated and the experimental images is excellent. Since the simulation did not include the tip, the experimental image exhibits a broader pattern with lower contrast. The simulation successfully reproduces the interference between the incident surface optical wave and the wave scattered by the Au particle. This confirms the relatively passive role of the probe tips even when scanning samples that sustain plasmon resonances.

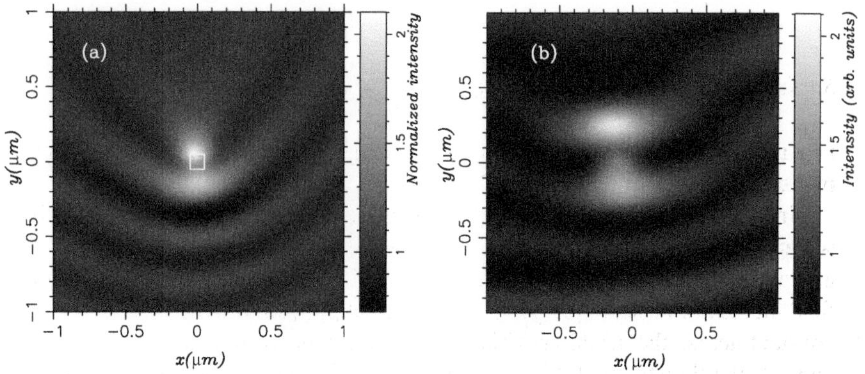

Figure 11.7. Theoretical distribution of the electric field associated with the optical near-field at 40 nm above an Au particle (100 × 100 × 40 nm^3) deposited on an ITO glass substrate (a) and the corresponding PSTM image (b).

In order to test the hypothesis of non-radiative coupling, small gold particles (100 nm × 100 nm × 40 nm) were aligned in a row with a spacing of 100 nm. The experimental result demonstrates that the plasmon coupling between the particles confines the electromagnetic field within the width of the chain (Fig. 11.8(a)).[22] In fact, the tip integrates the detection of the optical field at least over its own volume

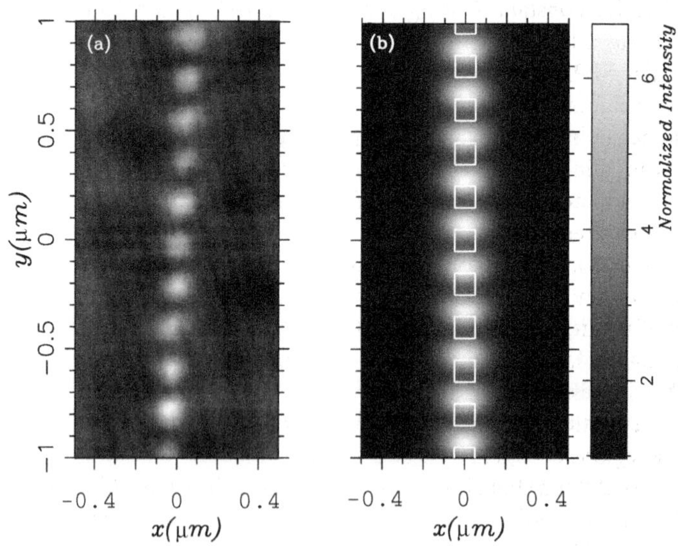

Figure 11.8. Constant height PSTM image (a) recorded above a section of a chain made of 10,000 Au particles deposited on an ITO glass substrate, compared to a numerical simulation (b) taking into account only a few tens of particles.

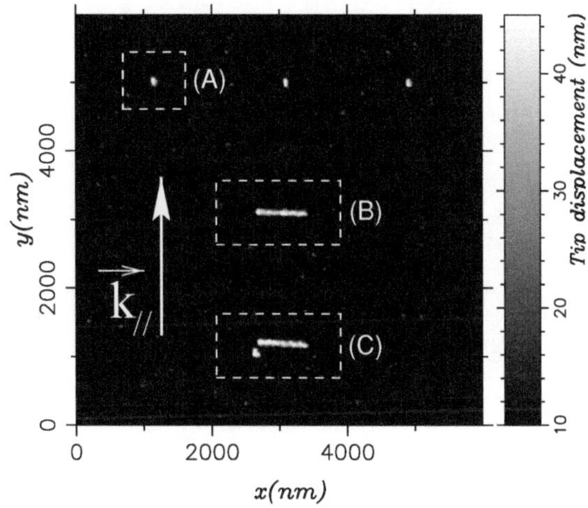

Figure 11.9. AFM image of the test sample used to demonstrate the control of plasmon coupling between nanostructures. The white arrow shows the direction of propagation of the surface wave obtained by total internal reflection.

(like in Fig. 11.7). Consequently, the field distribution in the absence of the tip is probably narrower.

The squeezing effect is related to the mutual coupling of the localized plasmons of each individual nanoparticle setting up a hybrid plasmon mode. The model calculation (Fig. 11.8(b)) confirms that the spots are narrower than in the case of the single Au particle, and that they are not centered on top of the particles. The effect observed in Fig. 11.8(a) is thus due to the plasmon coupling and not due to the underlying topography. However, the calculated squeezing is not as narrow as in the experiment, since the modeling involved only 30 particles instead of the 10,000 present in the experiment. The squeezing probably increases as the chain length grows.

11.6.2. Controlling the Coupling of Localized Plasmons

Finally, it was attempted to control the plasmon coupling between two Au nanostructures of different shapes.[23] In the sample of Fig. 11.9, all particles have a volume of $120 \times 60 \times 40$ nm^3 and all nanowires have a volume of $660 \times 60 \times 40$ nm^3. It was possible to switch on (Fig. 11.11) or off (Fig. 11.10) the excitation of the resonant mode of a Au nanowire by changing the position of a Au particle located at a subwavelength distance from the wire. In Fig. 11.10, the isolated particles (zone A) do not give rise to any significant signal, while the nanowires produce a signal proportional to their volumes. This explains the minor difference between the zone B (isolated nanowire) and the zone C (nanowire close to a particle).

In Fig. 11.11(a), the isolated particles (zone A) are excited resonantly, while the nanowire in zone B is not because of a selection rule involving the incident polarization. However, the nanowire in zone C is excited because it lies close to a resonant Au nanoparticle. This resonant nanoparticle scatters all kinds of possible

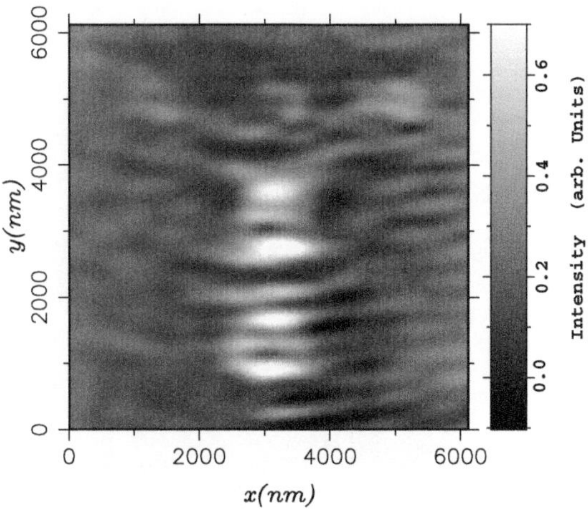

Figure 11.10. PSTM image above the sample shown in Fig. 11.9 obtained by shining a TM polarized laser beam (wavelength 633 nm). The angle of incidence is 60°. At this wavelength, the particles and the nanowires are not resonant.

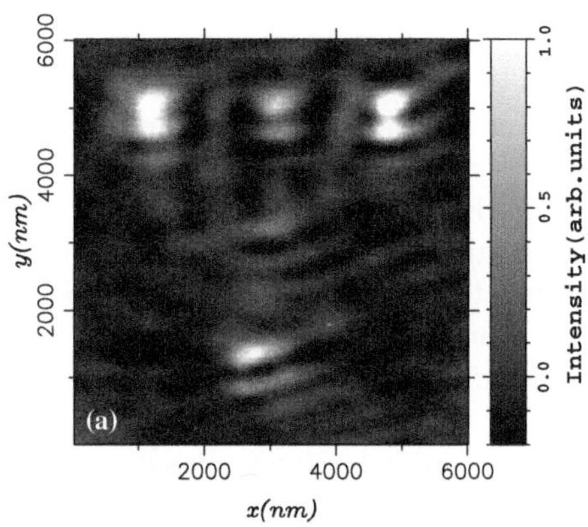

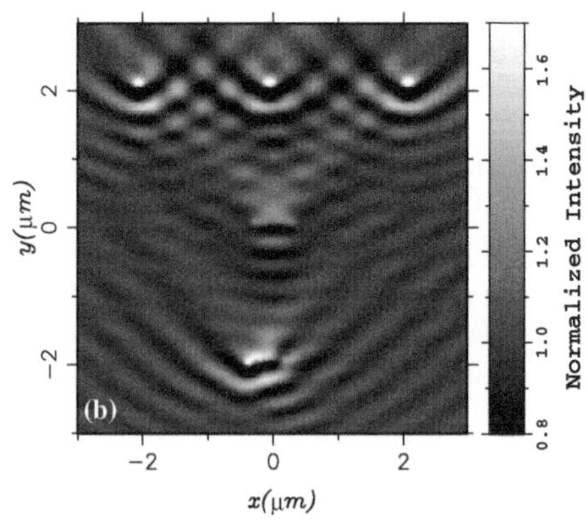

Figure 11.11. (a) PSTM image of the sample shown in Fig. 11.9 obtained by shining a TM polarized laser beam (wavelength 740 nm). (b) Theoretical distribution of the intensity of the electric field associated with the optical near-field (normalized relative to the intensity of the incident field, same illumination conditions as in the experiment).

polarizations, among which the polarization leading to the resonant excitation of the nanowire C. Let us note that, to design the sample which made this demonstration possible, the numerical simulations (Fig. 11.11(b)) were realized *before* the experiment.

ACKNOWLEDGMENT

This work was supported by the Regional Council of Burgundy (ARCEN project) and by the European Commission (Contracts NoE FP6-IST-2002-1-507879 & STRP FP6-NMP-2002-1-001601).

REFERENCES

1. D. Courjon, C. Bainier: Near field microscopy and near field optics, Rep. Prog. Phys. **57**, 989 (1994).
2. J. J. Greffet, R. Carminati: Image formation in near-field optics, Prog. Surf. Sci. **56**, 133 (1997).
3. B. Hecht *et al.*: Facts and artifacts in near-field optical microscopy, J. Appl. Phys. **81**, 2492 (1997).
4. A. Dereux, C. Girard, J. C. Weeber: Theoretical principles of near-field optical microscopies and spectroscopies, J. Chem. Phys. **112**, 7775 (2000).
5. A. Dereux *et al.*: Direct interpretation of near-field optical images, J. Microscopy **202**, 320 (2001).
6. R. G. Newton: ch. 4, *Scattering Theory of Waves and Particles* (McGraw-Hill, New York, 1966).
7. C. Girard, A. Dereux: Near-field optics theories, Rep. Prog. Phys. **59**, 657 (1996).
8. P. Morse, H. Feshbach: ch. 13, *Methods of Theoretical Physics* (McGraw-Hill, New York, 1953).
9. H. Levine, J. Schwinger: On the theory of electromagnetic wave diffraction by an aperture in an infinite plane conducting screen, Comm. Pure App. Math. **3**, 355 (1950).
10. G. S. Agarwal: Quantum electrodynamics in the presence of dielectrics and conductors, Phys. Rev. A **11**, 230 (1975).
11. H. Metiu: Surface enhanced spectroscopy, Prog. Surf. Sci. **17**, 153 (1984).
12. C. Girard, X. Bouju: Self-consistent study of dynamical and polarization effects in near-field optical microscopy, J. Opt. Soc. Am. B **9**, 298 (1992).
13. E. Devaux: Ph.D. thesis, Université de Bourgogne, Dijon, 2000.
14. J. C. Weeber *et al.*: Observation of light confinement effects with a near-field optical microscope, Phys. Rev. Lett. **77**, 5332 (1996).
15. E. Devaux *et al.*: Local detection of the optical magnetic field in the near zone of dielectric samples, Phys. Rev. B **62**, 10504 (2000).
16. U. Schröter, A. Dereux: Surface plasmon polaritons on metal cylinders with dielectric core, Phys. Rev. B **64**, 125420 (10 pages) (2001).
17. G. Colas des Francs *et al.*: Optical analogy to electronic quantum corrals, Phys. Rev. Lett. **86**, 4950 (2001).
18. G. Colas des Francs, C. Girard, J. C. Weeber, A. Dereux: Relationship between scanning near-field optical images and local density of photonic states, Chem. Phys. Lett. **345**, 512 (2001).
19. C. Chicanne *et al.*: Imaging the local density of states of optical corrals, Phys. Rev. Lett. **88**, 097402 (4 pages) (2002).
20. G. Colas des Francs, C. Girard, A. Dereux: Theory of near-field optical imaging with a single molecule as a light source, Single Molecules **3**, 311 (2002).
21. A. Dereux *et al.*: Subwavelength mapping of surface photonic states, Nanotechnology **14**, 935 (2003).
22. J. R. Krenn *et al.*: Squeezing the optical near-field zone by plasmon coupling of metallic nanoparticles, Phys. Rev. Lett. **82**, 2590 (1999).
23. J. R. Krenn *et al.*: Direct observation of localized surface plasmon coupling, Phys. Rev. B **60**, 5029 (1999).

CHAPTER TWELVE

OVERVIEW OF SIMULATION TECHNIQUES FOR PLASMONIC DEVICES

GEORGIOS VERONIS AND SHANHUI FAN
Ginzton Laboratory, Stanford University, Stanford, California 94305, USA

12.1. INTRODUCTION

Surface plasmons are electromagnetic waves that propagate along the interface of a metal and a dielectric. In a surface plasmon light interacts with the free electrons of the metal which oscillate collectively in response to the applied field. Recently, nanometer-scale metallic devices have shown the potential to manipulate light at the subwavelength scale using surface plasmons. This could lead to photonic circuits of nanoscale dimensions.

Surface plasmons can be described by macroscopic electromagnetic theory, i.e. Maxwell's equations, if the electron mean free path in the metal is much shorter than the plasmon wavelength.[1] This condition is usually fulfilled at optical frequencies.[1] We also note that in macroscopic electromagnetic theory, bulk material properties such as dielectric constants, are used to describe objects irrespective of their size. However, for particles of nanoscale dimensions, a more fundamental description of their optical and electronic properties may be required.[2]

In this chapter, we give an overview of simulation techniques for plasmonic devices. We focus our attention on techniques related to macroscopic electromagnetic theory. All materials are assumed to be non-magnetic ($\mu = \mu_0$) and are characterized by their bulk dielectric constant $\varepsilon(\mathrm{r}, \omega)$. We will focus on numerical simulation techniques and will not consider analytical methods, such as Mie theory.[3] Such methods can only be applied to planar geometries or to objects of specific shapes (spheres, cylinders) and have therefore limited importance in the analysis of plasmonic devices and structures.

Numerical modeling of plasmonic devices involves several challenges which need to be addressed. First, as mentioned above, plasmonic devices can have arbitrary geometries. Several techniques are specific for one type of geometrical configuration and are therefore not appropriate for modeling of arbitrary plasmonic devices.

M.L. Brongersma and P.G. Kik (eds.), Surface Plasmon Nanophotonics, 169–182.

Second, the dielectric constant of metals at optical wavelengths is complex, i.e. $\varepsilon_r(\omega) = \varepsilon_{\mathrm{Re}}(\omega) + i\varepsilon_{\mathrm{Im}}(\omega)$ and is a complicated function of frequency.[4] Thus, several simulation techniques which are limited to lossless, non-dispersive materials are not applicable to plasmonic devices. In addition, in time-domain methods the dispersion properties of metals have to be approximated by suitable analytical expressions.[5] In most cases the *Drude model* is invoked to characterize the frequency dependence of the metallic dielectric function

$$\varepsilon_{r,\mathrm{Drude}}(\omega) = 1 - \frac{\omega_p^2}{\omega(\omega + i\gamma)}, \tag{12.1}$$

where ω_p, γ are frequency-independent parameters.[6] However, the Drude model approximation is valid over a limited wavelength range.[6] The range of validity of the Drude model can be extended by adding Lorentzian terms to Eq. (12.1) to obtain the *Lorentz-Drude model*

$$\varepsilon_{r,\mathrm{LD}}(\omega) = \varepsilon_{r,\mathrm{Drude}}(\omega) + \sum_{j=1}^{k} \frac{f_j \omega_j^2}{(\omega_j^2 - \omega^2) - i\omega\gamma_j}, \tag{12.2}$$

where ω_j and γ_j stand for the oscillator resonant frequencies and bandwidths respectively, and f_j are weighting factors.[6] Physically, the Drude and Lorentzian terms are related to intraband (free-electron) and interband (bound-electron) transitions respectively.[6] Even though the Lorentz-Drude model extends the range of validity of analytical approximations to metallic dielectric constants, it is not suitable for the description of sharp absorption edges observed in some metals, unless a very large number of terms is used.[6] In particular, the Lorenz-Drude model cannot approximate well the onset of interband absorption in noble metals (Ag, Au, Cu) even if five Lorentzian terms are used.[6] In Fig. 12.1 we compare the Drude and Lorentz-Drude models with

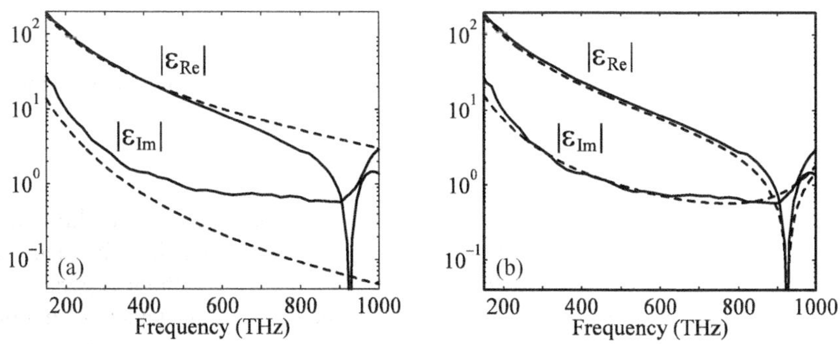

Figure 12.1. Real and imaginary part of the dielectric constant of silver at optical frequencies. The solid lines show experimental data.[4] The dashed lines show values calculated using (a) the Drude model, (b) the Lorentz-Drude model with five Lorentzian terms. The parameters of the models are optimized through an optimization procedure.[6]

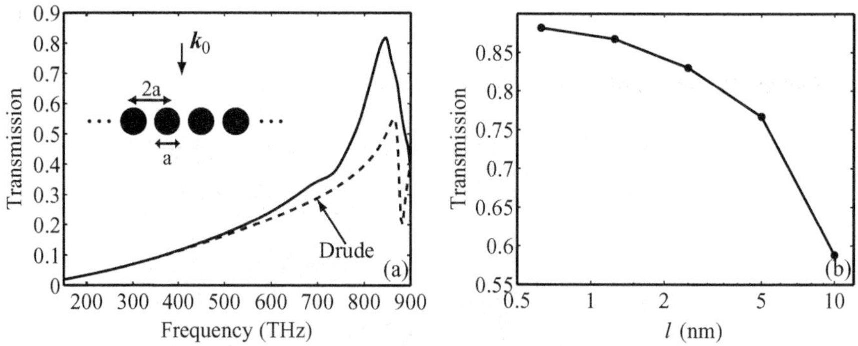

Figure 12.2. (a) Calculated transmission spectrum of an infinite array of silver cylinders (shown in the inset) for normal incidence and H polarization. Results are shown for a diameter $a = 100$ nm. The dashed line shows the transmission spectrum calculated using the Drude model with parameters $\omega_p = 1.37 \times 10^{16}\ \text{sec}^{-1}$ and $\gamma = 7.29 \times 10^{13}\ \text{sec}^{-1}$. (b) Calculated transmission at 855 THz as a function of the spatial grid size Δl.

experimental data for silver. We observe that even a five-term Lorentz-Drude model with optimal parameters results in a factor of two error at certain frequencies.

Third, in surface plasmons propagating along the interface of a metal and a dielectric, the field is concentrated at the interface, and decays exponentially away from the interface in both the metal and dielectric regions.[7] Thus, for numerical methods based on discretization of the fields on a numerical grid, a very fine grid resolution is required at the metal-dielectric interface to adequately resolve the local fields. In addition, several plasmonic devices are based on components of subwavelength dimensions.[7] In fact, most of the potential applications of surface plasmons are related to subwavelength optics. The nanoscale feature sizes of plasmonic devices pose an extra challenge to numerical simulation techniques.

We illustrate the challenges involved in modeling plasmonic devices using a simple example. We consider an infinite periodic array of silver cylinders illuminated by a plane wave at normal incidence (inset of Fig. 12.2(a)). We use the finite-difference frequency-domain (FDFD) method, described in more detail below, to calculate the transmission of the periodic array. This method allows us to directly use experimental data for the frequency-dependent dielectric constant of metals, including both the real and imaginary parts, with no further approximation. The fields are discretized on a uniform two-dimensional grid with grid size $\Delta x = \Delta y = \Delta l$. In Fig. 12.2(a) we show the calculated transmission as a function of frequency. We also show the transmission of the structure calculated with the Drude model of Eq. (12.1). We observe that the use of the Drude model results in substantial error. In general, the Drude model parameters are chosen to minimize the error in the dielectric function in a given frequency range.[8] However, this approach gives accurate results in a limited wavelength range, as illustrated in this example. In general the complicated dispersion properties of metals at optical frequencies pose a challenge in modeling of plasmonic devices not encountered in modeling of low- or high-index-contrast dielectric devices.

In Fig. 12.2(b) we show the calculated transmission at a specific wavelength as a function of the spatial grid size Δl. We observe that a grid size of $\Delta l \cong 1\,\text{nm}$ is required in this case to yield reasonably accurate results that no longer change substantially with decreasing grid size. The required grid size is directly related to the decay length of the fields at the metal–dielectric interface. In general, modeling of plasmonic devices requires much finer grid resolution than modeling of low- or high-index-contrast dielectric devices, due to the high localization of the field at metal–dielectric interfaces of plasmonic devices. The required grid size depends on the shape and feature size of the modeled plasmonic device, the metallic material used and the operating frequency.

In the following we examine several widely used simulation techniques for modeling of plasmonic devices. We also examine how they address the challenges mentioned above.

12.2. NUMERICAL SIMULATION TECHNIQUES

12.2.1. Green Dyadic Method

The Green dyadic method (GDM) is based on discretization of an integral equation to obtain a matrix equation.[9]

We consider an object with finite volume V embedded in a reference medium. The system is illuminated by an electric field with arbitrary spatial distribution and an $\exp(-i\omega t)$ harmonic time dependence. From Maxwell's equations

$$\nabla \times \mathbf{E}(\mathbf{r}) = i\omega\mu_0\mathbf{H}(\mathbf{r}), \tag{12.3}$$

$$\nabla \times \mathbf{H}(\mathbf{r}) = -i\omega\varepsilon_0\varepsilon_r(\mathbf{r})\mathbf{E}(\mathbf{r}), \tag{12.4}$$

we can eliminate $\mathbf{H(r)}$ and obtain the following wave equation for $\mathbf{E(r)}$

$$\nabla \times \nabla \times \mathbf{E}(\mathbf{r}) - \varepsilon_r(\mathbf{r})\frac{\omega^2}{c^2}\mathbf{E}(\mathbf{r}) = 0, \tag{12.5}$$

where the dielectric function $\varepsilon_r(\mathbf{r})$ is defined as $\varepsilon_r(\mathbf{r}) = \varepsilon_{r,\text{obj}}(\mathbf{r})$, if $\mathbf{r} \in V$, and $\varepsilon_r(\mathbf{r}) = \varepsilon_{r,\text{ref}}(\mathbf{r})$, if $\mathbf{r} \notin V$. Thus, Eq. (12.5) can be rewritten as[9]

$$\nabla \times \nabla \times \mathbf{E}(\mathbf{r}) - \varepsilon_{r,\text{ref}}(\mathbf{r})\frac{\omega^2}{c^2}\mathbf{E}(\mathbf{r}) = \frac{\omega^2}{c^2}[\varepsilon_{r,\text{ref}}(\mathbf{r}) - \varepsilon_{r,\text{obj}}(\mathbf{r})]\mathbf{E}(\mathbf{r}). \tag{12.6}$$

The right-hand side of Eq. (12.6), which is proportional to the electric field inside the object volume V, acts as a source. The solution of Eq. (12.6) can therefore be expressed as

$$\mathbf{E}(\mathbf{r}) = \mathbf{E}^i(\mathbf{r}) + \mathbf{E}^s(\mathbf{r}), \tag{12.7}$$

where the incident field $\mathbf{E}^i(\mathbf{r})$ is the homogeneous solution of Eq. (12.6) when the right-hand side source is zero. To determine the field scattered by the object $\mathbf{E}^s(\mathbf{r})$, one has to consider the response of the reference system to a point current source, the so-called *Green's function*. Since $\mathbf{E}(\mathbf{r})$ is a vector and the current source is also a vector, the Green's function must be a *dyad* $\bar{\mathbf{G}}(\mathbf{r}, \mathbf{r}')$ that operates on a vector giving rise to another vector.[10] The dyadic Green's function is therefore defined by

$$\nabla \times \nabla \times \bar{\mathbf{G}}(\mathbf{r}) - \varepsilon_{r,\mathrm{ref}}(\mathbf{r})\frac{\omega^2}{c^2}\bar{\mathbf{G}}(\mathbf{r}) = \bar{\mathbf{I}}\delta(\mathbf{r}-\mathbf{r}'). \tag{12.8}$$

Using Eqs. (12.6)–(12.8) we obtain the following integral equation for $\mathbf{E}(\mathbf{r})$

$$\mathbf{E}(\mathbf{r}) = \mathbf{E}^i(\mathbf{r}) + \frac{\omega^2}{c^2}\int_V \bar{\mathbf{G}}(\mathbf{r}, \mathbf{r}')[\varepsilon_{r,\mathrm{ref}}(\mathbf{r}') - \varepsilon_{r,\mathrm{obj}}(\mathbf{r}')]\mathbf{E}(\mathbf{r}')d\mathbf{r}'. \tag{12.9}$$

To solve Eq. (12.9), one needs to determine $\bar{\mathbf{G}}(\mathbf{r}, \mathbf{r}')$ first. If the reference medium in which the object is embedded is a homogeneous dielectric, i.e. $\varepsilon_{r,\mathrm{ref}}(\mathbf{r}) = \varepsilon_r$, it can be shown that

$$\bar{\mathbf{G}}(\mathbf{r}) = (\bar{\mathbf{I}} + \frac{1}{k^2}\nabla\nabla)\frac{e^{ik|\mathbf{r}-\mathbf{r}'|}}{4\pi\,|\mathbf{r}-\mathbf{r}'|}, \tag{12.10}$$

where $k^2 = \omega^2\varepsilon_r\varepsilon_0\mu_0$.[10] Of particular interest is also the Green's function of a layered reference system, since many practical plasmonic devices consist of metallic structures deposited on the interface between air and a multilayered substrate. The Green's function in this case is expressed in terms of Sommerfeld integrals which are evaluated numerically.[1] Once $\bar{\mathbf{G}}(\mathbf{r}, \mathbf{r}')$ is determined, the field inside the object can be computed numerically by discretizing the volume V into N cells and assuming that $\mathbf{E}(\mathbf{r})$ and $\varepsilon_{r,\mathrm{ref}}(\mathbf{r}) - \varepsilon_{r,\mathrm{obj}}(\mathbf{r})$ are constant within each cell.[9] If $\mathbf{r}_m$ is the center of the mth cell, Eq. (12.9) reduces to a dense system of linear equations

$$\begin{bmatrix} \bar{\mathbf{I}} - \xi_1\bar{\mathbf{G}}(\mathbf{r}_1, \mathbf{r}_1) & \cdots & \xi_N\bar{\mathbf{G}}(\mathbf{r}_1, \mathbf{r}_N) \\ \vdots & & \vdots \\ \xi_1\bar{\mathbf{G}}(\mathbf{r}_N, \mathbf{r}_1) & \cdots & \bar{\mathbf{I}} - \xi_N\bar{\mathbf{G}}(\mathbf{r}_N, \mathbf{r}_N) \end{bmatrix} \begin{bmatrix} \mathbf{E}(\mathbf{r}_1) \\ \vdots \\ \mathbf{E}(\mathbf{r}_N) \end{bmatrix} = \begin{bmatrix} \mathbf{E}^i(\mathbf{r}_1) \\ \vdots \\ \mathbf{E}^i(\mathbf{r}_N) \end{bmatrix}, \tag{12.11}$$

where $\xi_m = \omega^2/c^2[\varepsilon_{r,\mathrm{ref}}(\mathbf{r}_m) - \varepsilon_{r,\mathrm{obj}}(\mathbf{r}_m)]\delta V_m$ and δV_m is the volume of the mth cell. Special attention must be devoted to the calculation of the diagonal terms[11] due to the singularity of $\bar{\mathbf{G}}(\mathbf{r}, \mathbf{r}')$ for $\mathbf{r} = \mathbf{r}'$. Once the field inside the object is determined, the field at any other point can be calculated using the discretized form of Eq. (12.9)

$$\mathbf{E}(\mathbf{r}) = \mathbf{E}^i(\mathbf{r}) + \sum_{m=1}^{N}\xi_m\bar{\mathbf{G}}(\mathbf{r}, \mathbf{r}_m)\mathbf{E}(\mathbf{r}_m). \tag{12.12}$$

The GDM is particularly efficient in modeling scattering from metallic structures embedded in uniform or planarly layered media. This is due to the fact that Green's functions are available for such reference systems and only the metallic structure volume has to be discretized. However, it is hard to use the GDM in more general problems due to the difficulty in constructing the related Green's functions. The GDM is a frequency-domain technique, so it can treat arbitrary material dispersion. The GDM requires discretization of the metallic structure volume. Thus, a very fine grid resolution is required to accurately model the rapid field decay inside the metal.

In addition, assuming that an orthogonal grid is used, finer grid resolutions are required for curved metal–dielectric interfaces to avoid introducing large errors due to staircasing.[5] The efficiency of discretization can be increased by using nonuniform and/or nonorthogonal grids.[5,12] Non-uniform grids allow finer discretization in the vicinity of the metal–dielectric interface. Non-orthogonal grids allow better approximation of curved surfaces. These, however, are at the cost of more coding complexity.

12.2.2. Discrete Dipole Approximation

The discrete dipole approximation (DDA) is an approximation of a continuum target by a finite array of polarizable points which acquire dipole moments in response to the local electric field.[13,14]

We consider an electromagnetic wave with an electric field, $\mathbf{E}^i(\mathbf{r})$, incident on an object with dielectric function $\varepsilon(\mathbf{r})$. We approximate the object geometry with a finite number of polarizable elements. If $\mathbf{E}_m = \mathbf{E}(\mathbf{r}_m)$ is the electric field at the site of the mth element, the induced polarization, $\mathbf{P}_m$, at this site is

$$\mathbf{P}_m = \alpha_m \mathbf{E}_m, \tag{12.13}$$

where α_m is the dipole polarizability at $\mathbf{r}_m$. In most cases the Clausius-Mossotti equation[14] is used for α_m,

$$\alpha_m = \frac{3}{N}\frac{\varepsilon(\mathbf{r}_m) - 1}{\varepsilon(\mathbf{r}_m) + 2}, \tag{12.14}$$

where N is the number of polarizable elements per unit volume. For an infinite cubic lattice Eq. (12.14) is exact in the dc limit $kd \to 0$, where d is the dipole spacing. At finite wavelengths a radiative reaction correction[14] of $O[(kd)^3]$ should be included in Eq. (12.14). The contribution of the nth dipole to the electric field $\mathbf{E}_{mn}$ at site m is simply the electric field in vacuum of a dipole $\mathbf{P}_n$ at a distance $\mathbf{r}_{mn} \equiv \mathbf{r}_m - \mathbf{r}_n$. That is,[15]

$$\mathbf{E}_{mn} = \frac{\exp(ikr_{mn})}{4\pi\varepsilon_0 r_{mn}^3} \cdot \left\{k^2(\mathbf{r}_{mn} \times \mathbf{P}_n) \times \mathbf{r}_{mn} + [3(\mathbf{r}_{mn} \cdot \mathbf{P}_n)\mathbf{r}_{mn} - r_{mn}^2 \mathbf{P}_n]\frac{1 - ikr_{mn}}{\mathbf{r}_{mn}^2}\right\}, \tag{12.15}$$

where $r_{mn} \equiv |\mathbf{r}_m - \mathbf{r}_n|$. The total electric field $\mathbf{E}_m$ at site m is then

$$\mathbf{E}_m = \mathbf{E}_m^i + \sum_{n \neq m} \mathbf{E}_{mn}, \tag{12.16}$$

where $\mathbf{E}_m^i = \mathbf{E}^i(\mathbf{r}_m)$. If Eqs. (12.13)–(12.16) are applied at all N points, we obtain a system of $3N$ complex linear equations. Once the system is solved for the unknown fields $\mathbf{E}_m$, the electric field at any other position can be calculated by summing the contributions of the incident field and of the fields due to the N dipoles.[14]

If the structure dimensions are much smaller than the wavelength λ of the incident light, the quasistatic approximation can be used for the dipolar electric field[15]

$$\mathbf{E}_{mn} = \frac{1}{4\pi\varepsilon_0 r_{mn}^5}[3(\mathbf{r}_{mn}\cdot\mathbf{P}_n)\mathbf{r}_{mn} - r_{mn}^2\mathbf{P}_n]. \tag{12.17}$$

It has been shown that the DDA is equivalent to the GDM in the low-frequency limit.[14] The DDA can only be used for modeling scattering from metallic objects embedded in uniform media, since it is based on the field of a dipole in a homogeneous medium (Eq. 12.15). It is a frequency-domain technique, so it can treat arbitrary material dispersion. Similar to the GDM, the DDA is based on discretization of the metallic structure volume, so that nonuniform and/or nonorthogonal grids are required for efficient treatment of curved surfaces and rapid field decays at metal–dielectric interfaces.

12.2.3. Finite-Difference Frequency-Domain Method

In finite-difference methods, derivatives in differential equations are approximated by finite differences. To approximate the derivative $df/dx|_{x_0}$ we consider a Taylor's series expansion of $f(x)$ about the point x_0 to the points $x_0 + \Delta x$ and $x_0 - \Delta x$ and obtain[5]

$$\left.\frac{df}{dx}\right|_{x_0} = \frac{f(x_0+\Delta x) - f(x_0-\Delta x)}{2\Delta x} + O[(\Delta x)^2]. \tag{12.18}$$

Eq. (12.18) shows that a *central-difference* approximation of the first derivative is *second-order accurate*, meaning that the remainder term in Eq. (12.18) approaches zero as the square of Δx.

In finite-difference methods a continuous problem is approximated by a discrete one. Field quantities are defined on a discrete grid of nodes. The rectangular grid with node coordinates $\mathbf{r}_{ijk} = (x_i, y_j, z_k)$ is the simplest and most commonly-used. A field quantity at nodal location $\mathbf{r}_{ijk}$ is denoted for convenience as $f_{ijk} = f(\mathbf{r}_{ijk})$. Based on Eq. (12.18), the first derivative can be approximated by the following central-difference formula

$$\left.\frac{df}{dx}\right|_i \cong \frac{f_{i+1} - f_{i-1}}{2\Delta x}, \tag{12.19}$$

which is second-order accurate, based on the discussion above, if the rectangular gird is uniform, i.e. $x_i = i\Delta x$. Similarly, the second derivative can be approximated by the formula

$$\left.\frac{d^2f}{dx^2}\right|_i \cong \frac{f_{i+1} - 2f_i + f_{i-1}}{(\Delta x)^2}, \tag{12.20}$$

which is also second-order accurate on a uniform grid.[5]

By replacing derivatives in differential equations with their finite-difference approximations, we obtain algebraic equations which relate the value of the field at a specific node to the values at neighboring nodes. To solve Maxwell's equations with the FDFD method, we discretize the system of the three, coupled, scalar partial differential equations obtained from the wave equation for the electric field (Eq. 12.5). For simplicity we consider here two-dimensional problems with transverse electric (TE) polarization. For TE polarization we have $\mathbf{E} = E_z\hat{z}$ and the wave equation for the electric field becomes[16,17]

$$\left[\frac{\partial^2}{\partial x^2} + \frac{\partial^2}{\partial y^2} + k_0^2\varepsilon_r(x, y)\right]E_z(x, y) = -i\omega\mu_0 J_z(x, y), \qquad (12.21)$$

For simplicity we consider a uniform rectangular grid with $x_i = i\Delta x$, $y_j = j\Delta y$, and replace the derivatives in Eq. (12.21) with their finite-difference approximations of Eq. (12.20) to obtain

$$\frac{f_{i+1,j} - 2f_{i,j} + f_{i-1,j}}{(\Delta x)^2} + \frac{f_{i,j+1} - 2f_{i,j} + f_{i,j-1}}{(\Delta y)^2} + k_0^2\varepsilon_{ri,j} f_{i,j} = A_{i,j}, \qquad (12.22)$$

where $f = E_z$ and $A = -i\omega\mu_0 J_z$. Thus, application of finite-difference approximations at the node location $\mathbf{r}_{ij} = (x_i, y_j)$ results in a linear algebraic equation which relates the field f_{ij} to the fields at the four adjacent nodes $f_{i+1,j}, f_{i-1,j}, f_{i,j+1}, f_{i,j-1}$. By applying the finite-difference approximation to all nodes of the grid we obtain a system of linear equations of the form $\mathbf{Ax} = \mathbf{b}$, where $\mathbf{b}$ is determined by the source current $\mathbf{J}$. Since the equation for the field at each point involves only the fields at the four (six in three dimensions, two in one dimension) adjacent points, the resulting system matrix is extremely sparse.[17]

FDFD can be used to model plasmonic devices with arbitrary geometries. Compared to GDM, FDFD does not require construction of a Green's function and is therefore a more versatile method. For problems that both FDFD and GDM can be used, FDFD is in general less efficient because it requires discretization of both the metallic objects and the surrounding media. However, FDFD results in sparse matrix systems, while GDM results in dense matrix systems. Sparse problems can be solved more efficiently than dense problems with the same number of unknowns, if direct or iterative sparse matrix techniques are used.[16,17] An additional complexity in the use of FDFD, and of all other methods which are based on discretization of the differential form of Maxwell's equations in a finite volume, is the requirement of *absorbing boundary conditions (ABCs)* such that waves are not artificially reflected at the boundaries of the computational domain.[5,16] Very efficient and accurate ABCs such as the perfectly matched layer (PML) have recently been demonstrated.[18] FDFD is a frequency-domain technique and can thus treat arbitrary material dispersion. As with GDM, nonuniform and/or nonorthogonal grids are required in FDFD for efficient treatment of curved surfaces and rapid field decays at metal–dielectric interfaces.

Waveguiding Structures

In the previous section we described how Maxwell's equations are solved with the finite-difference method when the fields are excited by a current source $\mathbf{J}$. A slightly different approach is required if we are interested in determining the modes and the dispersion characteristics of waveguiding structures which are uniform in the z-direction. If we assume that all field components have a $\exp(-\gamma z)$ dependence, it can be shown that Maxwell's equations reduce to[19]

$$-\varepsilon_r k_0^2 h_x + \varepsilon_r \frac{\partial}{\partial y}\left[\varepsilon_r^{-1}\left(\frac{\partial h_y}{\partial x} - \frac{\partial h_x}{\partial y}\right)\right] - \frac{\partial}{\partial x}\left(\frac{\partial h_x}{\partial x} + \frac{\partial h_y}{\partial y}\right) = \gamma^2 h_x, \quad (12.23)$$

$$-\varepsilon_r k_0^2 h_y - \varepsilon_r \frac{\partial}{\partial x}\left[\varepsilon_r^{-1}\left(\frac{\partial h_y}{\partial x} - \frac{\partial h_x}{\partial y}\right)\right] - \frac{\partial}{\partial y}\left(\frac{\partial h_x}{\partial x} + \frac{\partial h_y}{\partial y}\right) = \gamma^2 h_y, \quad (12.24)$$

where $\mathbf{H}(x, y, z) = \mathbf{h}(x, y)\exp(-\gamma z)$. Equations (23) and (24) also satisfy the transversality constraint for the magnetic field. They are discretized by applying the finite-difference method on a numerical grid known as the Yee lattice which is described below. We then obtain a sparse matrix eigenvalue problem of the form $\mathbf{Ah} = \gamma^2\mathbf{h}$ which can be solved using iterative sparse eigenvalue techniques.[16] An important feature of this formulation is the absence of spurious modes in the solution[16] which is therefore robust.

Quasistatic Approximation

Several plasmonic structures have dimensions much smaller than the wavelength λ of the incident light. Under these conditions, retardation effects are negligible and the quasistatic approximation $\mathbf{E}(\mathbf{r}) = -\nabla\varphi(\mathbf{r})$ can be used for the scattered electric field, where $\varphi(\mathbf{r})$ is the electrostatic potential. The field distribution problem then reduces to solving

$$\nabla \cdot \{\varepsilon_r(\mathbf{r})[-\nabla\varphi(\mathbf{r}) + \mathbf{E}^i(\mathbf{r})]\} = 0, \quad (12.25)$$

which represents the current conservation law.[20] Equation (12.25) is discretized using the finite-difference method as described above. A sparse system of linear equations of the form $\mathbf{Ax} = \mathbf{b}$ is obtained, where $\mathbf{b}$ is determined by the incident field $\mathbf{E}^i(\mathbf{r})$.

12.2.4. Finite-Difference Time-Domain Method

The finite-difference time-domain (FDTD) method[5] solves directly Maxwell's time-dependent curl equations

$$\nabla \times \mathbf{E} = -\mu_0 \frac{\partial \mathbf{H}}{\partial t}, \quad (12.26)$$

$$\nabla \times \mathbf{H} = \varepsilon_0 \varepsilon_r \frac{\partial \mathbf{E}}{\partial t}, \quad (12.27)$$

so that both space and time have to be discretized. The standard FDTD is based on the Yee algorithm.[5] As we saw in the previous section, central-difference approximations are second-order accurate. To achieve second-order accuracy in time, the Yee algorithm uses a *leapfrog* arrangement.[5] $\mathbf{E}$ fields are calculated at $t = n\Delta t$ using previously calculated and stored $\mathbf{H}$ fields. Then $\mathbf{H}$ fields are calculated at $t = (n + 1/2)\Delta t$

using the previously calculated and stored **E** fields, and the process continues until time-stepping is concluded. Applying this scheme to Eq. (12.26) we obtain

$$\mathbf{H}|^{n+1/2} = \mathbf{H}|^{n-1/2} - \Delta t/\mu_0 \Delta \times \mathbf{E}|^n \tag{12.28}$$

We observe that the leapfrog scheme yields central-difference in time and therefore second-order accurate approximations. In addition, since **E(H)** fields are obtained from previously calculated and stored **H(E)** fields, the time-stepping is fully *explicit,* meaning that we do not have to solve a system of simultaneous equations.[5]

To achieve second-order accuracy in space, FDTD uses a special grid, known as the *Yee lattice*, where every **E** component is surrounded by four **H** components and every **H** component is surrounded by four **E** components.[5] Based on this arrangement, discretization of the x-component of Eq. (12.26) gives

$$H_x\Big|_{i,j,k}^{n+1/2} = H_x\Big|_{i,j,k}^{n-1/2} + \Delta t/\mu_0 \cdot$$

$$\left[\left(E_y\Big|_{i,j,k+1/2}^{n} - E_y\Big|_{i,j,k-1/2}^{n}\right)\Big/\Delta z - \left(E_z\Big|_{i,j+1/2,k}^{n} - E_y\Big|_{i,j-1/2,k}^{n}\right)\Big/\Delta y\right]. \tag{12.29}$$

We observe that, using the Yee lattice, all spatial finite-difference expressions are central and therefore second-order accurate. Similar finite-difference equations are obtained by discretizing the other components of Eqs. (12.26) and (12.27). In summary, FDTD is an explicit numerical scheme which is second-order accurate both in time and in space (in uniform media).

Treatment of Dispersive Media in FDTD

One of the major challenges in FDTD modeling of metals at optical frequencies is the treatment of the metallic dispersion properties. As mentioned above, in time-domain methods the dielectric constants of dispersive media have to be approximated by suitable analytical expressions. The most common algorithm for modeling dispersive materials with FDTD is the auxiliary differential equation (ADE) method.[5,21] In dispersive materials $\varepsilon(\omega)$ relates **E** and **D**

$$\mathbf{D} = \varepsilon(\omega)\mathbf{E}. \tag{12.30}$$

ADE is based on integrating an ordinary differential equation in time that relates $\mathbf{D}(t)$ to $\mathbf{E}(t)$, concurrently with Maxwell's equations. This equation is derived by taking the inverse Fourier transform of Eq. (12.30).

We consider here a simple example where the dielectric constant $\varepsilon_r(\omega)$ consists of a single Lorentzian term, i.e:

$$\varepsilon_r(\omega) = \frac{\omega_0^2}{(\omega_0^2 - \omega^2) - i\omega\gamma_0}. \tag{12.31}$$

If we substitute Eq. (12.31) into Eq. (12.30) and take the inverse Fourier transform, we obtain a second-order differential equation relating **D** and **E**

$$\omega_0^2\mathbf{D} + \gamma_0\frac{\partial \mathbf{D}}{\partial t} + \frac{\partial^2 \mathbf{D}}{\partial t^2} = \omega_0^2\varepsilon_0\mathbf{E}. \tag{12.32}$$

Equation (12.32) is discretized using a second-order accurate central-difference scheme similar to those described above. We note that, if the ADE method is used, **E** is obtained from **H** in two steps. First, **D** is obtained from **H** by solving the finite-difference approximation of

$$\nabla \times \mathbf{H} = \frac{\partial \mathbf{D}}{\partial t}. \tag{12.33}$$

Second, **E** is obtained from **D** by solving the finite-difference approximation of Eq. (12.32). Calculation of the finite-difference expressions of the first and second temporal derivatives of **D** in Eq. (12.32) requires storage of 2 previous values of **D**, in other words not only $\mathbf{D}|^{n+1}$ but also $\mathbf{D}|^{n}$ and $\mathbf{D}|^{n-1}$ are required to obtain **E** from **D**.

Another approach to model dispersive materials with FDTD is the recursive convolution method (RC).[5,21]

FDTD is a finite-difference method, so its performance in modeling plasmonic devices is similar to the performance of FDFD. However, there are some major differences. First, as already mentioned above, in time-domain methods the dispersion properties of metals have to be approximated by suitable analytical expressions which introduce substantial error in broadband calculations. In addition, the implementation of the ADE or RC methods requires additional computational cost and extra memory storage.[5,21] On the other hand, in FDTD it is possible to obtain the entire frequency response with a single simulation by exciting a broadband pulse and calculating the Fourier transform of both the excitation and the response.[5]

12.2.5. Other Numerical Methods

The finite-element frequency-domain (FEFD) method is a more powerful technique than FDFD, especially for problems with complex geometries. However, FDFD is conceptually simpler and easier to program. The main advantage of FEFD is that complex geometric structures can be discretized using a variety of elements of different shapes, while in FDFD a rectangular grid is typically used leading to staircase approximations of particle shapes.[16,22] In addition, in FEFD fields within elements are approximated by shape functions, typically polynomials, while in FDFD a simpler piecewise constant approximation is used.[22] In short, FEFD is more complicated than FDFD but achieves better accuracy for a given computational cost.[22]

The modes and dispersion characteristics of waveguiding structures which are uniform in the z-direction can also be calculated using the method of lines (MoL). In the MoL, the differential equations to be solved are discretized along one dimension only for a two-dimensional problem.[23] The equations are then solved analytically in the remaining dimension while applying continuity boundary conditions of tangential fields between consecutive layers. The main advantage of the MoL in modeling of plasmonic devices is that it can treat more efficiently the rapidly decaying fields at the metal–dielectric interfaces in the direction which is treated analytically. In general, the application of MoL is limited to planar structures and the method is not applicable to problems with complex geometries.[22]

For homogeneous scatterers an alternative to the Green dyadic method (GDM) is the boundary element method (BEM). In BEM Green's second identity is applied to derive boundary integral equations.[24] Compared to GDM, BEM has the advantage that it requires discretization of the scatterer's surface rather than the scatterer's volume.[24] However, its application is limited to homogeneous scatterers.

Another technique for modeling piecewise homogeneous material media is the multiple multipole (MMP) method. In the MMP method, the fields in every homogeneous region are expressed as a linear superposition of known basis functions which are analytical solutions of the Maxwell equations.[25] The most used basis functions are *multipolar* functions which represent harmonic monopolar, dipolar, (and so on) sources which are singular at the point where they are located.[25] The origins of multipolar functions used in a specific region are located outside that region. The coefficients of the multipole expansions in each region are determined by applying the boundary conditions at the interfaces. Thus, the MMP method is also a boundary method, since only the interfaces have to be discretized. As with BEM, application of the MMP method is limited to homogeneous scatterers.

As mentioned above, one of the major challenges in modeling plasmonic devices is the exponential decay of the field away from the metal–dielectric interface. The surface impedance boundary condition (SIBC) can be used to calculate the fields outside a metallic structure without having to model the interior of the structure.[5,26] Thus, the resolution of the rapidly decaying fields inside the metal is avoided. The SIBC relates the tangential electric and magnetic fields at the metal–dielectric interface

$$\mathbf{E}_{\text{tan}} = \sqrt{\frac{\mu_0}{\varepsilon_0 \varepsilon_r(\omega)}}\,\hat{\mathbf{n}} \times \mathbf{H}. \tag{12.34}$$

The SIBC is an approximate boundary condition which is quite accurate if the radius of curvature of the metallic object is much larger than the penetration depth of the field inside the metal[2.6] However, it can be used only if this condition is satisfied. If the field penetration in the metal is significant, the method fails to model the related physics.

12.3. CONCLUSIONS

The dispersion properties of metals at optical frequencies are a major challenge in modeling of plasmonic devices. Frequency-domain techniques can treat arbitrary material dispersion but require a large number of simulations to obtain the broadband response. One possible future research direction could be the development of suitable, fast, frequency-sweep methods.[16] Such methods allow the computation of the broadband response by simulating the device only at a limited number of frequencies. In time-domain techniques the entire frequency response can be obtained with a single simulation. However, the treatment of dispersion requires additional computational and storage cost and the analytical approximations used result in substantial error. A possible future research direction could be the development of analytical expressions which allow approximation of metallic dispersion with a small number of terms. Such expressions could increase the accuracy and decrease the computational and memory cost of broadband time-domain calculations. Another major challenge in modeling of plasmonic devices is the very fine discretization required to adequately resolve the rapid field decay away from metal–dielectric interfaces. Although nonuniform and/or nonorthogonal grids increase the efficiency of standard numerical methods for such problems, it would be interesting to investigate whether numerical grids could be optimized specifically for treatment of plasmonic devices. In conclusion, there are many exciting research opportunities in developing more accurate and more efficient methods for plasmonic devices.

REFERENCES

1. L. Novotny, B. Hecht, D.W. Pohl: Interference of locally excited surface plasmons, J. Appl. Phys. **81** (4), 1798–1806 (1997).
2. E. Prodan, P. Nordlander, N.J. Halas: Effects of dielectric screening on the optical properties of metallic nanoshells, Chem. Phys. Lett. **368** (1–2), 94–101 (2003).
3. C.F. Bohren, D.R. Huffman: *Absorption and Scattering of Light by Small Particles* (Wiley, New York, 1983).
4. E.D. Palik ed: *Handbook of Optical Constants of Solids* (Academic, New York, 1985).
5. A. Taflove: *Computational Electrodynamics* (Artech House, Boston, 1995).
6. A.D. Rakic, A.B. Djurisic, J.M. Elazar, M.L. Majewski: Optical properties of metallic films for vertical-cavity optoelectronic devices, Appl. Opt. **37** (22) 5271–5283 (1998).
7. W.L. Barnes, A. Dereux, T.W. Ebbesen: Surface plasmon subwavelength optics, Nature **424**, 824–830 (2003).
8. A. Vial, A.S. Grimault, D. Macias, D. Barchiesi, M.L. de la Chapelle: Improved analytical fit of gold dispersion: application to the modeling of extinction spectra with a finite-difference time-domain method, Phys. Rev. B **71** (8), 85416 (2005).
9. J.C. Weeber, A. Dereux, C. Girard, J.R. Krenn, J.P. Goudonnet: Plasmon polaritons of metallic nanowires for controlling submicron propagation of light, Phys. Rev. B **60** (12), 9061–9068 (1999).
10. J.A. Kong: *Electromagnetic Wave Theory* (Wiley, New York, 1990).
11. A.D. Yaghjian: Electric dyadic Green's functions in the source region. Proc. IEEE **68** (2), 248–263 (1980).
12. J.P. Kottmann, O.J.F. Martin: Accurate solution of the volume integral equation for high-permittivity scatterers, IEEE Trans. Antennas Propagation **48** (11), 1719–1726 (2000).
13. E.M. Purcell, C.R. Pennypacker: Scattering and absorption of light by nonspherical dielectric grains, Astrophys. J. **186** (2), 705–714 (1973).
14. B.T. Draine, P.J. Flatau: Discrete-dipole approximation for scattering calculations, J. Opt. Soc. Am. A **11** (4), 1491–1499 (1994).
15. J.D. Jackson: *Classical Electrodynamics* (Wiley, New York, 1999).
16. J. Jin: *The Finite Element Method in Electromagnetics* (Wiley, New York, 2002).
17. G. Veronis, R.W. Dutton, S. Fan: Method for sensitivity analysis of photonic crystal devices, Opt. Lett. **29** (19), 2288–2290 (2004).
18. J.P. Berenger: A perfectly matched layer for the absorption of electromagnetic waves, J. Comput. Phys. **114** (2), 185–200 (1994).
19. J.A. Pereda, A. Vegas, A. Prieto: An improved compact 2D fullwave FDFD method for general guided wave structures, Microwave Opt. Technol. Lett. **38** (4), 331–335 (2003).
20. D.A. Genov, A.K. Sarychev, V.M. Shalaev: Plasmon localization and local field distribution in metal-dielectric films, Phys. Rev. E **67** (5), 56611 (2003).
21. J.L. Young, R.O. Nelson: A summary and systematic analysis of FDTD algorithms for linearly dispersive media. IEEE Antennas Propagation Mag. **43** (1), 61–77 (2001).

22. M.N.O. Sadiku: *Numerical Techniques in Electromagnetics* (CRC Press, Boca Raton, 2001).
23. P. Berini, K. Wu: Modeling lossy anisotropic dielectric waveguides with the method of lines, IEEE Trans. Microwave Theory Tech. **44** (5), 749–759 (1996).
24. C. Rockstuhl, M.G. Salt, H.P. Herzig: Application of the boundary-element method to the interaction of light with single and coupled metallic nanoparticles, J. Opt. Soc. Am. A **20** (10), 1969–1973 (2003).
25. E. Moreno, D. Erni, C. Hafner, R. Vahldieck: Multiple multipole method with automatic multipole setting applied to the simulation of surface plasmons in metallic nanostructures, J. Opt. Soc. Am. A **19** (1), 101–111 (2002).
26. D.M. Pozar: *Microwave Engineering* (Wiley, New York, 1998).

CHAPTER THIRTEEN

PLASMON HYBRIDIZATION IN COMPLEX NANOSTRUCTURES

J.M. STEELE,[1] N.K. GRADY,[2] P. NORDLANDER[3] AND N.J. HALAS[2,4]
[1]Department of Physics, Trinity University, San Antonio, Texas USA
[2]Department of Electrical and Computer Engineering, Rice University, Houston, Texas, USA
[3]Department of Physics and Astronomy, Rice University, Houston Texas,USA
[4]Department of Chemistry, Rice University, Houston, Texas, USA

13.1. INTRODUCTION

The recent development of a multitude of different metal-based nanoparticles and nanostructures has been fueled by a variety of uses for these structures in spectroscopic,[1,2] biomedical,[3,4] and photonic applications.[5] Examples include biosensing applications such as surface plasmon resonance (SPR) sensing,[6,7] Raman spectroscopy,[8–10] whole blood immunoassays,[11] and in vivo optical contrast agents.[12] In addition to sensing, medical applications include drug delivery materials[4] and photothermal cancer therapy.[13] New synthesis procedures have produced nanoparticle morphologies such as rods,[14] shells,[15,16] cups,[17,18,10] rings,[19] and cubes.[20] Complementary to nanoparticle chemistry, new planar fabrication methods have produced a variety of nanopatterned metal films that can support both propagating and localized surface plasmons.[1,21,22] The applications of these metal nanostructures take advantage of the enhancement of the local electromagnetic field associated with their plasmon resonances. In general, the frequency at which these plasmon resonances occurs is determined both by the dielectric properties of the materials composing the nanoparticles and the geometry of the nanoparticles.

The usefulness of these structures relies on the ability of nanoscientists to fabricate metallic nanoparticles and nanostructures with plasmon resonances in regions of the electromagnetic spectrum appropriate to a particular application. In order to predict the spectral locations of plasmon resonances of nanoparticles, early theoretical efforts focused on solutions of Maxwell's equations for highly symmetric nanoparticles[23–25] The plasmon resonances appear as one or more singularities in the frequency dependent polarizibility. However, even for simple particles the absorption cross section can

M.L. Brongersma and P.G. Kik (eds.), Surface Plasmon Nanophotonics, 183–196.

become very complicated. Computational methods such as the Discrete Dipole Approximation (DDA)[26] and the finite difference time domain (FDTD) method[27,28] have enabled calculations of the plasmonic properties for more complicated nanoparticles and nanostructures.

Although such brute force computational methods produce the optical response of arbitrary nanoparticle geometries, they provide very little insight into the nature of plasmon resonances. More importantly, these methods do not provide scientists with the means to intuitively predict the plasmon resonances of new nanostructures. Plasmon hybridization theory provides a conceptually enlightening method for calculating the plasmon resonance of complex nanostructures.[29–33] This method is a mesoscale electromagnetic analog of how atomic orbitals interact to form molecular orbitals in electronic structure theory. Plasmon hybridization theory deconstructs a nanoparticle or composite structure into more elementary shapes, and then calculates how the plasmon resonances of the composite particles interact or hybridize with each other. This theory enables scientists to draw on decades of intuition from molecular orbital theory to predict the plasmonic response of complicated nanostructures. In this chapter, the general formalisms of plasmon hybridization will be outlined, and then applied to more complex nanostructures such as multilayer concentric gold nanoshells and multi-particle geometries.

13.2. PLASMON HYBRIDIZATION OF NANOSHELLS

Hollow metallic spheres support plasmon resonances that are highly sensitive to the inner and outer radius of their shell.[23] Experimentally realizable nanoshells consist of a thin noble metal shell grown onto a functionalized silica core.[15,34] Nanoshells exhibit a broad spectral range of geometrically tunable plasmon resonances that far exceed the spectral range of solid nanoparticles (see Fig. 13.1.). Additionally, nanoshell plasmon resonances are very sensitive to the embedding dielectric medium.[35] An intuitive understanding for this tunability clearly arises from plasmon hybridization theory. In this formalism, the tunable nanoshell plasmon is viewed as a hybridization of the essentially fixed-frequency plasmon resonances of a nanosphere and a nanocavity.

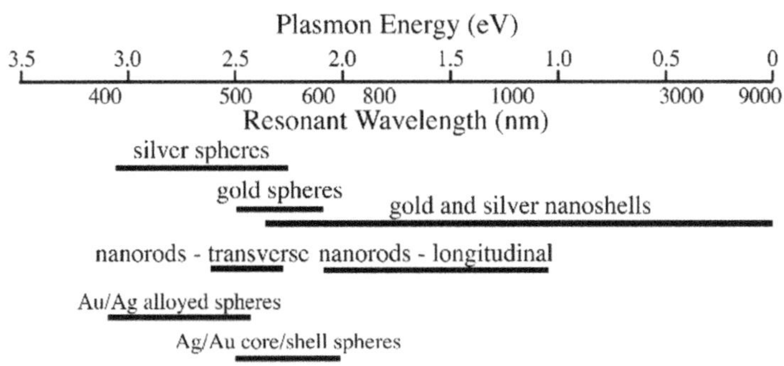

Figure 13.1. Range of plasmon resonances for a variety of particle morphologies.

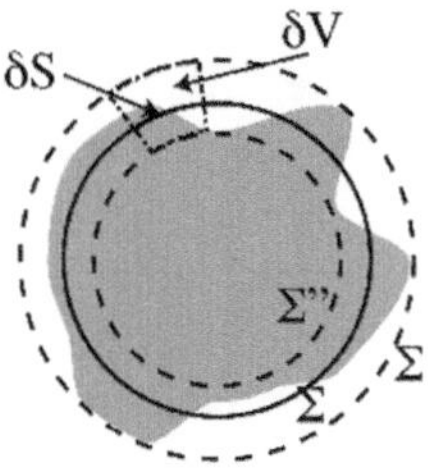

Figure 13.2. Configuration of a fluid of conduction electrons. The surfaces Σ' and Σ'' are the maximum and minimum boundaries of the fluid, and Σ denotes the nanoparticle boundary. In the limit of small deformations, surfaces Σ' and Σ'' coalesce with Σ, and the charge fluctuations can be treated as changes in the surface charge.

13.2.1. The Incompressible Fluid Model

Plasmons can be thought of as incompressible irrotational deformations of the conduction electron gas in metallic structures.[31] For simplicity, we will assume the electrons form a liquid of uniform electron density on the ion cores that form a rigid uniform background charge ρ_0. The system is assumed to be charge neutral. The current $\mathbf{j}$ and electron charge density ρ must satisfy the general equations

$$\begin{aligned} &\frac{\partial \rho}{\partial t} + \nabla \cdot \mathbf{j} = 0 \\ &\nabla \times \mathbf{j} = 0 \end{aligned} \tag{13.1}$$

A constant electron charge density inside the fluid of $n_0 e$ is assumed and the current is related to the density by $\mathbf{j} = \rho \mathbf{v}$, where $\mathbf{v}$ is the velocity of the fluid at a given position. We define a scalar potential η that satisfies the Laplace equation $\nabla^2 \eta = 0$. It then follows that the fluid velocity can be written as $\mathbf{v} = \nabla \eta$ and $\mathbf{j} = n_0 e\, \nabla \eta$, where n_0 is the electron density of the conduction electrons and e the unit charge. A net charge distribution of ρ–ρ_0 appears on the metallic surface for small deformations of the fluid. For infinitesimal deformations, this spill out charge, σ, can be defined as a superficial charge distribution:

$$\sigma = \lim_{\delta S \to 0} \frac{1}{\delta S} \int_{\delta V} (\rho - \rho_0) dV \tag{13.2}$$

where δV is a volume centered on a surface, S, as illustrated in Fig. 13.2. From the continuity equation we can derive a relationship between η and the surface charge over the volume δV. The continuity equation is given by:

$$\frac{\mathrm{d}}{\mathrm{dt}} \int \rho dV + \int \mathbf{j} \cdot d\mathbf{S} = 0. \tag{13.3}$$

With no current flowing through the boundaries of δV and where δS goes to zero, the above equation reduces to

$$\frac{\partial \sigma}{\partial t} - \mathbf{n} \cdot \mathbf{j} = 0 \Leftrightarrow \partial_t \sigma = n_0 e \frac{\partial \eta}{\partial \mathbf{n}} \tag{13.4}$$

where **n** is the surface normal vector. The dynamics of small deformations in this fluid will be determined by the Lagrangian of the system[31]

$$L = \frac{n_0 m_e}{2} \int \eta \dot{\sigma} dS - \frac{1}{2} \int \frac{\sigma(\mathbf{r})\sigma(\mathbf{r}')}{|\mathbf{r} - \mathbf{r}'|} dS_r dS_{r'}, \tag{13.5}$$

where m_e is the electron mass at $\sigma(\mathrm{r})$ is the superficial charge density at r.

13.2.2. Cavity and Solid Sphere Plasmons

To apply this formalism to calculate the plasmon resonances of a nanoshell, we must first calculate the plasmon resonances of the constituent particles: a solid metallic sphere and a spherical cavity inside a bulk metal. The metal is modeled using an electron liquid of uniform density n_0 which will give rise to a Drude dielectric function, $\varepsilon(\omega) = 1 - \omega_B^2/\omega^2$ where ω_B is the bulk plasma frequency.

$$\omega_B = \sqrt{\frac{4\pi e^2 n_0}{m_e}} \tag{13.6}$$

For brevity, the effects of the dielectric polarizibilities of the metal due to the ion cores and of the medium are ignored. Including these effects is straightforward and yields results in good agreement with experimental measurements and other theories.[31]

For a cavity of radius a, the solution of the Laplace equation can be written as a sum of spherical harmonics, $Y_{lm}(\Omega)$,

$$\eta(r, \Omega) = \sum_{l,m} \sqrt{\frac{a^{2l+1}}{l+1}} \dot{C}_{lm}(t) r^{-l-1} Y_{lm}(\Omega) \tag{13.7}$$

where the normalization constants, C_{lm}, were chosen for convenience, l is the angular momentum (multipolar order), and m is the azimuthal index. The expressions for the current and cavity surface charge, σ_c, are then[31]

$$\begin{aligned} j(t, r, \Omega) &= \sum_{l,m} n_0 e \sqrt{\frac{a^{2l+1}}{l+1}} \dot{C}_{lm}(t) \nabla \left(r^{-l-1} Y_{lm}(\Omega) \right) \\ \sigma_C(t, \Omega) &= n_0 e \sum \sqrt{\frac{l+1}{a^3}} C_{lm}(t) Y_{lm}(\Omega) \end{aligned} \tag{13.8}$$

A straightforward calculation of the kinetic and electrostatic energy of the void leads to the following Lagrangian for the electron gas:[31]

$$L_C = \frac{n_0 m_e}{2} \sum_{l,m} \left[\dot{C}_{lm}^2 - \omega_{C,l}^2 C_{lm}^2\right]. \tag{13.9}$$

This Lagrangian reduces to a set of decoupled oscillators with frequency $\omega_{C,l}$, which is the plasmon frequency predicted by classical Mie scattering theory

$$\omega_{C,l} = \omega_B \sqrt{\frac{l+1}{2l+1}}. \tag{13.10}$$

A similar procedure can be followed for a solid metallic sphere of radius b.[31] The solution of the Laplace equation now becomes

$$\eta(r, \Omega) = \sum_{l,m} \sqrt{\frac{1}{lb^{2l+1}}} \dot{S}_{lm}(t) r^l Y_{lm}(\Omega), \tag{13.11}$$

where S_{lm} are normalization constants for a solid sphere. The expressions for the current and surface charge of the sphere, σ_s, are then

$$\begin{aligned} j(t, r, \Omega) &= n_0 e \sum_{l,m} \sqrt{\frac{1}{lb^{2l+1}}} \nabla(\dot{S}_{lm}(t) r^l Y_{lm}(\Omega)) \\ \sigma_S(t, \Omega) &= n_0 e \sum_{l,m} \sqrt{\frac{1}{b^3}} S_{lm}(t) Y_{lm}(\Omega) \end{aligned} . \tag{13.12}$$

The Lagrangian for the sphere is[31]

$$L_S = \frac{n_0 m_e}{2} \sum_{l,m} \left[\dot{S}_{lm}^2 - \omega_{S,l}^2 S_{lm}^2\right]. \tag{13.13}$$

Similar to the cavity modes, this Lagrangian reduces to a set of oscillators with frequencies equal to the plasmon frequency predicted by Mie scattering theory for a solid sphere

$$\omega_{S,l} = \omega_B \sqrt{\frac{l}{2l+1}}. \tag{13.14}$$

13.2.3. Hybridization of Metallic Nanoshells

The geometry of a nanoshell is illustrated in Fig. 13.3. We will define the nanoshell aspect ratio x as the ratio of the inner to outer radius, $x = a/b$. For a shell geometry, the general solution for η is[31]

$$\eta(r, \Omega) = \sum_{l,m} \left(\sqrt{\frac{a^{2l+1}}{l+1}} \dot{C}_{lm}(t) r^{-l-1} + \sqrt{\frac{1}{lb^{2l+1}}} \dot{S}_{lm}(t) r^l \right) Y_{lm}(\Omega) \tag{13.15}$$

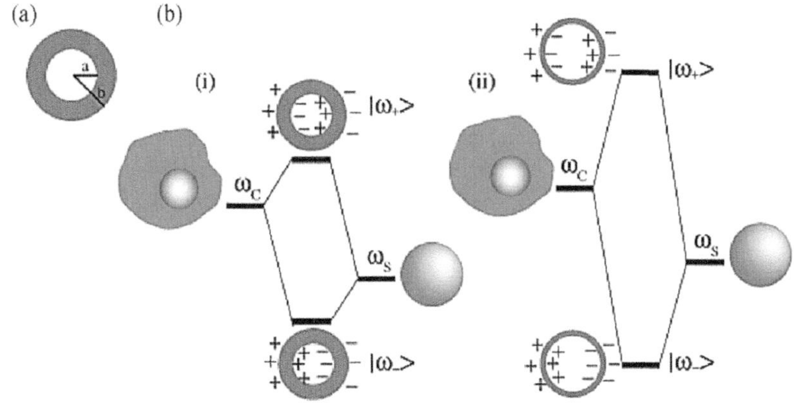

Figure 13.3. Diagram of (a) the nanoshell geometry of core radius a and total radius b. (b) shows the energy diagram for the hybridization model of a nanoshell showing both a weak (i) and strong (ii) plasmon interaction.

which leads to the following expression for the surface charges on the inner or cavity surface:

$$\sigma_C(t,\Omega) = n_0 e \sum_{l,m} \left[\sqrt{\frac{l+1}{a^3}} C_{lm} - x^{l-1} \sqrt{\frac{1}{b^3}} S_{lm}(t) \right] Y_{lm}(\Omega) \tag{13.16}$$

and on the outer surface,

$$\sigma_S(t,\Omega) = n_0 e \sum_{l,m} \left[-x^{l+2} \sqrt{\frac{l+1}{a^3}} C_{lm}(t) + \sqrt{\frac{1}{b^3}} S_{lm}(t) \right] Y_{lm}(\Omega). \tag{13.17}$$

Following the procedure outlined in the last section, the Lagrangian of the system becomes[31]

$$L_{NS} = (1 - x^{2l+1}) \left[L_C + L_S + n_0 m_e \omega_{C,l} \omega_{S,l} x^{l+(1/2)} C_{lm} S_{lm} \right]. \tag{13.18}$$

For each angular number l, this system has two fundamental frequencies

$$\omega_{l\pm}^2 = \frac{\omega_B^2}{2} \left[1 \pm \frac{1}{2l+1} \sqrt{1 + 4l(l+1)x^{2l+1}} \right]. \tag{13.19}$$

The frequencies obtained from Eq. (13.19) for nanoshells agree with plasmon frequencies calculated from classical Mie scattering theory as well as quantum mechanical *ab initio* time-dependent local density approximation (TDLDA) calculations.[36] The plasmon resonances of nanoshells can now be understood as the interaction of the sphere and cavity plasmons. This interaction or hybridization results in a splitting of the plasmon resonances into a lower energy symmetric (bonding) plasmon and a higher energy antisymmetric (antibonding) plasmon, shown schematically in Fig. 13.3(b) for both a weak (i) and strong (ii) interaction. The bonding plasmon has a larger dipole moment than the antibonding plasmon, and therefore couples more easily to incident electromagnetic radiation. The bonding plasmon is therefore the more prevalent feature in experimentally obtained extinction spectrum.

The strength of this interaction depends on the aspect ratio of the shell, x, to the power $2l + 1$ as seen in Eq. (13.19). The energy of the plasmon resonance of a nanoshell will therefore depend on both the plasmon energies of the constituent sphere and cavity, but also on the aspect ratio, or thickness of the shell. This explains the unique tunability of nanoshells compared to solid nanoparticles. If the metal shell is thin, the interaction will be strong and the splitting of the plasmons will be large, as seen in Fig. 13.3 (b). For a thick shell, the interaction will be weak and therefore a small splitting of the plasmons will occur.

13.3. HYBRIDIZATION IN MORE COMPLICATED STRUCTURES

By using the above formalisms and the calculated plasmon resonances for elementary geometries, we can now understand the plasmonic response of complex nanostructures by considering their more elementary constituent parts. In this section, we will use plasmon hybridization to calculate the plasmon resonance of concentric metallic shells and metallic dimer particles.

13.3.1. Multiconcentric Metallic Shells

Recently, the tuning range of nanoshells has been expanded into the infrared by the fabrication of multiconcentric gold nanoshells.[32,33] These 'nano-matryushkas' are composed of a silica core, a gold shell, a silica spacer layer, and finally a second gold shell. Nano-matryushka plasmons arise from the hybridization of the two individual metal shell plasmons, with the interaction strength determined in part by the thickness of the spacer layer.

To calculate the plasmon response of N concentric nanoshells, we define the nanoshells with inner radius a_j and outer radius b_j, where the jth shell is inside the $j + 1$ nanoshell. The individual nanoshell surface charge expressions will be the same as derived in sect. 13.2.3, and the kinetic energy of the system is simply the sum of the kinetic energies of the individual nanoshells. The electrostatic energy system is the sum of the interactions between the different nanoshells and can be written as[31]

$$V_{lm}^{ij} = 2\pi e n_j \frac{1 - x_j^{2l+1}}{2l+1} \begin{cases} \sqrt{l+1}\, a_j^{-l-(1/2)} \left[a_i^{l+2} \sigma_{C,lm}^i + b_i^{l+2} \sigma_{S,lm}^i \right] C_{lm}^j & \text{for } i < j \\ \sqrt{(l+1) a_i^3}\, \sigma_{C,lm}^i C_{lm}^i + \sqrt{l b_i^3}\, \sigma_{S,lm}^i S_{lm}^i & \text{for } i = j \\ \sqrt{l}\, b_j^{l+(1/2)} \left[a_i^{-l+1} \sigma_{C,lm}^i + b_i^{-l+1} \sigma_{S,lm}^i \right] S_{lm}^j & \text{for } i > j \end{cases} \tag{13.20}$$

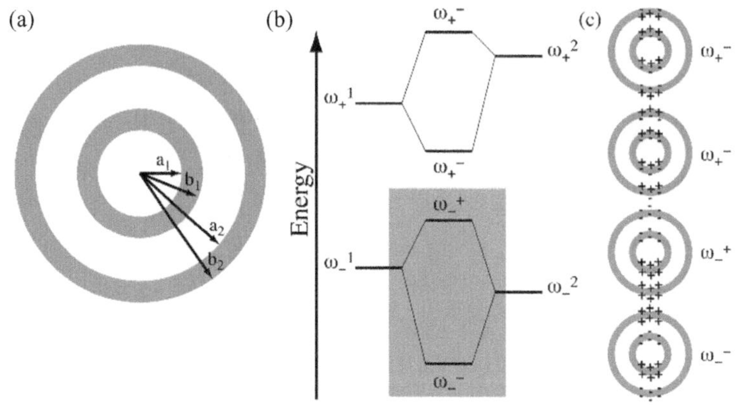

Figure 13.4. (a) Schematic of a concentric nanoshell where the gray regions are the metallic phases and the white regions are the dielectric phases. (b) Energy diagram of the plasmon modes excited in a concentric nanoshell. (c) Induced surface charge for each plasmon mode.

For multiconcentric nanoshells, the Lagrangian is simply

$$L_{CS} = \sum_i L^i_{NS} - \sum_{i \neq i'} V^{ii'}. \tag{13.21}$$

In the case of two concentric shells made of the same metal, shown schematically in Fig. 13.4 (a), the Lagrangian reduces to[31]

$$L_{CS} = L^1_{NS} + L^2_{NS} - n_0 m_e (1 - x_1^{2l+1})(1 - x_2^{2l+1}) \omega^2_{C,l} \omega^1_{S,l} \left(\frac{b_1}{a_2}\right)^{l+(1/2)} S^1_{ml} C^2_{lm}. \tag{13.22}$$

This system can now be described as the interaction of four linearly independent charge deformations, or plasmons. The interaction strength is determined by the thickness of the dielectric spacer layer $|a_2 - b_1|$. Although the symmetric and antisymmetric plasmons on a given shell can interact with both plasmons on the other shell, the large energy separation between the ω_+ and ω_- plasmons will cause the hybridization to occur predominantly between two plasmons of the same symmetry. The resulting energy diagram for the composite plasmon resonances are shown in Fig. 13.4 (b), and the induced surface charge for each mode in (c). Experimentally accessible modes are highlighted in gray. The hybridization of the two symmetric modes yields an antisymmetric (or bonding) mode, ω_-^- and a symmetric (or antibonding) mode, ω_-^+. The antisymmetric mode has a smaller dipole moment and is of lower energy. The symmetric mode has a higher energy due to the electrostatic repulsion at the internal interfaces of the metal shells. For the interactions of the antisymmetric modes, the symmetric alignment is favored because the interaction of the two ω_+ modes has an opposite sign compared to the symmetric case.

Concentric nanoshells have recently been fabricated and experimental measurements of their plasmon resonances match well with theory.[32,33] Figure 13.5 shows

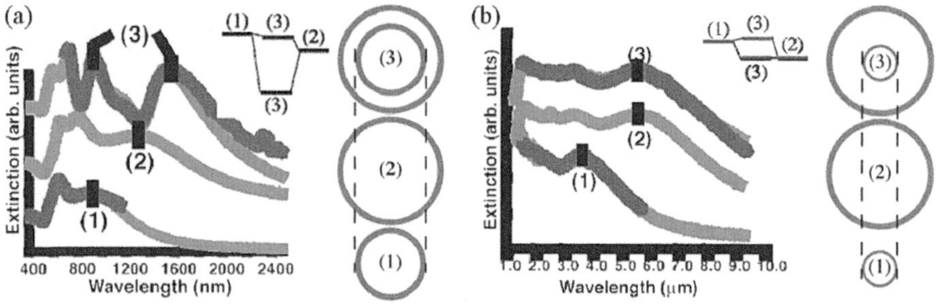

Figure 13.5. Experimental and theoretical extinction spectra of concentric nanoshells (3) along with their constituent inner (1) and outer (2) single nanoshells. Concentric nanoshell with dimensions $a_1/b_1/a_2/b_2$ of (a) 80/107/135/157 nm resulting in a strong coupling and (b) 396/418/654/693 nm showing the weak coupling case.

the optical response of concentric gold nanoshells with a silica core and spacer layer for both strong (a) and weak (b) coupling cases. In this figure, the spectra labeled (1) show the experimental and theoretical extinction spectra for the isolated inner shell $\omega_{-\mathrm{NS1}}$. The theoretical spectra for the isolated outer shell, $\omega_{-\mathrm{NS2}}$, are labeled (2), and the theoretical and experimental spectra for the composite concentric nanoshell are labeled (3). In spectra (3), both the $\omega^{+}_{-\mathrm{CS}}$ and $\omega^{-}_{-\mathrm{CS}}$ plasmons are clearly apparent. In these plots, the theoretical extinction spectrum is calculated using Mie scattering theory.

The concentric nanoshell depicted in Fig. 13.5 (a) shows strong coupling because the inter-shell spacing is small (28 nm) and the inner and outer nanoshell plasmons are nearly resonant with each other. The asymmetry of the splitting can be attributed to retardation effects and interactions with modes of higher l. Figure 13.5 (b) shows the case where the plasmon modes are fully decoupled because the inter-shell spacing is very large (236 nm). In this case, the outer shell feels little effect of the inner shell even being present.

13.3.2. Nanoparticle Dimers

In this section, the plasmon hybridization method will be applied to a system of nanoparticle dimers. The massive field enhancements occurring in the junction between two nanoparticles make nanoparticle dimers greatly important for applications such as Surface Enhanced Raman Spectroscopy (SERS).[37,38] While this has sparked a number of theoretical and experimental studies,[27,39–41] there is a lack of consensus about the effect of interparticle spacing on the dimer plasmon resonances. In two studies considering nanoparticle dimers with a high aspect ratio, one finds that the energy shifts of the dimer plasmon energies can be described as a simple dipolar interaction,[40] while another reports an exponential dependence on the interparticle spacing.[41]

To illustrate the plasmon hybridization method, two isolated sphere plasmons will be mixed to calculate the corresponding dimer plasmon energies. For nanospheres of radius b and large interparticle spacing D, the shifts in the plasmon energy should follow the interaction between classical dipoles $(1/D^3)$. However, as D becomes smaller the single particle plasmons start mixing with higher order l plasmons in the other particle, causing the dimer plasmon energy to shift more strongly and much faster than $1/D^3$. To calculate the dimer plasmon energies, the same surface charge expressions and Lagrangian derived in sect. 13.2.2 will be used. Dielectric effects will again be neglected to elucidate the nature of the plasmon resonances of dimers, however they can be included if experimentally accurate results are desired.[31]

Retardation effects can be neglected if the overall size of the nanoparticle dimer is significantly smaller than the plasmon wavelength $(D + 2b < \lambda/4)$. In this limit, the instantaneous Coulomb interaction between the surface charges determines the potential energy of the system[29]

$$V(D) = \int b_1^2 d\Omega_1 \int b_2^2 d\Omega_2 \frac{\sigma^1(\Omega_1)\sigma^2(\Omega_2)}{|\mathbf{r}_1 - \mathbf{r}_2|}. \tag{13.23}$$

This interaction is diagonal in the azimuthal quantum number, m, if the polar axis lies along the dimer axis, indicating that the plasmon modes of different m are decoupled. To simplify the notation, for the following equations i will denote noninteracting plasmon modes, with energy ω_i and angular momentum l_i on sphere N_i with radius b_i. For each number m, the Lagrangian is[29]

$$L^{(m)} = \frac{n_0 m_e}{2} \sum_{i,j} \left[(\dot{S}_i^2 - \omega_i^2 S_i^2)\delta_{ij} - \frac{\omega_B^2}{4\pi} V_{ij}^{(m)}(D) S_i S_j \right] \tag{13.24}$$

The interaction $V_{ij}^{(m)}$ is zero if i and j refer to the same sphere. Otherwise the interaction is[29]

$$V_{ij}^{(m)}(D) = 4\pi \sqrt{l_i l_j b_i^{2l_i+1} b_j} \int d\theta_j \sin\theta_j \frac{P_{l_i}^m(\cos\Theta_i(\theta_j))}{(2l_i + 1) X_i(\theta_j)^{l_i+1}} P_{l_j}^m(\cos\theta_j) \tag{13.25}$$

For the above equation, the integration is taken over the surface of the particle N_i and the interaction matrix is symmetric. Fig. 13.6 (a) illustrates the geometry of the problem. At this point its useful to note that the interaction between two sphere plasmon modes of angular momentum l and l' will vanish as $D^{-(l+l'+1)}$ or faster. The Euler–Lagrange equations results in the follow eigenvalue problem

$$\det\left[A_{ij}^{(m)} - \omega^2\right] = 0 \tag{13.26}$$

where the matrix A is defined by:

$$A_{ij}^{(m)} = \omega_i^2 \delta_{ij} + \frac{\omega_B^2}{8\pi} \left(V_{ij}^{(m)} - V_{ji}^{(m)} \right) \tag{13.27}$$

Figure 13.6 (b) shows the calculations for a dimer made of nanospheres with a radius of 15 nm, $\omega_B = 9$ eV, and an electron density corresponding to $r_s = 3.0$. The

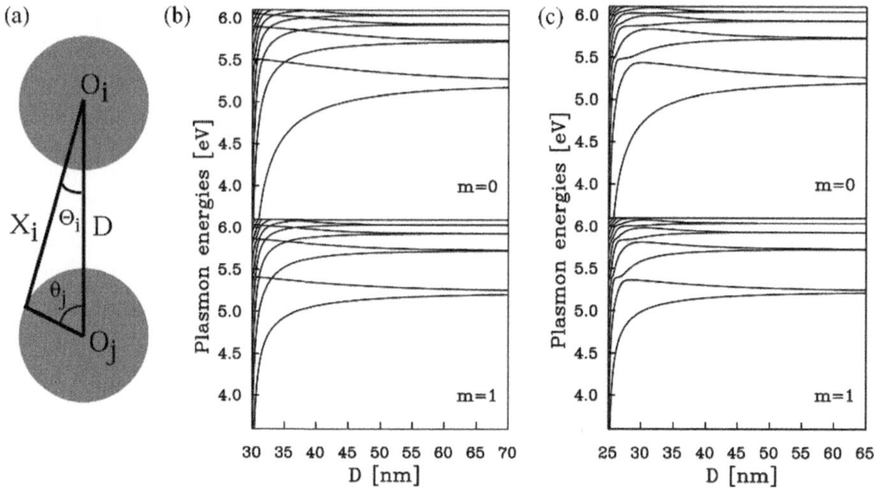

Figure 13.6. (a) Schematic picture of the dimmer geometry described in the text with the origin of the particles i and j at O_i and O_j, respectively. (b) Calculated plasmon energies of a nanosphere dimer as a function of interparticle spacing for 2 nanospheres of radius 15 nm. (c) Calculated plasmon energies for a heterodimer composed of nanoshperes of 10 nm and 15 nm radii as a function of interparticle distance. For both (b) and (c), the top panel is for $m = 0$ and the bottom panel is for $m = 1$. The lowest two energy curves correspond to the $l = 1$ plasmons.

upper graph shows the dimer plasmon energies as a function of sphere separation for plasmons polarized along the dimer axis ($m = 0$). Concentrating on the lowest two plasmons modes ($l = 1$), the bonding configuration corresponds to the two dipoles oscillating in phase (symmetric electric fields or positive parity of dipole moments). The antibonding configuration corresponds to the negative parity of the dipoles (antisymmetric electric fields). Because the net dipole moments of the bonding configuration are large, they will couple efficiently with light and will therefore be referred to as bright plasmons. The net dipole moment of the antibonding plasmons is zero, and therefore not easily excited by light. These plasmons will be referred as dark plasmons. For large separations, the splitting increases as the interaction increases, i.e., as $1/D^3$. As the separation decreases, the split becomes asymmetric as higher multipoles come into play. The bonding plasmon then shifts downward in energy much faster than the antibonding plasmon shifts upward. The bottom graph shows the $m = \pm 1$ plasmons for polarizations perpendicular to the dimer axis. Overall the dimer plasmon energies versus distance is similar to the $m = 0$ case. However, since the dipole coupling for this case has the opposite sign, the bright and dark plasmon assignments are reversed: the dark plasmons are the bonding (antisymmetric) modes and the bright plasmons are the antibonding (symmetric) modes.

In Fig. 13.6 (c), the dimer energies are calculated for two nanospheres with radii of 10 and 15 nm. For a heterodimer, parity is broken and the dimer energies therefore exhibit avoided crossings, i.e., the dimer plasmon modes repel each other. This results in particularly strong interactions as the distance between the spheres is decreased.

For $|m| < 1$, all plasmons have a net dipole moment and multiple peaks are expected in the optical absorption spectra, or a broad band absorption line if the resonances overlap.

13.4. CONCLUSIONS

In this chapter we have reviewed the plasmon hybridization method for calculating the plasmonic response of nanostructures with complicated geometries. By considering the electrons as an incompressible fluid, the optical response of a complex nanoparticle can now be modeled as a system of interacting plasmons on each of the constituent surfaces, in direct analogy with molecular orbital theory. This analogy provides for a simple intuitive understanding of plasmon resonances in complex systems. Applications of this method is currently ongoing, with recent efforts focusing on nanoparticles in close proximity to metallic surfaces[30] and larger multi-particle systems. It provides a general and powerful tool for scientists to use when developing nanoparticles and nanostructures for use in various applications.

REFERENCES

1. S. Ekgasit, C. Thammacharoen, F. Yu, W. Knoll: Evanescent field in surface plasmon resonance and surface plasmon field-enhanced fluorescence spectroscopies, Anal. Chem. **76** (8), 2210–2219 (2004).
2. R.P. Van Duyne: Molecular plasmonics, Science **306** (5698), 985–986 (2004).
3. A.J. Haes, L. Chang, W.L. Klein, R.P. Van Duyne: Detection of a biomarker for Alzheimer's disease from synthetic and clinical samples using a nanoscale optical biosensor, J. Am. Chem. Soc. **127** (7), 2264–2271 (2005).
4. S.R. Sershen, S.L. Westcott, N.J. Halas, J.L. West: Temperature-sensitive polymer-nanoshell composites for photothermally modulated drug delivery, J. Biomed. Mater. Res. **51** (3), 293–298 (2000).
5. S.A. Maier, M.L. Brongersma, P.G. Kik, S. Meltzer, A.A.G. Requicha, B.E. Koel, H.A. Atwater: Plasmonics—A route to nanoscale optical devices, Adv. Mater. **15** (7–8), 562–562 (2003).
6. Y.G. Sun, Y.N. Xia,: Increased sensitivity of surface plasmon resonance of gold nanoshells compared to that of gold solid colloids in response to environmental changes, Anal. Chem. **74** (20), 5297–5305 (2002).
7. F. Tam, C. Moran, N. Halas: Geometrical parameters controlling sensitivity of nanoshell plasmon resonances to changes in dielectric environment, J. Phys. Chem. B **108** (45), 17290–17294 (2004).
8. J.B. Jackson, N.J. Halas: Surface-enhanced Raman scattering on tunable plasmonic nanoparticle substrates, Proc. Natl. Acad. Sci. U S A **101** (52), 17930–17935 (2004).
9. K. Kneipp, H. Kneipp, I. Itzkan, R.R. Dasari, M.S. Feld: Ultrasensitive chemical analysis by Raman spectroscopy, Chem. Rev. **99** (10), 2957–+ (1999).
10. Y. Lu, G.L. Liu, J. Kim, Y.X. Mejia, L.P. Lee: Nanophotonic crescent moon structures with sharp edge for ultrasensitive biomolecular detection by local electromagnetic field enhancement effect, Nano Lett. **5** (1), 119–124 (2005).
11. L.R. Hirsch, J.B. Jackson, A. Lee, N.J. Halas, J. West: A whole blood immunoassay using gold nanoshells, Anal. Chem. **75** (10), 2377–2381 (2003).
12. C. Loo, A. Lin, L. Hirsch, M.H. Lee, J. Barton, N. Halas, J. West, R. Drezek: Nanoshell-enabled photonics-based imaging and therapy of cancer, Technol. Cancer Res. Treat. **3** (1), 33–40 (2004).
13. D.P.O'Neal, L.R. Hirsch, N.J. Halas, J.D. Payne, J.L. West: Photo-thermal tumor ablation in mice using near infrared-absorbing nanoparticles, Cancer Lett. **209** (2), 171–176 (2004).

14. B. Nikoobakht, M.A. El-Sayed: Preparation and growth mechanism of gold nanorods (NRs) using seed-mediated growth method, Chem Mater. **15** (10), 1957–1962 (2003).
15. S.J. Oldenburg, R.D. Averitt, S.L. Westcott, N.J. Halas: Nanoengineering of optical resonances, Chem. Phys. Lett. **288** (2–4), 243–247 (1998).
16. Y.G. Sun, Y.N. Xia: Gold and silver nanoparticles: A class of chromophores with colors tunable in the range from 400 to 750 nm, Analyst **128** (6), 686–691 (2003).
17. C. Charnay, A. Lee, S.Q. Man, C.E. Moran, C. Radloff, R.K. Bradley, N.J. Halas: Reduced symmetry metallodielectric nanoparticles: Chemical synthesis and plasmonic properties, J. Phys. Chem. B **107** (30), 7327–7333 (2003).
18. J.C. Love, B.D. Gates, D.B. Wolfe, K.E. Paul, G.M. Whitesides: Fabrication and wetting properties of metallic half-shells with submicron diameters, Nano Lett. **2** (8), 891–894 (2002).
19. J. Aizpurua, P. Hanarp, D.S. Sutherland, M. Kall, G.W. Bryant, F.J.G. de Abajo: Optical properties of gold nanorings, Phys. Rev. Lett. **90** (5), (2003).
20. Y. Sun, Y. Xia: Shape-controlled synthesis of gold and silver Nanoparticles, Science **5601**, 2176–2179 (2002).
21. C.L. Haynes, R.P. Van Duyne: Nanosphere lithography: A versatile nanofabrication tool for studies of size-dependent nanoparticle optics, J. Phys. Chem. B **105** (24), 5599–5611 (2001).
22. C.E. Moran, J.M. Steele, N.J. Halas: Chemical and dielectric manipulation of the plasmonic band gap of metallodielectric arrays, Nano Lett. **4** (8), 1497–1500 (2004).
23. A.L. Aden, M. Kerker: Scattering of electromagnetic waves from 2 concentric spheres, J. Appl. Phys. **22** (10), 1242–1246 (1951).
24. K.L. Kelly, C. Eduardo, L.L. Zhao, G.C. Schatz: The optical properties of metal nanoparticles: the influence of size, shape, and dielectric environment, J. Phys. Chem. B **107** (3), 668–677 (2003).
25. G. Mie: Articles on the optical characteristics of turbid tubes, especially colloidal metal solutions, Annalen Der Physik **25** (3), 377–445 (1908).
26. B.T. Draine, P.J. Flatau: Discrete-dipole approximation for scattering calculations, J. Opt. Soc. Am. A **11** (4), 1491–1499 (1994).
27. M. Futamata, Y. Maruyama, M. Ishikawa: Local electric field and scattering cross section of Ag nanoparticles under surface plasmon resonance by finite difference time domain method, J. Phys. Chem. B **10 7**(31), 7607–7617 (2003).
28. C. Oubre, P. Nordlander: Optical properties of metallodielectric nanostructures calculated using the finite difference time domain method, J. Phys. Chem. B **108** (46), 17740–17747 (2004).
29. P. Nordlander, C. Oubre, E. Prodan, K. Li, M.I. Stockman: Plasmon hybridizaton in nanoparticle dimers, Nano Lett. **4** (5), 899–903 (2004).
30. P. Nordlander, E. Prodan: Plasmon hybridization in nanoparticles near metallic surfaces, Nano Lett. **4** (11), 2209–2213 (2004).
31. E. Prodan, P. Nordlander: Plasmon hybridization in spherical nanoparticles, J. Chem. Phys. **120** (11), 5444–5454 (2004).
32. E. Prodan, C. Radloff, N.J. Halas, P. Nordlander: A hybridization model for the plasmon response of complex nanostructures, Science **302** (5644), 419–422 (2003).
33. C. Radloff, N.J. Halas: Plasmonic properties of concentric nanoshells, Nano Lett. **4** (7), 1323–1327 (2004).
34. S.J. Oldenburg, J.B. Jackson, S.L. Westcott, N.J. Halas: Infrared extinction properties of gold nanoshells, Appl. Phys. Lett. **75** (19), 2897–2899 (1999).
35. E. Prodan, A. Lee, P. Nordlander: The effect of a dielectric core and embedding medium on the polarizability of metallic nanoshells, Chem. Phys. Lett. **360** (3–4), 325–332 (2002).
36. E. Prodan, P. Nordlander: Electronic structure and polarizability of metallic nanoshells, Chem. Phys. Lett. **352** (3–4), 140–146 (2002).
37. K. Kneipp, H. Kneipp, I. Itzkan, R.R. Dasari, M.S. Feld: Ultrasensitive chemical analysis by Raman spectroscopy, Chem. Rev. **99** (10), 2957–2975 (1999).
38. H.X. Xu, E.J. Bjerneld, M. Kall, L. Borjesson: Spectroscopy of single hemoglobin molecules by surface enhanced Raman scattering, Phys. Rev. Lett. **83** (21), 4357–4360 (1999).

39. J. Prikulis, F. Svedberg, M. Kall, J. Enger, K. Ramser, M. Goksor, D. Hanstorp: Optical spectroscopy of single trapped metal nanoparticles in solution, Nano Lett. **4** (1), 115–118 (2004).
40. W. Rechberger, A. Hohenau, A. Leitner, J.R. Krenn, B. Lamprecht, F.R. Aussenegg: Optical properties of two interacting gold nanoparticles, Opt. Commun. **220** (1–3), 137–141 (2003).
41. K.H. Su, Q.H. Wei, X. Zhang, J.J. Mock, D.R. Smith, S. Schultz: Interparticle coupling effects on plasmon resonances of nanogold particles, Nano Lett. **3** (8), 1087–1090 (2003).

CHAPTER FOURTEEN

SENSING PROTEINS WITH ADAPTIVE METAL NANOSTRUCTURES

VLADIMIR P. DRACHEV, MARK D. THORESON AND VLADIMIR M. SHALAEV

School of Electrical and Computer Engineering and Birck Nanotechnology Center, Purdue University, West Lafayette, IN 47907, USA

14.1. INTRODUCTION

Raman scattering spectra enable molecular "fingerprinting", which is of particular interest for molecular sensing and bio-applications. Surface enhanced Raman scattering (SERS) provides greater detection sensitivity than conventional Raman spectroscopy,[1–3] and it is quickly gaining traction in the study of biological molecules adsorbed on a metal surface.[4–12] SERS spectroscopy allows for the detection and analysis of minute quantities of analytes because it is possible to obtain high-quality SERS spectra at sub-monolayer molecular coverage as a result of the large scattering enhancements. SERS has also been shown to be sensitive to molecular orientation and the distance of the molecule to the metal surface.[13]

The SERS enhancement mechanism originates in part from the large local electromagnetic fields caused by resonant surface plasmons that can be optically excited at certain wavelengths for metal particles of different shapes or closely spaced groups of particles.[14–21] For aggregates of interacting particles, which are often structured as fractals, plasmon resonances can be excited in a very broad spectral range.[22] In addition to electromagnetic field enhancement, metal nanostructures and molecules can form charge-transfer complexes that provide further enhancement for SERS.[23–29] The resulting overall enhancement depends critically on the particle or aggregate nanostructure morphology.[22,30–36] Enhancement factors can be as high as 10^5 to 10^9 for the area-averaged macroscopic signal and 10^{10} to 10^{15} within the local resonant nanostructures.

A variety of structures have been found to be appropriate for SERS, including roughened metal electrodes,[1–3] aggregated films,[15] metal islands of different morphologies,[14,15,17–20] and semicontinuous metal films near the percolation

M.L. Brongersma and P.G. Kik (eds.), Surface Plasmon Nanophotonics, 197–215.

threshold.[37–39] Among SERS-active substrates, vacuum-evaporated nanostructured metal films are well suited for SERS studies and have a high potential for applications.[14–20,24,39–43] The effect on the metal film due to deposition rate, mass thickness, and thermal annealing were previously studied in detail.[18,40–43]

The most advanced tendency in SERS substrate preparation is to engineer optimal nanostructures with a controlled particle shape, such as triangles,[35,44] nanoshells,[36,45] or with a regular arrangement achieved by nanosphere lithography,[35] electron beam lithography,[46] or through metal coating of dielectric spheres.[4,42] It has been shown that the maximum local field can be obtained between a pair of particles with an appropriate shape like a bow-tie structure[47,48] fabricated with electron beam lithography or nanosphere lithography. In particular, the largest macroscopic enhancement factor related to unit area covered with metal was reported with an array of triangular particles.[35] Strong SERS enhancement can also occur in a periodic array of metal nanoparticles.[80]

We found recently that vacuum evaporated silver films fabricated within a certain range of evaporation parameters allow fine rearrangement of the local structure under protein deposition.[49–53] Such a substrate, which is referred to as *adaptive* enables adsorption of proteins without significant changes in their conformational states and provides large SERS enhancement to allow protein sensing at monolayer protein surface density. This adaptive property appears to be especially important for the sensing of large molecules whose size is comparable with the typical sizes of the metal film nanostructure (particle size, inter-particle spacing). In the case of proteins in the nanometer size range, it is still an issue to match an optimal design of engineered nanostructures with a molecule of particular size and shape. This issue can be addressed with adaptive nanostructures where protein-mediated restructuring forms groups of metal particles naturally covered and matched with the molecules of interest.

In this chapter we discuss specific properties of adaptive silver films (ASFs) and provide several examples of their uses for protein sensing and potential applications in protein microarrays. An example of SERS spectral sensitivity to protein conformational state is presented in the case of insulin. We also describe further development of this type of substrate to improve the SERS signal by addition of a sublayer of bulk metal.

14.2. BASIC FORMULAS FOR SERS ENHANCEMENT FACTOR

The concept of the enhanced local field in aggregated particle films based on the Maxwell-Garnett approach was applied to SERS in the 1980s.[14–15,17–19] These theories account for particle shape and, in a strongly simplified manner, particle interactions. The early theories also emphasize that strong local fields can only result from large field fluctuations.[16]

In many cases, aggregates of metal particles form fractal structures, which are characterized by a scale-invariant distribution such that they look similar at different length scales. A special type of fractal structure is represented by a percolation film (also referred to as a semicontinuous metal film). A percolation film is formed by fractal metal clusters of different sizes, from a cluster of a single or just few particles up to the "infinite" fractal cluster that spans over the whole film and provides a conducting (percolating) path for the film. Below the percolation threshold a metal-dielectric film is an insulator whereas above the threshold it is a conductor. Although macroscopically a percolation film is homogeneous, the metal clusters can still be characterized by a scale-invariant distribution.

For fractal systems, the powerful scaling approach can be applied. A theory describing optical properties of fractal aggregates was developed by Shalaev and Stockman,[22,39] and an optical theory for percolation composites was put forward by Sarychev and Shalaev.[38,39] There is much in common between fractal aggregates and percolation films; in both cases, the electromagnetic excitations are localized in small, nanometer-sized areas which are referred to as "hot spots". The local-field intensity in the hot spots can exceed the intensity of an incident wave by three to five orders of magnitude, resulting in a dramatic Raman scattering enhancement of up to twelve orders of magnitude under optimum conditions.

This enhancement results from the excitation of local plasmon modes in aggregates (clusters) of metal nanoparticles. Because of scale-invariance, metal clusters in fractal and percolation systems are characterized by a large variety of shapes and sizes. Different local structures formed by metal particles resonate at different frequencies so that all together they provide a broad range of enhancement for Raman scattering. Because different metal clusters resonate at different frequencies, the hot spots at different frequencies are spatially separated. A typical size of the hot spots is between 10 and 100 nm. Light at different frequencies excite different distributions of hot spots on a fractal or percolation system. The resonant plasmon modes in such systems cover a very broad spectral range from the near-UV to the mid-IR.

In many cases, random aggregates of particles do not form fractal or percolation systems (for example, an island film well below the percolation). Still, hot spots are excited in various small groups of metal particles and the optical properties of such random systems are often similar to those obtained for fractal and percolation systems. In fact, it has been shown that the optical scaling theory developed for percolation systems works well in a rather broad range of metal concentrations around the percolation threshold, including those well below and above the percolation threshold.[38,39] For optical frequencies, the scaling formulas of the Shalaev–Sarychev theory remain valid in a broad range of metal filling factors, roughly from 0.3 to 0.7. The fractal theory is also very robust and describes random aggregates of particles (even if they are not fractal), provided that there is frequency-spatial selectivity so that hot spots at different frequencies are located in different locations, resulting in an inhomogeneously-broadened distribution of plasmon modes (see Sect. 3.6.1 of Ref. 39).

Below we present formulas describing SERS enhancement in random metal-dielectric composites, including the important case of metal-dielectric films.

14.2.1. Enhancement Factor for Raman Scattering

Let's assume that Raman-active molecules are placed on a metal-dielectric film, where the plasmon modes can be excited resulting in local-field enhancement. Raman scattering (RS) of an incident wave at frequency ω results in a scattered field at a shifted frequency ω_s.The shift ω–ω_s is equal to one of the molecule's vibrational frequencies, and the combination of all such vibrational frequencies represents the molecule's "fingerprint". Spontaneous Raman scattering is a linear, incoherent optical process. Because the plasmon modes cover a very broad spectral range, the fields at frequencies ω and ω_s can both experience the enhancement caused by the resonant plasmon modes.[16] As a result, the enhancement factor for surface-enhanced Raman scattering, G_{RS}, is given by the product of the enhancements for the two fields at frequencies ω and ω_s:[39]

$$G_{RS} \sim \left\langle \left| \frac{E_\omega}{E^0_\omega} \right|^2 \left| \frac{E_{\omega_s}}{E^0_{\omega_s}} \right|^2 \right\rangle, \tag{14.1}$$

where the angular brackets stand for spatial averaging over the random film. In Eq. (14.1), E_ω and $E_{\omega s}$ are the local enhanced fields at frequencies ω and ω_s, respectively, whereas E^0_ω and $E^0_{\omega s}$ are the probe, non-enhanced fields at these frequencies. Below, for simplicity, we set these non-enhanced amplitudes both equal to unity so that $|E_\omega|^2$ and $|E_{\omega s}|^2$ represent the local enhancements for the field intensities at frequencies ω and ω_s. Then, the enhancement factor for RS in Eq. (14.1) can be rewritten as

$$G_{RS} \sim \left\langle \left| E_\omega E_{\omega_s} \right|^2 \right\rangle. \tag{14.2}$$

There are two important limiting cases for Eq. (14.2) that we consider below. If the Stokes frequency shift is smaller than the plasmon resonance width Γ, then the hot spots at the two frequencies occur in the same spatial positions so that the enhancement can be written as

$$G_{RS} \sim \left\langle |E|^4 \right\rangle, \tag{14.3}$$

where the ω and ω_s fields have been set equal to each other because the two enhancements are fully correlated. Thus under these optimum conditions, the SERS enhancement is proportional to the fourth power of the enhanced local field averaged over the film. Such nonlinear dependence for the *enhancement* should not be confused with the Raman signal itself, which is still proportional to the field intensity because it is a linear process.

A more conventional situation, however, is represented by the case when the shift ω—ω_s exceeds the plasmon resonance width Γ. For large shifts, the positions of the hot spots at the two frequencies can be approximated as statistically independent. Then, we can decouple the averaging for the two fields in Eq. (14.2) and present the enhancement as

$$G_{RS} \sim \left\langle |E_\omega|^2 \right\rangle \left\langle \left| E_{\omega_s} \right|^2 \right\rangle. \tag{14.4}$$

Equation (14.4) is in agreement with experiments[15,18] and theory[17–18] showing that the enhanced Raman intensity is proportional to the product of the absorptions at the laser and scattered frequencies, i.e. to $A(\omega)A(\omega_s)$[15,17–19] (note that the absorption A

is proportional to $\langle |E|^2 \rangle$). This fact helps in evaluating various substrates for their possible use as SERS substrates.

For noble and coinage metals, the dielectric permittivity, ε, can be well-described by the Drude formula

$$\varepsilon = \varepsilon_0 - \frac{\omega_p^2}{\omega\left(\omega + i\tau^{-1}\right)}, \tag{14.5}$$

where $\tau = \Gamma^{-1}$ is the relaxation time for plasmon oscillations, ε_0 is the interband contribution to the permittivity and ω_p is the bulk plasma frequency. For effective SERS metals the parameters above have the following magnitudes:[54,55] Ag ($\omega_p = 9.1$ eV, $\tau_{eff}^{-1} = 0.021$ eV), Au ($\omega_p = 9.0$ eV, $\tau_{eff}^{-1} = 0.066$ eV), and Cu ($\omega_p = 8.8$ eV, $\tau_{eff}^{-1} = 0.09$ eV).

14.2.2. Relationship between Electromagnetic Enhancement Factor and Effective Optical Properties

Below we estimate the SERS enhancement factor, using formulas obtained for fractal and percolation systems.[22,38,39] As mentioned, these results remain approximately valid for a rather broad class of metal-dielectric composites and films.

For fractals, the enhancement for a nonlinear process which is proportional to the nth power of the local field is estimated as[39]

$$\left\langle |E|^n \right\rangle_{frac} \sim \left(\frac{\left|\varepsilon'\right|^2}{\varepsilon'' \varepsilon_h} \right)^{n-1} \sim \left(\frac{\omega_p^2}{\omega} \tau \right)^{n-1}. \tag{14.6}$$

For the latter estimate we used the Drude formula given by Eq. (14.5) and the dielectric permittivity of the host materials ε_h was estimated as $\varepsilon_h \sim 1$. There is also a frequency-independent pre-factor in Eq. (14.6) which depends on the specific geometry of the system; for the sake of the simplicity we omit it here.

For a percolation system, the enhancement factor is given by

$$\left\langle |E|^n \right\rangle_{perc} \sim \left[\left(\frac{\left|\varepsilon'\right|}{\varepsilon_h} \right)^{\frac{\nu(n-2)+s}{t+s}} \right] \left(\frac{\left|\varepsilon'\right|}{\varepsilon''} \right)^{(n-1)(1-2\gamma)+\gamma}, \tag{14.7}$$

where ν, t, s, and γ are the critical exponents of the percolation theory. We use the fact that for 2D systems $\nu = t = s = 4/3$ and set $\gamma = 0$ for simplicity (γ takes into account the presence of delocalized modes[56]). Then, by applying the Drude formula (Eq. (14.6)) and taking $\varepsilon_h \sim 1$, we have

$$\left\langle |E|^n \right\rangle_{perc} \sim \left[\frac{\left|\varepsilon'\right|^{3/2}}{\varepsilon_h \varepsilon''} \right]^{n-1} \sim \left(\omega_p \tau\right)^{n-1}. \tag{14.8}$$

By using Eqs. (14.3)–(14.6), we obtain the SERS enhancement in fractal systems for the two limiting cases of small and large Stokes shifts,

$$G_{\mathrm{RS}}^{(\mathrm{frac})}(\omega-\omega_{\mathrm{s}}\lesssim\Gamma)\sim c\left(\frac{|\varepsilon'|^{2}}{\varepsilon''\varepsilon_{\mathrm{h}}}\right)^{3}\sim c\left(\frac{\omega_{\mathrm{p}}^{2}}{\omega}\tau\right)^{3} \tag{14.9}$$

and

$$G_{\mathrm{RS}}^{(\mathrm{frac})}(\omega-\omega_{\mathrm{s}}\gg\Gamma)\sim c\left(\frac{|\varepsilon'|^{2}}{\varepsilon''\varepsilon_{\mathrm{h}}}\right)^{2}\sim c\left(\frac{\omega_{\mathrm{p}}^{2}}{\omega}\tau\right)^{2}, \tag{14.10}$$

where we added the geometry-dependent pre-factor c, which is estimated to be in the range of 10^{-1} to 10^{-2} (it can have different values in Eqs. (14.9) and (14.10)).

For percolation systems, Eqs. (14.3)–(14.5) and (14.8) result in the following SERS enhancement factors for large and small Stokes shifts, respectively,

$$G_{\mathrm{RS}}^{(\mathrm{perc})}(\omega-\omega_{\mathrm{s}}\lesssim\Gamma)\sim c\left(\frac{|\varepsilon'|^{3/2}}{\varepsilon_{\mathrm{h}}\varepsilon''}\right)^{3}\sim c\left(\omega_{\mathrm{p}}\tau\right)^{3} \tag{14.11}$$

and

$$G_{\mathrm{RS}}^{(\mathrm{perc})}(\omega-\omega_{\mathrm{s}}\gg\Gamma)\sim c\left(\frac{|\varepsilon'|^{3/2}}{\varepsilon_{\mathrm{h}}\varepsilon''}\right)^{2}\sim c\left(\omega_{\mathrm{p}}\tau\right)^{2}, \tag{14.12}$$

where the geometry factor c has been included.[39] Although the values of factor c can be different in Eqs. (14.9)–(14.12), in all cases it is frequency independent and estimated to be in the range of 1 to 10^{-3}.

The enhancement factors for fractal and percolation systems are similar and differ by the factor $\omega_{\mathrm{p}}/\omega$, which is about 4 for the visible range. Although these estimates were obtained for the special case of scale-invariant systems such as fractal and percolation systems, they remain approximately valid for many systems of randomly distributed clusters of metal particles including island films or colloidal aggregates. In all of these systems, the enhanced local fields are concentrated in randomly distributed hot spots whose spatial positions depend on the frequency of the driving field.

It is also important to mention that the macroscopic, average electromagnetic enhancement of Raman scattering is larger for Ag than for Au or Cu, which follows from the formulas above and the optical constants for these metals. According to Eqs. (14.9)–(14.12) and the optical constants given above, the macroscopic SERS enhancement can range between 10^4 and 10^9. The local enhancement in the hot spots can be one to three orders of magnitude larger and thus reaches 10^{10} or even 10^{12} for optimal conditions. SERS enhancement factor with larger than these magnitudes can occur when a chemical enhancement mechanism is also present, which can provide an additional "boost" to the overall SERS enhancement. The chemical enhancement factor ranges typically from 10 to 10^3. Therefore, under optimal conditions when both the electromagnetic and chemical mechanisms contribute, the SERS enhancement can be as high as 10^{15}.

14.3. ADAPTIVE PROPERTY OF SILVER FILMS

As we mentioned above, vacuum evaporated silver films fabricated at a certain range of evaporation parameters allow fine rearrangement of their nanostructure under protein deposition in buffer solution. Although all of the physical and chemical processes resulting in the restructuring are not yet fully established, we have used several methods to study the restructuring mechanism. The changes in the film nanostructure, color, and other properties have been studied by optical methods, including UV–Vis spectrophotometry and Raman spectroscopy. We have also employed such analysis as field emission scanning electron microscopy (FE SEM), adhesion testing, atomic force microscopy, X-ray photoemission spectroscopy, and X-ray diffractometry.

The adaptive silver films are typically formed on a dielectric substrate under vacuum evaporation with an electron beam at an initial pressure inside the system of approximately 10^{-7} Torr.[49] The dielectric (glass) slides were covered first by a layer of 10 nm of SiO_2 followed by an 8–13 nm Ag layer deposited at a rate of 0.05 nm s^{-1}. During the silver deposition process, small isolated metal granules are formed first on the dielectric substrate. As the silver coverage increases, the granules coalesce, resulting in various sizes of silver particles and aggregates. Qualitative estimates of adhesion performed with the common tape test[57] show good adhesion for the Ag/SiO_2/glass substrate structure. Absorption and reflection measurements before and after the tape test indicate the relative level of adhesion of the film to the glass substrate. The changes in absorption/reflection spectra after the tape test were less than 5% for the Ag/SiO_2/glass substrate. A comparison of silver substrates with and without SiO_2 shows that adhesion of silver on glass is poor. In the case of very high adhesion (such as when an adhesion-promoting titanium adhesion layer is included), the deposition of the protein solution does not lead to spot color changes nor structural modifications, and typically little or no SERS is observed.

Excitation of the collective electron oscillations (plasmons) in a metal nanostructure results in strong light and metal particle interactions and eventually in increased absorption relative to a thick metal film. Typical absorbance and reflectance spectra of an ASF substrate are similar in shape and have a maximum at around 500 nm with a broad wing into the longer wavelengths (see Fig. 14.1, solid line). Reflection is typically comparable or slightly larger than absorption by a factor of roughly 1–1.4 when both spectra are expressed in percent. Both the visible color of the film and the extinction spectrum change after protein deposition, as is shown in Fig. 14–1 (dashed line) for an insulin spot. The spectrum inside the analyte spot typically shows a blue-shifted maximum, reduced slope of the long wavelength wing, and reduced extinction integrated over the 300–1100 nm spectral range. A FE SEM image of the same insulin spot (Fig. 14.2(b)) clearly shows nanoscale restructuring, where groups of closely spaced metal nanoparticles are formed. This is in contrast to the film outside the spot, where rather disintegrated particles are typical (Fig. 14.2(a)).

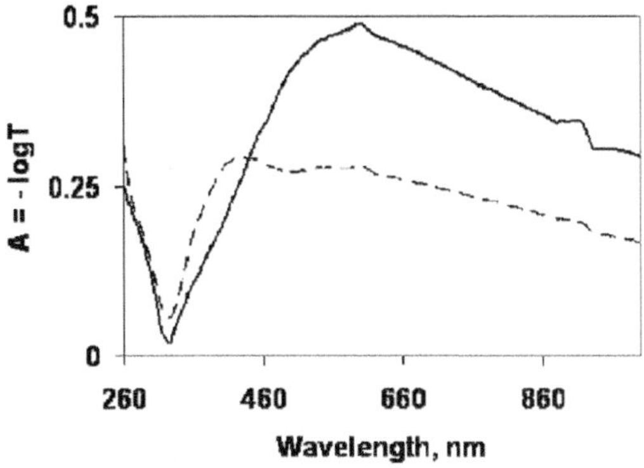

Figure 14.1. Absorption (extinction) spectra for a typical ASF substrate. The solid line indicates spectra of bare ASF substrate and the dashed line is spectra of insulin spot on an ASF substrate.

A representative view of an ASF substrate after protein deposition and drying is shown in Fig. 14.3(a). Note that after washing with a Tris-buffered saline (TBS) solution containing 0.5% Tween-20, the silver coating has been removed everywhere except the areas under the analyte spots, as clearly seen in Fig. 14.3(b). From this one can conclude that the proteins stabilize the silver film, allowing the silver film to remain in place even through washing procedures. This indicates that the biomolecules themselves play a key role in forming a stable complex with the silver particles. By varying the biomolecule and buffer concentrations, we observe that both factors are important in the formation of uniform, stabilized analyte spots. A lower protein concentration (by roughly a factor of 10–20) in a deposition solution results in an almost transparent spot and hence no metal particles. The solvent may etch the metal particles through the oxidation and reduction reactions, leaving merely silver salt on the substrate. X-ray diffraction measurements show AgCl crystals in the transparent areas after silver film treatment with either TBS containing NaCl and KCl or HCl 0.1–1 mM solution.[53] This redox process may affect the interface between the silver particles and the silicon dioxide, decreasing adhesion or even removing

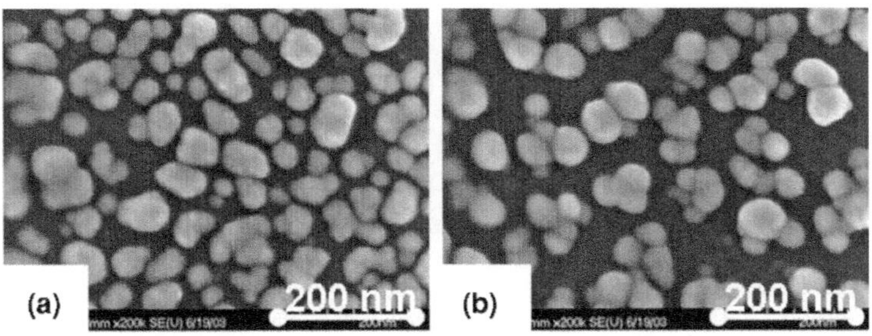

Figure 14.2. FE SEM images of an ASF substrate outside (a) and inside (b) an insulin spot.

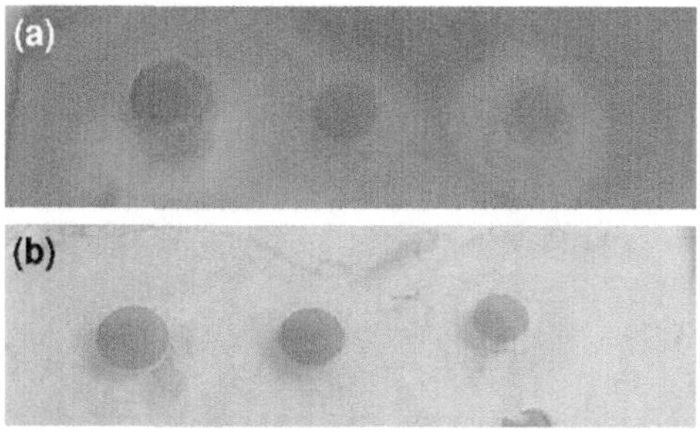

Figure 14.3. Photos of ASF substrate with insulin spots before (a) and after (b) washing with a Tris-buffered saline solution.

particles in solution. Deposition of the protein solution without buffer reveals no visible changes of the film surface in some cases, e.g. insulin, and might result in restructuring for other proteins, e.g. M2 monoclonal antibody. The specific dependence on protein properties is not established yet. Elemental analysis with X-ray photoelectron spectroscopy[52] show that the silver on the substrate is in the metal state (Ag 3d 5/2 peak at 368.5 eV) without oxidation during the first 2–3 weeks after fabrication while the metal is in an oxidized state about eight weeks after fabrication (the peak is shifted to 367.0–367.4 eV). Since many successful experiments were done with month-old ASFs, we conclude that the silver particles are covered with an oxide layer initially and then deoxidize under deposition of protein molecules in buffer solution.

One more example of the restructuring mediated by M2 monoclonal antibody in TBS buffer solution is shown in Fig. 14.4(b) and by antigen in Fig. 14.4(c). A lower concentration of protein results in lower metal coverage (the ratio of white area to total area in the FE SEM images). Fig. 14.4(b) and (c) show that a decrease of metal coverage correlates with the decreasing optical absorption (transmission $T = 0.25$ for Fig. 14.4(b) and $T = 0.4$ for Fig. 14.4(c)).

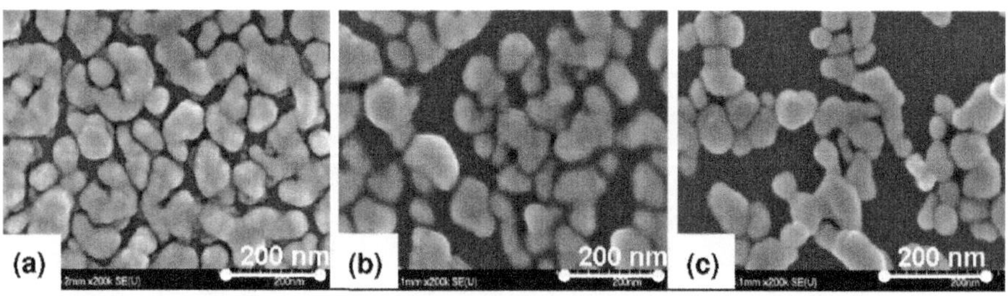

Figure 14.4. FE SEM images of two substrates from one batch: (a) 12 nm silver film (transmission at 568 nm $T = 0.27$); (b) 12 nm silver film inside Ab spot ($T = 0.23$–0.27); (c) same substrate as (b) but inside antigen (FLAG-BAP) spot ($T = 0.4$). Images were collected eight weeks after fabrication; proteins were deposited one week after fabrication.

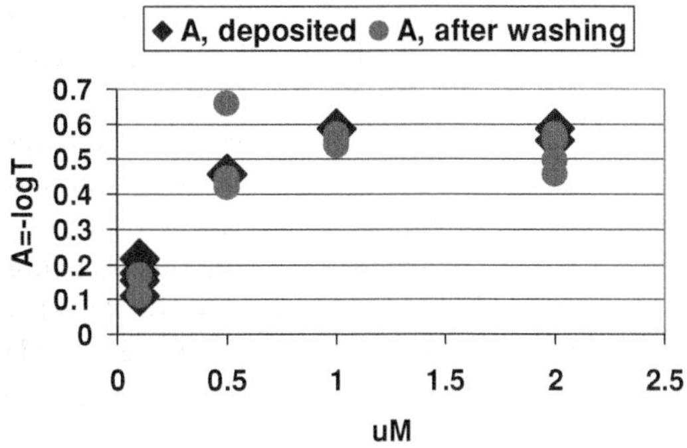

Figure 14.5. Absorbance at 568 nm inside protein spot (anti-interleukin 10) versus protein concentration.

Typically, protein solutions contain buffer. The effect of the protein concentration on silver film stabilization is illustrated in Fig. 14.5. A lower concentration of protein results in a lower metal particle surface density. The absorbance of the silver film is governed mostly by silver particles in the metal state. The absorbance increases with protein concentration and then saturates above a certain concentration which can be considered as optimal. The concentration dependence shows almost no change after 30 min of washing in TBS/Tween-20 solution (the circles in Fig. 14.5), which confirms the stabilization of the film by the proteins.

14.4. SERS ENHANCEMENT

Depending on the mass thickness of the initial film, small or large fractal-like aggregates can be formed. The analyte SERS signal, normalized per metal mass coverage, is comparable for both the small (Fig. 14.3b) and large aggregates (Fig. 14.4b–c) that we examined. SERS enhancement is high enough to detect a monolayer of analyte.[49] Note that an aggregated structure provides conditions for both electromagnetic and chemical SERS enhancements. Even small aggregates provide strong electromagnetic enhancement in the visible and near infrared, as has been shown for polarization nonlinearities[58] and SERS.[31] Large aggregates typically have a fractal morphology, which is known to provide a particularly strong SERS signal.[22,39] In addition, the first molecular layer may also produce conditions to make an optical tunnel current possible, either through the molecule at the point of nearest approach between two particles,[59] or through a system operating as a molecular tunnel junction between particles across a vibrating molecular bridge.[60] Thus, the adaptive feature of our films produces cavity sites enclosed by two or more particles, which are optimal for enhanced Raman. It is even more important that the cavity sites are naturally filled with proteins as a result of restructuring.

The measured macroscopic enhancement factor for insulin on our ASF substrate[49] relative to normal Raman of insulin on quartz is about 3×10^6. This is among the largest observed for random metal-dielectric films: 10^5 for nitrobenzonate[19] and 5.3×10^5 for *trans*-1,2-bis(4-pyridyl)-ethylene.[42]

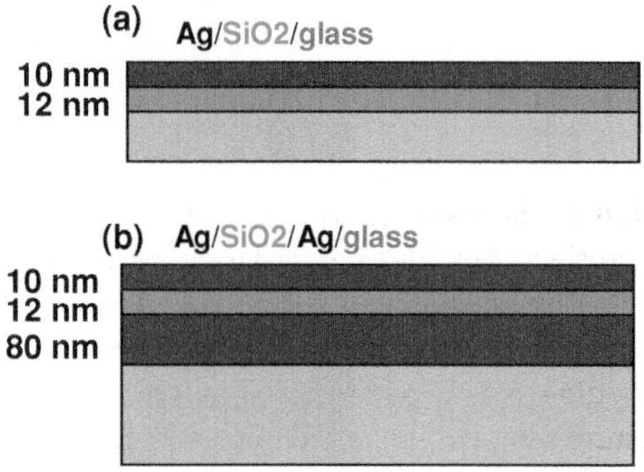

Figure 14.6. Vertical layer structure for ASF substrates: (a) two-layer structure, (b) sandwich structure with an additional thick metal sublayer.

A metal nanostructured film positioned near a mirror-like metal surface with a sandwiched dielectric layer has been employed to further increase SERS signal from ASF based biosensors.[52] Such a sandwich structure can show dramatic change in the film's optical properties.[61–63] The sandwich structure (Fig. 14.6(b)) contains a bulk silver layer (80 nm) deposited on glass, then a layer of SiO_2 (10 nm), and finally a 12 nm nano-structured silver film. So, relative to the usual ASF (Fig. 14.6(a)) it has an additional sublayer of bulk metal. The bulk silver layer provides an additional enhancement of the local fields caused by interaction between particles and their images in the bulk layer and far-field interactions between particles. Test experiments with three analytes (human insulin, anti-human interleukin 10, and anti-human interleukin 10 incubated with 1 nM R6G) show that the multi-layer sandwich structure provides a signal increase of roughly 4–5 times relative to the usual ASF.

To summarize the above discussion, adaptive silver films allow protein/buffer-mediated restructuring of the metal nanoparticles, which makes it possible to address three issues. Specifically, proteins adsorb on the metal surface without significant structural changes (*soft*-adsorption), the silver film stabilizes which makes the analyte/metal combinations resistant to washing, and the SERS signal improves for given set of particles.

14.5. SERS OF INSULIN AND ANTIBODY–ANTIGEN BINDING DETECTION

We first discuss ASF experiments involving the detection of insulin analogs. Next we cover the detection of antibody–antigen binding events using ASF substrates.

In general, protein sensing using Raman spectroscopy provides important structural information on conformational changes. Changes between native and denatured insulin in the solid form as well as the spectral features of proinsulin and insulin fibrils were studied previously.[64–66] Signatures of allosteric conformation changes in hexameric insulin have been assessed using Raman difference spectroscopy.[67]

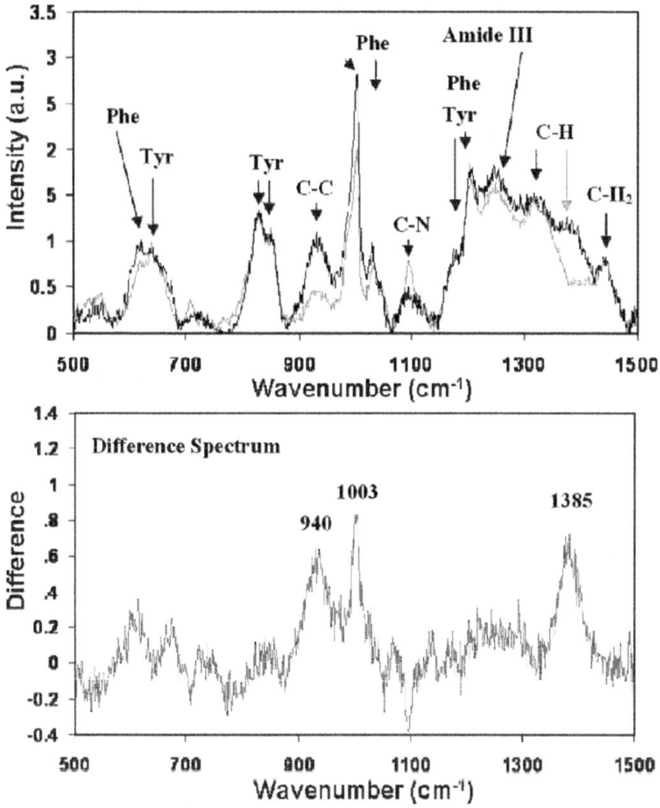

Figure 14.7. SERS spectra collected at 568 nm incident laser wavelength for human insulin (black) and insulin lispro (gray) on ASF substrates.The SERS difference spectrum (insulin–lispro) in the lower panel clearly shows the observed spectral differences between the two isomers.

Insulin is a protein consisting of 51 amino acids split into two chains (called A and B) and is a glucose regulation agent in the bloodstream. The experiments with adaptive substrates[49] examined the differences in Raman spectra of two insulin isomers, human insulin and its analog insulin lispro. These two insulins differ only in the interchange of two neighboring amino acids; specifically, the propyl-lysyl sequence at the C-terminus of the B-chain in insulin lispro is inverted as compared to human insulin. This propyl-lysyl switch leads to conformational changes at the C- and N-termini and has an important clinical effect for diabetes treatment. The difference in SERS spectra for the two insulins (Fig. 14.7) was detected at a sub-monolayer density of 80 fmol mm^{-2}, with only 25 amol in the probed area.

A comparison of our SERS and normal Raman spectra for insulin on a quartz surface[50] and for Zn-insulin in solution[67] suggests that the SERS spectra reveal all known Raman fingerprint spectral peaks for insulin. The Raman peaks are assigned mainly to amide I and amide III bands of peptide backbone vibrations, to vibrational modes of Phe (located at the B1, B24, and B25 residues of the B-chain) and tyrosine (Tyr) (A14, A19, B16, and B26).[64]

The SERS difference between the two insulins can be attributed, in part, to i) Phe(B1) displacement, and ii) the α-helical N-terminus of the B-chain in human insulin, which is a feature of the R-sate conformation. As mentioned, insulin is adsorbed on the surface primarily through the N-terminus. Because of the excess negative charge, metal particles attract the positively charged N-terminus of the B-chain and thus move Phe(B1) closer to the surface. Depending on the conformational state of the displaced Phe(B1), it can be at different distances and thus have different orientations with respect to the metal surface, enabling the observed increase in Phe peak intensities by a factor of 1.4 for human insulin as compared to insulin lispro. The CH deformation band at 1385 cm^{-1} and the C-C skeletal band at 940 cm^{-1} are stronger in human insulin than in insulin lispro. The 890–945 cm^{-1} band is a characteristic spectral line for an α-helix and is known to be sensitive to structural changes.[68–71] This spectral line is typically centered at 940 cm^{-1} and disappears or displays weak intensity upon conversion to β-sheet or random coil structures. It is also known that the C–H deformation band at 1371 cm^{-1} appears for the R_6 conformation of hexameric human insulin, which has the longest sequence (B1–B19) of α-helix.[67] We note that this band appears in R_6-T_6 Raman difference spectra and disappears in $T_3R_3^f$- T_6 spectra.[67] This points out the critical contribution of the α-helical residues Phe(B1), Val(2), and Ala(B3) at N-termini of R_6 hexamers in the C-H deformation band at 1370–1385 cm^{-1}. The observed differences in the SERS spectra suggest that human insulin and insulin lispro have different conformational states on the surface. Specific orientations of molecular bonds on the silver surface emphasize the SERS spectral difference between the two insulins, making the differences much stronger than for conventional Raman. The observed SERS spectral differences are in agreement with X-ray crystallographic studies of hexameric human insulin and insulin lispro.[72]

Since human insulin and its analog have the same set of side chains and differ only in conformational states, the observed difference reveals Raman features of the conformational state with the use of ASF substrates. In this study we used Raman difference spectroscopy, which is a general method of probing protein structure for comparison between closely related proteins,[73] which we extended to SERS in order to study the spectral features of insulin conformation. In our experiments, all insulin vibration modes are enhanced by approximately the same factor. This makes the SERS spectra similar to the conventional Raman spectra in liquid and solid forms and simplifies the analysis.

We have also studied ASF substrates for use in antibody/antigen binding research. The detection of protein binding by optical means is of critical importance to current protein analysis, and promises to lead to a number of exciting future applications in biomedical diagnostics, research and discovery. The results show that SERS substrates based on ASFs allow direct, label-free SERS detection of antibody–antigen binding at a monolayer level. In experiments we used the following protocol. Antibodies (anti-FLAG M2 monoclonal antibody) were immobilized on ASF substrates to form a monolayer array, followed by incubation with an antigen solution at 1 nM

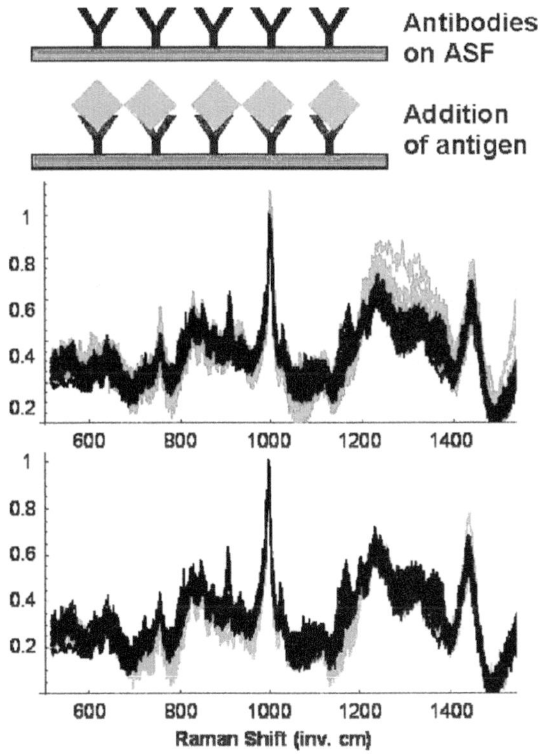

Figure 14.8. Antibody (anti-FLAG M2 monoclonal) and antigen (C-terminal FLAG-BAP) binding events. Spectra were taken before antigen incubation (black) and after antigen incubation (gray), nine spectra each. As a control experiment, non-binding BAP without FLAG was incubated on an antibody array, and the resulting SERS spectra (gray, lower panel) reveal no spectral changes.

concentration (C-terminal FLAG-BAP (bacterial alkaline phosphatase)). In each experiment, after the immobilization of antibodies and incubation with antigens the film was washed for about 20 min in a TBS/Tween-20 (as mentioned in Sect. 3) and then rinsed five times with deionized water. After washing, SERS spectra were collected to observe spectral changes following the incubation of antibodies with the antigen. Antigen–antibody binding events result in distinct SERS spectral changes as shown in Fig. 14.8 (top panel black—before incubation, gray—after). In a control experiment incubation with BAP containing no FLAG reveals no spectral changes (Fig. 14–8, lower panel, gray curves). It is important to note that ASF substrates allow independent *in situ* binding activity validation using traditional chemiluminescence and fluorescence methods. Such validation has been performed and confirms that antigens and antibodies retain their binding properties on our SERS-active substrate. As with insulin, we find that the deposited biomolecules (antibodies or antigens in this case) restructure and stabilize the ASFs so that the protein binding activities are preserved and, in parallel, SERS is improved.

Label-free detection using the ASF method produces unique advantages relative to prior optical binding detection methods (typically based on different types of labels) such as scintillation counting,[74] electrochemical,[75] enzymatic,[76] fluorescence,[76–78] and chemiluminescence methods.[79] An additional feature of ASFs is an ability to employ various detection methods on the same substrate, such as label-free SERS, chemiluminescence, and fluorescence.

ASF substrates used for protein microarrays reveal a promising opportunity to detect SERS spectra along with a fluorescence signal. Results from a microarray prepared using a quill-type spotter show that ASF substrates enable both fluorescence (with excitation at 633 nm) and SERS with no fluorescence (excitation at 568 nm) for the streptavidinCy5 fluorescence reporter.[52] SERS spectra of streptavidinCy5 can be used to distinguish between desirable and undesirable binding events. Since the interaction of biotin with streptavidin forms the basis of several widely used detection methods in bio-array technology, this application of SERS detection could be very important.

14.6. SUMMARY

Adaptive silver films allow protein sensing at monolayer surface densities. The adaptive property of these substrates enables the adsorption of proteins without significant alteration in their conformational state. In an example using human insulin and insulin lispro, the results show that the SERS spectra reveal unique features attributable to distinct conformational states, which is in agreement with X-ray crystallographic studies. The experiments on protein array applications show evidence of distinct SERS spectral changes upon antigen–antibody binding using direct, label-free detection. Independent immunochemical assay validation confirms that the antibodies retain binding properties on ASFs. Experiments with a sandwich structure including a bulk metal layer below the ASF structure reveal a promising way to further improve the sensitivity of SERS-based biosensors.

ACKNOWLEDGMENTS

We thank our colleagues E.N. Khaliullin, V.C. Nashine, D. Ben-Amotz, and V.J. Davisson for their contributions to the described work. This research was sponsored in a part by a grant from Inproteo.

REFERENCES

1. M. Fleischmann, P.J. Hendra, A.J. McQuillan: Raman spectra of pyridine adsorbed at a silver electrode, Chem. Phys. Lett. **26**, 163–166 (1974).
2. D.J. Jeanmaire, R.P. Van Duyne: Surface Raman spectroelectrochemistry Part I. Heterocyclic, aromatic, and aliphatic amines adsorbed on the anodized silver electrode, J. Electroanal. Chem. **84**, 1–20 (1977).
3. M.G. Albrecht, J.A. Creighton: Anomalously intense Raman spectra of pyridine at a silver electrode, J. Am. Chem. Soc. **99,** 5215–5217 (1977).

4. T. Vo-Dinh: Surface-enhanced Raman spectroscopy using metallic nanostructures, Trends Anal. Chem. **17,** 557–582 (1998).
5. G. Bauer, N. Stich, T.G.M. Schalkhammer. In: *Methods and Tools in Biosciences and Medicine: Analytical Biotechnology*, ed by T.G.M. Schalkhammer (Birkhauser Verlag Basel, Switzerland, 2002).
6. M.S. Sibbald, G. Chumanov, T.M. Cotton: Reductive properties of iodide-modified silver nanoparticles, J. Electroanal. Chem. **438,** 179–185 (1997).
7. T. Vo-Dinh, D.L. Stokes, G.D. Griffin, M. Volkan, U.J. Kim, M.I. Simon: Surface-enhanced Raman scattering (SERS) method and instrumentation for genomics and biomedical analysis, J. Raman Spec. **30,** 785–793 (1999).
8. K.R. Brown, A.P. Fox, M.J. Natan: Morphology-dependent electrochemistry of cytochrome *c* at Au colloid-modified SnO_2 electrodes, J. Am. Chem. Soc. **118**, 1154–1157 (1996).
9. K.E. Shafer-Peltier, C.L. Haynes, M.R. Glucksberg, R.P. Van Duyne: Toward a glucose biosensor based on surface-enhanced Raman scattering, J. Am. Chem. Soc. **125,** 588–593 (2003).
10. Y.W.C. Cao, R. Jin, C.A. Mirkin: Nanoparticles with Raman spectroscopic fingerprints for DNA and RNA detection, Science **297,** 1536–1540 (2002).
11. Y.C. Cao, R. Jin, J.M. Nam, C.S. Thaxton, C.A. Mirkin: Raman dye-labeled nanoparticle probes for proteins, J. Am. Chem. Soc. **125,** 14676–14677 (2003).
12. D.S. Grubisha, R.J. Lipert, H.Y. Park, J. Driskell, M.D. Porter: Femtomolar detection of prostate-specific antigen: An immunoassay based on surface-enhanced Raman scattering and immunogold labels, Anal. Chem. **75,** 5936–5943 (2003).
13. M. Moskovits: Surface-enhanced spectroscopy, Rev. Mod. Phys. **57**, 783–826 (1985).
14. M. Moskovits: Surface roughness and the enhanced intensity of Raman scattering by molecules adsorbed on metals, J. Chem. Phys. **69**, 4159–4161 (1978).
15. C.Y. Chen, E. Burstein, S. Lundquist: Giant Raman scattering by pyridine and Cn-adsorbed on silver, Solid State Commun. **32,** 63–66 (1979).
16. S.L. McCall, P.M. Platzman, P.A. Wolff: Surface enhanced Raman scattering, Phys. Lett. **77A**, 381–383 (1980).
17. C.Y. Chen, E. Burstein: Giant Raman scattering by molecules at metal island films, Phys. Rev. Lett. **45**, 1287–1291 (1980).
18. J.G. Bergman, D.S. Chemla, P.F. Liao, A.M. Glass, A. Pinczuk, R.M. Hart, D.H. Olson: Relationship between surface-enhanced Raman scattering and the dielectric properties of aggregated silver films, Opt. Lett. **6**, 33–35 (1981).
19. D.A. Weitz, S. Garoff, T.J. Gramila: Excitation spectra of surface-enhanced Raman scattering on silver island films, Opt. Lett. 7, 168–170 (1982).
20. G. Ritchie, C.Y. Chen. In: *Surface Enhanced Raman Scattering*, ed by P.K. Chang, T.E. Furtak (Plenum, New York, 1982), p. 361.
21. G.C. Schatz. In: *Fundamentals and Applications of Surface Raman Spectroscopy*, ed by R.L. Garrell, J.E. Pemberton, T.M. Cotton (VCH Publishers, Deerfield Beach, FL, 1993).
22. M.I. Stockman, V.M. Shalaev, M. Moskovits, R. Botet, T.F. George: Enahnced Raman scattering by fractal clusters—scale invariant theory, Phys. Rev. B **46,** 2821–2830 (1992).
23. A. Otto: Raman-spectra of (CN)- adsorbed at a silver surface, Surf. Sci. **75**, L392–L396 (1978).
24. I. Pockrand, A. Otto: Coverage dependence of Raman scattering from pyridine adsorbed to silver-vacuum interfaces, Solid State Commun. **35**, 861–865 (1980).
25. B.N.J. Persson: On the theory of surface-enhanced Raman scattering, Chem. Phys. Lett. **82,** 561–565 (1981).
26. F.J. Adrian: Charge transfer effects in surface-enhanced Raman scattering, J. Chem. Phys. **77,** 5302–5314 (1982).
27. P.K.K. Pandey, G.C. Schatz: A detailed analysis of the Raman enhancement mechanisms associated with the interaction of a Raman scatterer with a resonant metal cluster: results for Li_n–H_2, J. Chem. Phys. **80,** 2959–2972 (1984).
28. J.R. Lombardi, R.L. Birke, T. Lu, J. Xu: Charge-transfer theory of surface enhanced Raman spectroscopy: Herzberg–Teller contributions, J. Chem. Phys. **84,** 4174–4180 (1986).

29. A. Campion, P. Kambhampati: Surface-enhanced Raman scattering, Chem. Soc. Rev. **27**, 241–250 (1998).
30. S. Nie, S.R. Emory: Probing single molecules and single nanoparticles by surface-enhanced Raman scattering, Science **275,** 1102–1106 (1997); Screening and enrichment of metal nanoparticles with novel optical properties, J. Phys. Chem. B **102,** 493–497 (1998).
31. K. Kneipp, Y. Wang, H. Kneipp, L.T. Perelman, I. Itzkan, R.R. Dasari, M. Feld: Single molecule detection using surface-enhanced Raman scattering (SERS), Phys. Rev. Lett. **78**, 1667–1670 (1997).
32. G.C. Schatz, R.P. Van Duyne. In: *Handbook of Vibrational Spectroscopy*, ed by J.M. Chalmers, R.P. Griffiths (Wiley, New York, 2002), pp. 759–744.
33. M.D. Musick, C.D. Keating, M.H. Keefe, M.J. Natan: Stepwise construction of conductive Au colloid multilayers from solution, Chem. Mater. **9**, 1499–1501 (1997).
34. A.M. Michaels, M. Nirmal, L.E. Brus: Surface enhanced Raman spectroscopy of individual rhodamine 6G molecules on large Ag nanocrystals, J. Am. Chem. Soc. **121**, 9932–9939 (1999).
35. C.L. Haynes, R.P. Van Duyne: Plasmon-sampled surface-enhanced Raman excitation spectroscopy, J. Phys. Chem. B **107,** 7426–7433 (2003).
36. E. Prodan, C. Radloff, N.J. Halas, P. Nordlander: A hybridization model for the plasmon response of complex nanostructures, Science **302,** 419–422 (2003).
37. P. Gadenne, D. Gagnot, M. Masson: Surface enhanced resonant Raman scattering induced by silver thin films close to the percolation threshold, Physica A **241,** 161–165 (1997).
38. A.K. Sarychev, V.M. Shalaev: Electromagnetic field fluctuations and optical nonlinearities in metal-dielectric composites, Phys. Rep. **335,** 275–371 (2000).
39. V.M. Shalaev: *Nonlinear Optics of Random Media: Fractal Composites and Metal-Dielectric Films*, STMP v.158 (Springer, Heidelberg, 2000).
40. J.P. Davies, S.J. Pachuta, R.G. Cooks, M. Weaver: Surface-enhanced Raman-scattering from sputter-deposited silver surface, J. Anal. Chem. **58**, 1290 (1986).
41. V.L. Schlegel, T.M. Cotton: Silver island films as substrates for enhanced Raman-scattering—effect of deposition rate on intensity, Anal. Chem. **63,** 241–247 (1991).
42. R.P. Van Duyne, J.C. Hulteen, D.A. Treichel: Atomic force microscopy and surface-enhanced Raman spectroscopy. I. Ag island films and Ag film over polymer nanosphere surfaces supported on glass, J. Chem. Phys. **99,** 2101–2115 (1993).
43. E. Vogel, W. Kiefer, V. Deckert, D. Zeisel: Laser-deposited silver island films: an investigation of their structure, optical properties and SERS activity, J. Raman Spec. **29,** 693–702 (1998).
44. J.J. Mock, M. Barbic, D.R. Smith, D.A. Schultz, S. Schultz, Shape effects in plasmon resonance of individual colloidal silver nanoparticles, J. Chem. Phys. **116**, 6755–6759 (2002).
45. M. Kerker: Electromagnetic model for surface-enhanced Raman scattering (SERS) on metal colloids, Acc. Chem. Res. **17**, 271–277 (1984).
46. J.R. Krenn, A. Dereux, J.C. Weeber, E. Bourillot, Y lacroute, J.P. Goudonnet, G. Schider, W. Gotschy, A. Leitner, F.R. Aussenegg, C. Girard: Squeezing the optical near-field zone by plasmon coupling of metallic nanoparticles, Phys. Rev. Lett. **82**, 2590–2593 (1999).
47. E. Hao, G.C. Schatz: Electromagnetic fields around silver nanoparticles and dimmers, J. Chem. Phys. **120**, 357–366 (2004).
48. D.P. Fromm, A. Sundaramurthy, P.J. Schuck, G. Kino, W.E. Moerner: Gap-dependent optical coupling of single "bowtie" nanoantennas resonant in the visible, Nano Lett., **4**, 957–961 (2004).
49. V.P. Drachev, M.D. Thoreson, E.N. Khaliullin, V.J. Davisson, V.M. Shalaev: Surface-enhanced Raman difference between human insulin and insulin lispro detected with adaptive nanostructures, J. Phys. Chem. A, **108**, 18046–18052 (2004).
50. V.P. Drachev, M.D. Thoreson, E.N. Khaliullin, A.K. Sarychev, D. Zhang, D. Ben-Amotz, V.M. Shalaev: Semicontinuous silver films for protein sensing with SERS, SPIE Proc., **5221**, 76–81 (2003).
51. V.P. Drachev, V.C. Nashine, M.D. Thoreson, D. Ben-Amotz, V.J. Davisson, V.M. Shalaev: Adaptive silver films for detection of antibody-antigen binding, Langmuir, **21**(18), 8368–8373 (2005).

52. V.P. Drachev, M.D. Thoreson, V.C. Nashine, E.N. Khaliullin, D. Ben-Amotz, V.J. Davisson, V.M. Shalaev: Adaptive silver films for surface-enhanced Raman spectroscopy of biomolecules, J. Raman Spectrose, **36**(6–7), 648–656 (2005).
53. V.P. Drachev, M.L. Narasimhan, H.-K. Yuan, M.D. Thoreson, Y. Xie, V.J. Davisson, V.M. Shalaev: Adaptive silver films towards bio-array applications, SPIE Proc., **5703,** 13 (2005).
54. P.B. Johnson, R.W. Christy: Optical constants of the noble metals, Phys. Rev. B **6**, 4370–4379 (1972).
55. C. Kittel: *Introduction to Solid State Physics* (Wiley, New York, 1995).
56. D.A. Genov, A.K. Sarychev, V.M. Shalaev: Surface plasmons excitation in semicontinuous metal films. In: *Progress in Condensed Matter Physics*, ed by F. Columbus (Nova Science Publishers, Hauppauge, New York, 2005).
57. L. Eckertova: *Physics of Thin Films* (Plenum Press, New York and London, 1977).
58. V.P. Drachev, S.V. Perminov, S.G. Rautian, V.P. Safonov. In: *Optical Properties of Nanostructured Random Media*, Topics in Applied Physics v. 82, ed by V.M. Shalaev (Springer Verlag, Berlin, 2001), pp. 113–148.
59. A.M. Michaels, J.J. Jiang, L.E. Brus: Ag nanocrystal junctions as the site for surface-enhanced Raman scattering of single Rhodamine 6G molecules, J. Phys. Chem. B **104,** 11965–11971 (2000).
60. S. Holzapfel, W. Akemann, D. Schummacher, A. Otto: Variations of dc-resistance and SERS intensity during exposure of cold-deposited silver films, Surf Sci. **227**, 123–128 (1990).
61. W.R. Holland, D.G. Hall: Frequency shifts of an electric dipole resonance near a conducting surface, Phys. Rev. Lett. **52,** 1041–1044 (1984).
62. A. Leitner, Z. Zhao, H. Brunner, F.R. Aussenegg, A. Wokaun: Optical properties of a metal island film close to a smooth metal surface, Appl. Opt. **32,** 102–110 (1993).
63. H.R. Stuart, D.G. Hall: Enhanced dipole–dipole interaction between elementary radiators near a surface, Phys. Rev. Lett. **80,** 5663–5666 (1998).
64. N.-T. Yu, C.S. Liu: Laser Raman spectra of native and denatured insulin in the solid state, J. Am. Chem. Soc. **94**, 3250–3251 (1972.).
65. N.-T. Yu, C.S. Liu, D.C O'Shea: Laser Raman spectroscopy and the conformation of insulin and proinsulin, J. Mol. Biol. **70,** 117–132 (1972).
66. N.-T. Yu, B.H. Jo, R.C.C. Chang, J.D. Huber: Single-crystal Raman spectra of native insulin: structures of insulin fibrils, glucagon fibrils, and intact calf lens, Arch. Biochem. Biophys. **160,** 614–622 (1974).
67. D. Ferrari, J.R. Diers, D.F. Bocian, N.C. Kaarsholm, M.F. Dunn: Raman signature of ligand binding and allosteric conformation change in hexameric insulin,Biopolymers (Biospectroscopy), **62**, 249–260 (2001).
68. T.-J. Yu, J.L. Lippert, W.L. Peticolas: Laser Raman studies of conformational variations of poly-l-lysine, Biopolymers **12**, 2161–2176 (1973).
69. M.C. Chen, R.C. Lord, R. Mendelson: Laser-excited Raman spectroscopy of biomolecules 4: thermal denaturation of aqueous lysozyme, Biochem. Biophys. Acta **328**, 252–260 (1973).
70. B.G. Frushour, J.L. Koenig: Raman spectroscopy study of tropomyosin denaturation, Biopolymers **13**, 1809–1819 (1974).
71. M.C. Chen, R.C. Lord, R. Mendelson: Laser-excited Raman spectroscopy of biomolecules 5: conformational changes associated with chemical denaturation of lysozyme, J. Am. Chem. Soc. **96**, 3038–3042 (1976).
72. E. Ciszak, J.M. Beals, B.H. Frank, J.C. Baker, N.D. Carter, G.D. Smith: Role of C-terminal B-chain residues in insulin assembly: the structure of hexameric $Lys^{B28}Pro^{B29}$- human insulin, Structure **3**, 615–622 (1995).
73. R. Callender, H. Deng, R. Gilmanshin: Raman difference studies of protein structure and folding, enzymatic catalysis and ligand binding, J. Raman Spec. **29,** 15–21 (1998).
74. S. Gutcho, L. Mansbach: Simultaneous radioassay of serum vitamin-B12 and folic acid, Clin. Chem. **23,** 1609–1614 (1977).
75. F.J. Hayes, H.B. Halsall, W.R. Heineman: Simultaneous immunoassay using electrochemical detection of metal-ion labels, Anal. Chem. **66,** 1860–1865 (1994).
76. J.E. Butler: Enzyme-linked immunosorbent assay, J. Immunoassay **21,** 165–209 (2000).

77. J. Vuori, S. Rasi, T. Takala, K. Vaananen: Dual-label time-resolved fluoroimmunoassay for simultaneous detection of myoglobin and carbonic acnhydrase-III in serum, Clin. Chem. **37,** 2087–2092 (1991).
78. Y.Y. Xu, K. Pettersson, K. Blomberg, I. Hemmila, H. Mikola, T. Lovgren: Simultaneous quadruple-label fluorometric immunoassay of thyroid-stimulating hormone, 17-alpha-hydroxyprogresterone, immunoreactive trypsin, and creatine-kinase MM isoenzyme in dried blood spots, Clin. Chem. **38,** 2038–2043 (1992).
79. C.R. Brown, K.W. Higgins, K. Frazer, L.K. Schoelz, J.W. Dyminski, V.A. Marinkovich, S.P. Miller, J.F. Burd: Simultaneous determination of total IgE and allergen-specific IgE in serum by the MAST chemi-luminescent assay system, Clin. Chem. **31,** 1500–1505 (1985).
80. D.A. Genov, A.K. Sarychev, V.M. Shalaev, A. Wei: Resonant field enhancement from metal nanoparticle arrays, Nano Lett., **4**, NL-0343710 (2004).

CHAPTER FIFTEEN

INTEGRATED OPTICS BASED ON LONG-RANGE SURFACE PLASMON POLARITONS

PIERRE BERINI
School of Information Technology and Engineering, Universtiy of Ottawa, 161 Louis Pasteur, Ottawa, ON, K1N 6N5, Canada, and Spectalis Corporation, 610 Bronson, Ottawa, Canada

15.1. INTRODUCTION

The bound optical modes propagating along a metal film of thickness t and width w embedded in an optically infinite homogeneous background dielectric, as shown in cross-sectional view as the inset to Fig. 15.1, have recently been computed and discussed.[1,2] As pointed out in Ref. 2, this structure can be used as the foundation waveguide for a new integrated optics technology, due to the convergence of some desirable attributes, including of course, the ability of the waveguide to provide confinement along both dimensions transverse to the direction of propagation. This waveguide and the passive integrated optics technology that emerges from its use are the subjects of this chapter.

In subsequent sections, experimental and theoretical work conducted on passive elements implemented using this waveguide is reviewed and discussed, including work on straight and curved waveguides, s-bends, four-port couplers, y-junctions, Mach-Zehnder interferometers and Bragg gratings. Most experiments reported to date have been conducted near 1550 nm, reflecting current interests for applications within the research community, rather than a limitation to free-space wavelengths in this region.

15.2. STRAIGHT WAVEGUIDES

15.2.1. Modes of the Generic Straight Waveguide

Reviewing the salient features of the modes supported by the straight symmetric structure shown as the inset to Fig. 15.1, and describing the rationale for its use in integrated optics, seems like a natural place to start. Only the salient features are reviewed, the details being readily available.[2]

M.L. Brongersma and P.G. Kik (eds.), Surface Plasmon Nanophotonics, 217–233.

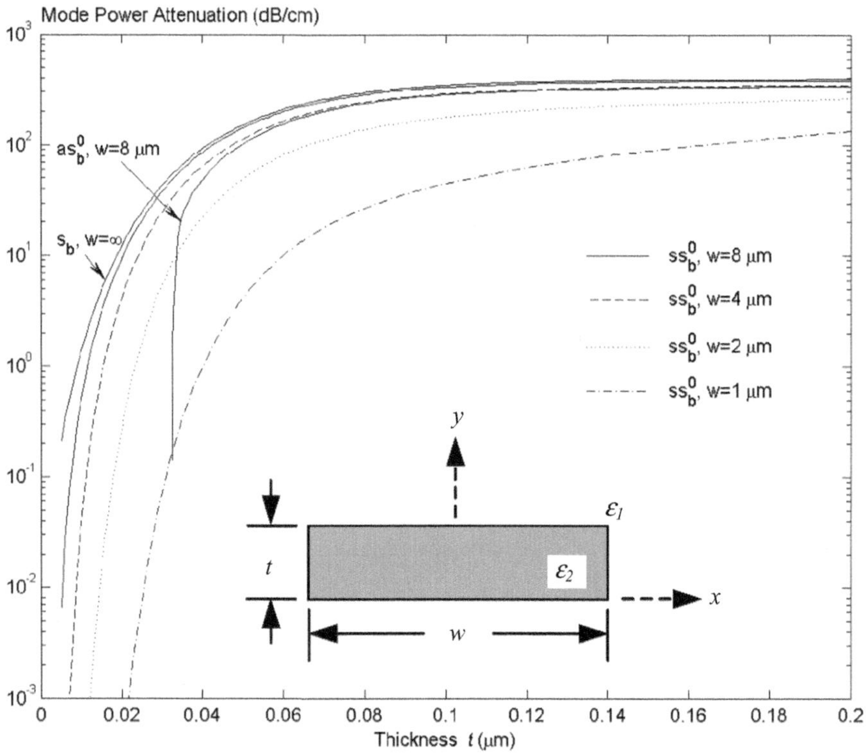

Figure 15.1. Attenuation of the ss_b^0 mode supported by an Au film in SiO_2 at $\lambda_0 = 1550$ nm for various film dimensions. The waveguide is shown in cross-sectional view as the inset.

The structure shown is obtained from the (infinitely wide) metal slab[3–6] by simply limiting its width. This leads to a number of changes, including the emergence of a new mode spectrum, the creation of lateral confinement and a significant reduction in loss for some of the modes compared to the slab.[3–6] Another significant change pertains to the manner by which modal solutions to Maxwell's equations are obtained: the modes of the slab can be derived analytically in a rather straightforward manner, whereas the modes of the finite width structure are obtained numerically, considerably increasing the analysis effort. Nonetheless, the finite width structure can be handled by well-established numerical techniques if appropriate care is taken, as discussed below.

There are four fundamental modes supported by the generic structure shown in inset to Fig. 15.1, labelled aa_b^0, as_b^0, sa_b^0, and ss_b^0.* Given the finite width, higher order

*The mode nomenclature describes the E_y field component of the mode and is an extension[1] of that used to identify the modes of the slab;[5] *a* and *s* refer to asymmetric and symmetric, respectively, the first position modes displaying extrema along x in their field distribution can also be supported. The E_y field component dominates for all modes when the aspect ratio w/t of the film is greater than 1, so the modes are mainly TM in nature, though not purely TM since H_z is always non-zero.

The evolution of modes with metal dimensions (w, t), material parameters $(\varepsilon_1, \varepsilon_2)$ and operating free-space wavelength (λ_0) can be complex, and for asymmetric structures (differing substrate and cover) the trends difficult to generalize.[7,8] The complexity is due to all modes in fact being supermodes created from the coupling of "interface" and "corner" modes,* with their inclusion into a supermode depending on the similarity in their phase constants.

Some trends, however, can be generalized for the symmetric structure considered herein. For instance, the evolution of the modes as $t \to 0$ resembles the evolution of the s_b and a_b modes of the slab in that all modes eventually become partitioned into either lower loss (ss_b^m and as_b^m are s_b-like) or higher loss (aa_b^m, sa_b^m are a_b-like) modes with vanishing t, as determined by the symmetry or asymmetry of E_y along y.

One of the fundamental modes, the ss_b^0 mode, evolves smoothly and predictably as the metal film vanishes $(t, w \to 0)$ into the TEM wave supported by the background dielectric. Its E_y and H_x fields evolve from being highly localized near the metal corners to being distributed in a Gaussian-like fashion over the waveguide cross-section. This transformation is accompanied by a significant reduction in attenuation, often of a few orders of magnitude, due to reduced field penetration into the metal. The resulting Gaussian-like field distribution can be well-matched with modes of dielectric waveguides like single mode fibre, leading to efficient excitation via butt-coupling (end-fire coupling). The ss_b^0 mode thus evolves into the fundamental long-range mode in symmetric structures, and following convention one can refer to it as a Long-Range Surface Plasmon-Polariton (LRSPP). It is important for the structure to be symmetric (within tolerance) in order for the ss_b^0 mode to remain purely bound and long-ranging.[8]

One of the other fundamental modes, the as_b^0 mode, evolves in a similar manner as the metal vanishes, except that its E_y field develops two extrema along x, and the mode becomes unguided below cut-off dimensions (w_c, t_c). This mode, which is *fundamental* for a large metal cross-section, evolves into the first long-range, *higher-order* mode as $t \to 0$. Long-range modes of order higher than the as_b^0 mode may also exist, originating from the mode families ss_b^m $(m > 0$, odd) and as_b^m $(m > 0$, even). They have cut-off dimensions that increase with the mode order m.

All modes of the aa_b^m and sa_b^m families exhibit an increasing loss as $t \to 0$ and do not butt-couple efficiently with a Gaussian field. This is due to the asymmetry in the distribution of E_y along y for all of these modes.

---

being associated with the horizontal dimension and the second with the vertical one, b signifies *bound* and the superscript identifies the number of extrema in the horizontal distribution of E_y not counting the corner peaks.

*An interface can support a surface plasmon, as can a corner and an edge (i.e.: an interface of finite width with corners).

In summary, the metal film can be dimensioned such that: (i) the ss_b^0 mode propagates with low loss, (ii) the ss_b^0 mode butt-couples efficiently to the TM mode propagating in a single mode dielectric waveguide, (iii) lateral and vertical confinement is provided, (iv) all long-range, higher-order modes are cut off, and (v) any remaining modes are unexcited, or at best excited with very low efficiency and rapidly absorbed. This situation corresponds to aforementioned "convergence of some desirable attributes," and leads to a new integrated optics technology based on propagating the long-range ss_b^0 mode along thin metal films of finite width surrounded by a homogeneous dielectric. In the next sub-section, some performance metrics for this mode are given for an example waveguide structure.

15.2.2. Performance Characteristics of the ss_b^0 Mode

The mode power attenuation at $\lambda_0 = 1550$ nm of the ss_b^0 mode supported by Au films ($\varepsilon_{r,2} = -131.95 - j12.65$)[9] of various dimensions in SiO_2 ($\varepsilon_{r,1} = 1.444^2$)[9] is shown in Fig. 15.1. One notes that a reduction in attenuation of more than two orders of magnitude is possible compared to the slab by dimensioning the film properly, with values in the range of 0.1 to 1 dB/mm being readily achievable for $w \sim 4$ μm and $t \sim 20$ nm. The curve of the as_b^0 mode for $w = 8$ μm shows it cutting off at $t = 33$ nm so the 8 μm wide waveguide supports a single long-range mode below this thickness. The cut-off thickness of the as_b^0 mode moves to larger values as the width narrows.

Parts (a), (b) and (c) of Fig. 15.2 show contours of the real part of the Poynting vector components S_x, S_y and S_z of the ss_b^0 mode for $w = 4$ μm and $t = 20$ nm, all other parameters being the same as in Fig. 15.1. The contours are normalized such that $\max|\text{Re}\{S_z\}| = 1$ and the relative magnitude of S_x, S_y, and S_z are conserved. It is noted that even though $\text{Re}\{S_x\}$ and $\text{Re}\{S_y\}$ are much smaller than $\text{Re}\{S_z\}$, they are non-zero since the waveguide is absorbing. Indeed, it is observed from Part (a) that $\text{Re}\{S_x\} > 0$ to the left of the $x = 0$ axis and $\text{Re}\{S_x\} < 0$ to its right, while from Part (b) $\text{Re}\{S_y\} > 0$ below the metal and $\text{Re}\{S_y\} < 0$ above, indicating that power is flowing into the metal from all directions to compensate for the power dissipated therein. Parts (d) and (e) give contours of the angle between the S_y, S_x components and the S_z component, respectively, indicating that the Poynting vector has a slight tilt of about 0.04° towards the metal.

Though not visible in Part (c), the z-directed power flow in the metal is actually in the opposite direction to that in the dielectric since $\text{Re}\{\varepsilon_{r,2}\} < 0$. The total mode power, however, flows in the same direction as the power in the dielectric which is along the direction of a constant phase plane.

For comparison, Part (f) shows normalized intensity contours for standard single mode fibre at the same wavelength (core radius of 4.1 μm, numerical aperture of 0.12). By comparing with Part (c), it is observed that the contours are quite similar. A power coupling efficiency of ~90% is computed in this case.

Trends for the mode are as follows: decreasing λ_0 or increasing $\varepsilon_{r,1}$ leads to greater confinement and attenuation for the same metal dimensions, changing the metal to Al

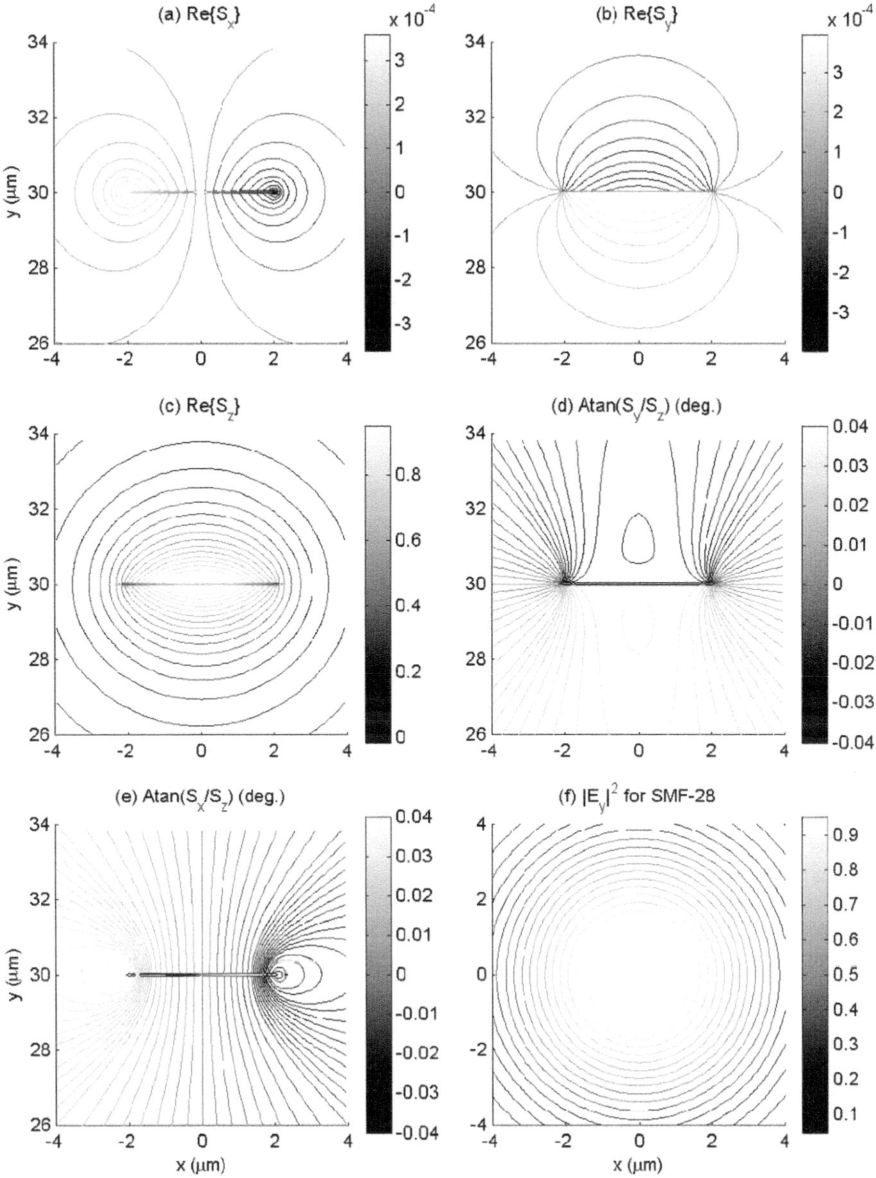

Figure 15.2. Parts (a) through (e) show contours associated with the Poynting vector of the ss_b^0 mode. Part (f) shows normalized intensity contours of standard single mode fibre.

leads to an increase in loss by a factor of about two for the same confinement, while changing to Ag leads to a reduction in loss for the same confinement.

The results given in Fig. 15.1 and Fig. 15.2 are theoretical, obtained by modeling the waveguide using the method of lines (MoL) formulated in rectangular coordinates.[2] The MoL is a very accurate and efficient frequency domain numerical technique;[10] it is vectorial and ideally suited to the divergent dimensional scales present in this structure (t ~nm's, $w \sim$ μm's). Open structures are modelled using an absorbing boundary condition, which is desirable for modeling weak waveguides ($w, t \rightarrow 0$) and to determine the cut-off dimensions of higher order long-range modes.

Vectorial finite element methods (FEMs) have also been used for modeling these waveguides,[11,12] however, the FEM requires significantly more numerical effort to approach the accuracy of the MoL.[11] This is due to the divergent dimensional scales present in the structure and the fact that both dimensions are discretised in the FEM versus only one in the MoL. The same comment applies to the use of the vectorial finite difference method for the same reasons.[13] An equivalent dielectric waveguide has also been proposed to model this structure and comparisons with published modal computations are favourable.[14]

The existence of the ss_b^0 mode was established experimentally using Au in SiO_2[15] and the cutback technique was used to measure the attenuation and coupling loss of the mode.[16] Attenuation and coupling loss measurements were also reported for Au in polymer.[17] The lowest attenuations reported to date for the ss_b^0 mode are 0.3 dB/mm and 0.4 dB/mm using Ag and Au films, respectively, with the metal evaporated directly onto an SiO_2 lower cladding and an index-matched polymer used as the uppercladding.[18] Power coupling efficiencies from single mode fibre into the ss_b^0 mode of greater than 95% have also been achieved.[18] The experimental results reported in Ref. 18 are in very good quantitative agreement with those computed with the MoL using established optical parameters for the materials.[9] All of this experimentation[15–18] was conducted near $\lambda_0 = 1550$ nm.

As noted earlier, it is important for the structure to be symmetric in order for the ss_b^0 mode to propagate properly.[8] As the asymmetry increases (via differing substrate and cover indices) the mode becomes distorted with the fields extending deeper into the high index side while remaining localized on the low index side, and eventually the mode cuts off becoming radiative. A model based on modal decomposition, capable of describing a mode as it transitions through and beyond cut-off, was proposed[19,20] and validated experimentally[20] for the ss_b^0 mode using Au on SiO_2 and a thermally tunable polymer upper cladding. Leaky modes in asymmetric finite-width structures have also been predicted.[21]

Square cross-section metal films ($w = t$) have been proposed in lieu of the rectangular structure ($w \gg t$) in order to create a waveguide that does not exhibit polarisation sensitivity.[22] Indeed, it was found that the symmetric structure supports a fundamental TE-like LRSPP mode, denoted $ss_{b,x}^0$, which is degenerate with the fundamental TM-like LRSPP mode, denoted the $ss_{b,y}^0$ mode.[22] These long-range modes were recently observed experimentally using a square cross-section Au film in polymer at $\lambda_0 = 1550$ nm.[23]

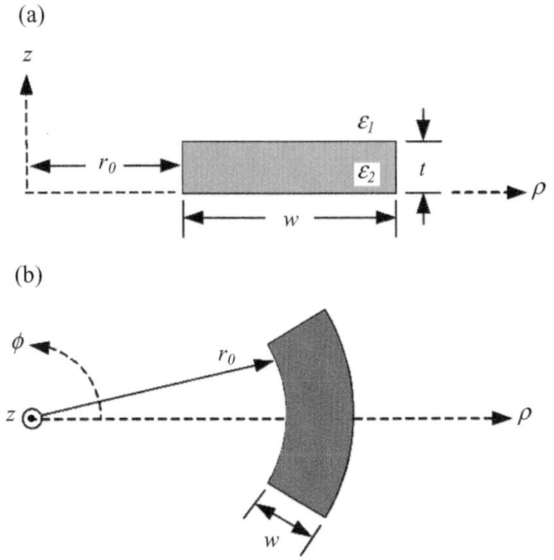

Figure 15.3. Curved waveguide sketched in cross-sectional (a) and top (b) views.

15.3. CURVED WAVEGUIDES

The design of curved waveguides is of central importance in integrated optics, and particularly so in this technology, since a waveguide that is comprised of a vanishing metal supports the ss_b^0 mode with less loss but also with less confinement, potentially compromising the ability of the mode to round bends due to excessive radiation.

Figure 15.3 shows a curved waveguide in cross-sectional view in Part (a), and in top view in Part (b), along with the cylindrical coordinate system used to analyse the structure. In this structure, the bend modes propagate along the ϕ direction and radiation occurs along the outside of the bend (right side in Part (b)) when the radius of curvature r_0 is small.

Figure 15.4 shows in Part (a) the total insertion loss of a 90° bend for a 1 μm wide, 15 nm thick Ag film ($\varepsilon_{r,2} = -19 - j0.53$)[1] in Si_3N_4 ($\varepsilon_{r,1} = 4$) at $\lambda_0 = 633$ nm as a function of the radius of curvature r_0 of the bend.[24] The total insertion loss is comprised of the propagation and radiation losses.

One immediately notes the existence of an optimum radius of curvature $r_{0,\text{opt}}$ where the insertion loss is minimized, occurring in this situation at $r_{0,\text{opt}} \sim 130$ μm. An optimum radius exists because the waveguide includes an absorbing medium (the metal). For $r_0 > r_{0,\text{opt}}$ the insertion loss is dominated by the propagation loss, which increases with r_0 due to the increasing arc length of a fixed angle bend (say 90°), while for $r_0 < r_{0,\text{opt}}$ radiation loss dominates and increases with decreasing r_0.

Part (b) gives the radiation loss component, which is obtained by setting $\text{Im}\{\varepsilon_{r,2}\} = 0$. It is noted that some radiation loss indeed occurs at the optimum radius of $r_{0,\text{opt}} \sim 130$ μm.

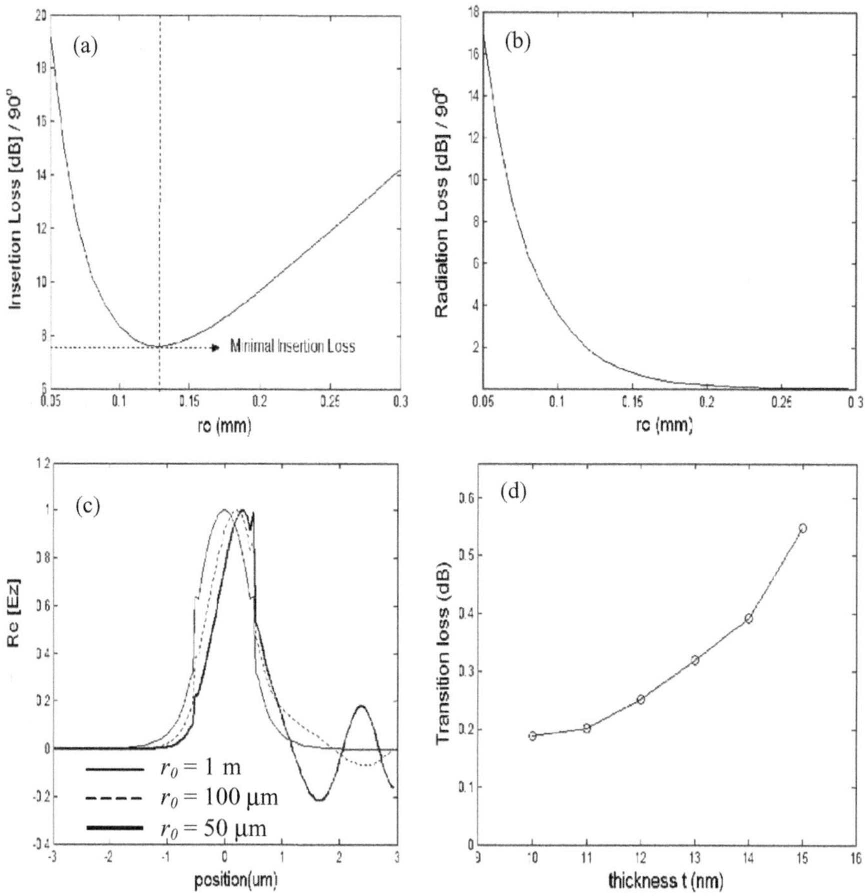

Figure 15.4. Parts (a) and (b) show total and radiation losses, respectively, for a 90° bend; Part (c) horizontal field distributions; Part (d) transition losses. (Adapted from 24.)

Part (c) shows normalized distributions of the $\mathrm{Re}\{E_z\}$* along a horizontal cut immediately above the metal for three radii of curvature, including: $r_0 = 1$ m, 100 and 50 μm, the largest radius (1 m) being used to model the waveguide as straight ($r_0 \rightarrow \infty$). Two principal effects are noted as r_0 decreases: (i) the field becomes oscillatory along the outside of the bend, indicating radiation, and (ii) the mode peak shifts from the center towards the outside of the bend. This latter effect causes a coupling (transition) loss to occur at the junction between curved and straight sections,

* E_z in the cylindrical coordinate system is normal to the metal film and parallel to the E_y field component of the straight waveguide in the rectangular coordinate system; compare the inset of Fig. 15.2 with Fig. 15.3 (a).

as can be appreciated qualitatively by comparing the field for $r_0 = 1$ m with that for $r_0 = 50$ μm. Part (d) gives the transition loss of bends designed at $r_{0,\text{opt}}$ as a function of the metal thickness.

The results shown in Fig. 15.4 are theoretical, obtained by modeling the bend using the MoL formulated in cylindrical coordinates.[24,25] An absorbing boundary condition positioned on the radiating side of the bend (right side in Fig. 15.3) is required and used. The MoL for bends provides the same computational advantages as the MoL for straight waveguides and is also vectorial. The ability of the ss_b^0 mode to round bends was demonstrated experimentally using Au in SiO_2 at $\lambda_0 = 1550$ nm.[26]

15.4. PASSIVE DEVICES

15.4.1. Experimental Demonstrations

Straight and curved waveguides are the main building blocks from which more sophisticated devices can be constructed, such as s-bends, four-port couplers, y-junctions and Mach-Zehnder interferometers. The operation of such devices was demonstrated experimentally using Au in SiO_2 at $\lambda_0 = 1550$ nm,[16] and more recently using Au on SiO_2 with an index matched polymer uppercladding, also at $\lambda_0 = 1550$ nm.[27] Figure 15.5 shows a set of measured outputs associated with these passive elements, along with the output of a Bragg Grating (extreme right—to be discussed in the next section). A mosaic of outputs for a series of four-port couplers is shown on the extreme left, the separation between the straight parallel segments being the only difference between the couplers. This separation varies in steps of 1 μm, from 8 μm in the case of the top image of the mosaic, down to 2 μm in the case of the bottom image, and light is always launched into the left input waveguide. Experimental results for similar passive elements have been published using Au in polymer at $\lambda_0 \sim 1550$ nm, where

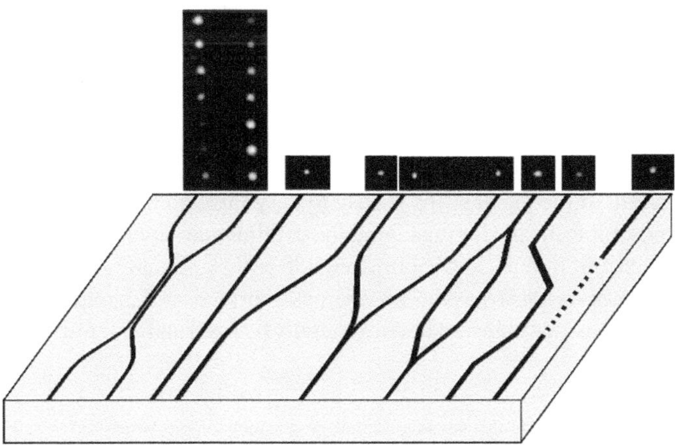

Figure 15.5. Measured outputs for a collection of passive elements. (Adapted from 27.)

the authors also fit simple models used in conventional integrated optics to some of their experimental results.[28] Broadside (vertical) coupling between a pair of stacked waveguides has recently been demonstrated using Au in polymer at $\lambda_0 \sim 1550$ nm.[29] From a qualitative standpoint, these experimental results[16,27–29] generally agree with each other.

15.4.2. Thermo-Optic Devices

Thermo-optic devices using the metal film as both the waveguiding strip and an Ohmic heating element driven by the passage of current have been reported.[30–32] Such devices are capable of operating at high efficiency due to the excellent overlap that is achieved between the heated region and the optical mode. Optically non-invasive electrical contacts to the metal waveguide film are typically achieved through patterning such that a few tabs extend from the metal film along the perpendicular direction and in the same plane. Knowing the electromigration current density limit of the metal and operating well away from this limit is essential if burn-out of the thin film is not to occur.

Thermo-optic variable optical attenuators (VOAs) based on mode cut-off induced by index asymmetry in a straight waveguide section[8,20] were demonstrated using Au on SiO_2 with an index matched polymer uppercladding at $\lambda_0 = 1550$ nm;[30] the metal film in these devices was also used as a temperature monitor by simultaneously measuring its resistivity. A thermo-optic switch implemented as a four-port coupler,[31] a thermo-optic VOA implemented as a Mach-Zehnder interferometer[31] and a thermo-optic VOA implemented as a straight waveguide section[32] were demonstrated using Au in polymer near the same wavelength. An optical power monitor was also demonstrated by monitoring the resistivity of the metal film which changes with the power carried by the ss_b^0 mode due to heating caused by absorption in the metal.[33]

15.4.3. Modelling and Design Considerations

Device designs published to date have generally not been optimized, where an optimal design, for instance, is one that minimizes the total fibre to fibre insertion loss. Achieving an optimal device requires a design strategy that trades-off confinement against attenuation at a holistic level.

The width of the metal strips is easy to vary through lithographic patterning, thus providing simple means for managing the confinement-attenuation trade-off. Designs might benefit from the use of optimized adiabatic tapers for a change in width.* The metal thickness is not so easy to vary across a surface, as this requires multiple lithography and deposition steps, so it will generally be fixed and thus must be chosen wisely.

*It should be borne in mind that tapers introduce propagation lengths which add loss. The corresponding step-in-width might in fact introduce less loss than the taper, presenting a potential alternative if stray radiation is not a concern.

Changing the width affects the mode size along *both* transverse dimensions[27]; *i.e.*: the horizontal *and* vertical extents of the mode are changed. Using an adiabatic taper, the mode profile can be modified to achieve high coupling efficiencies with the chosen input and output means. Adiabatic tapers would also be used in a y-junction for example, widening the input mode to better match those supported by the waveguides at the split of the output branches.

The confinement of a guide might be increased by widening prior to entering a curved section in order to benefit from a smaller radius of curvature potentially achieving a lower loss. Bends can be offset from the end-coupled input and output straight sections in order to better align the modes thus reducing the transition loss. Bends can also be designed using a varying radius of curvature as in traditional integrated optics.

Clearly, accurate and efficient design tools based on electromagnetic field theory are essential if optimized designs are to be achieved. Modelling difficulties posed by 3-D structures include: the presence of divergent length scales ($t \sim$ nm, $w \sim$ μm, L ~mm), the large dielectric contrast between the metal and background dielectric, and the strong frequency dispersion and negative real part of permittivity of the metal.

Given that the structures of interest are generally adiabatic (or close to it), a modeling framework based on decomposition into local modes can be used.[34] In this approach, three-dimensional structures are partitioned into straight or curved waveguide sections, each section supporting a local mode, with forward coupling between sections determined via overlap integrals. The approach relies on having accurate mode solvers for constitutive straight and curved sections, yielding complex propagation constants and 2-D mode field distributions. The mode solvers used were principally the MoL for straight and curved waveguides, providing accurate descriptions of the mode fields everywhere over the cross-section, including inside the metal and within the tail of bend modes.

Figure 15.6 shows a Mach-Zehnder interferometer constructed by joining identical y-junctions. A model constructed for this structure according to the framework described above, is comprised of the segments and loss contributions highlighted on the Figure. C_1 is the coupling loss between the input/output means and the input/output straight waveguides, C_2 is the coupling loss at the output of the tapered section and the input of the curved branches forming the arms and C_3 is the coupling loss between bends having opposite radii of curvature. The tapers are formed from the overlap of two oppositely curved sections as shown, a taper being modeled crudely as a straight section of width equal to the average of its input and output widths.

Measurements conducted over the range from $\lambda_0 = 1525$ nm to $\lambda_0 = 1620$ nm on Mach-Zehnder interferometers fabricated from Au in SiO_2 were compared with results obtained at the same wavelengths using the theoretical model described above (and shown in Fig. 15.6) and found to agree within a few percent. Models constructed in a similar manner for s-bends, y-junctions and four-port couplers were likewise validated, with measurement and theory also agreeing to within a few percent.[34]

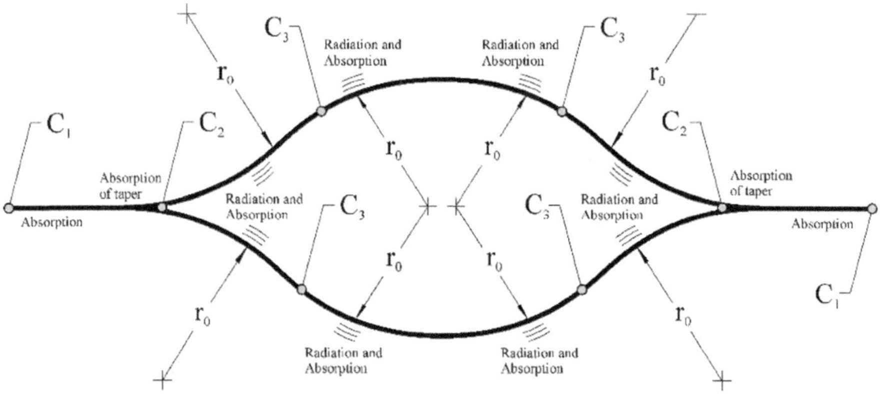

Figure 15.6. Model of a Mach-Zehnder interferometer highlighting all loss contributions. (Adapted from Ref. 34.)

The modeling framework described is clearly accurate enough to reliably design and optimize passive elements.

15.5. BRAGG GRATINGS

The inset along the bottom of Fig. 15.7 shows an example of a uniform periodic Bragg grating obtained by stepping the width of the metal film over a prescribed length.[35] Gratings of this architecture can be implemented through lithographic means by simply patterning the metal film, the only limitation on the design being that imposed by the resolution limit of the selected lithography process. The operation of such gratings was demonstrated near $\lambda_0 = 1550$ nm using Au on SiO_2 with an index matched polymer uppercladding.[36]

These gratings propagating the long-range ss_b^0 mode operate in a manner analogous to gratings in traditional integrated optics, in that the reflected wave originates from the multiplicity of small perturbations in effective index associated with the steps in width. Based on this observation, a simple model for the gratings was proposed consisting of cascaded dielectric slabs, each slab taking on the complex effective index of the ss_b^0 mode supported by the associated waveguide segment.[37] The complex effective index of the ss_b^0 mode in a segment is determined using an accurate mode solver such as the MoL for straight waveguides discussed earlier. The model must include the loss associated with each segment if accurate responses are to be generated.

Figure 15.7 compares the experimental and theoretical wavelength responses of a 3rd order uniform periodic Bragg grating designed as a 3 mm long, 22 nm thick, and 8 μm wide metal film periodically stepped in width to 2 μm and fabricated from Au on SiO_2 with an index matched polymer uppercladding.[37] The patterning was achieved using a contact lithography process. The reflection strength of this design is about 40% at the centre (Bragg) wavelength of $\lambda_B = 1544.22$ nm. Its bandwidth is about 0.3 nm full width at half maximum. The agreement between the experimental

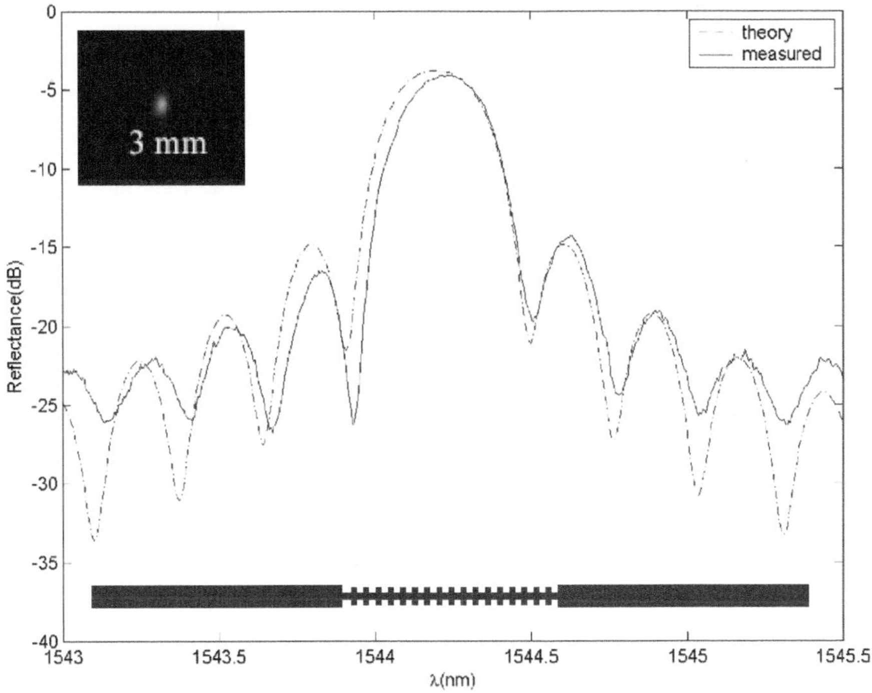

Figure 15.7. Theoretical and experimental reflection responses of a uniform Bragg Grating.

response and the theoretical one obtained using the model described above is very good.*

The inset located within the top left portion of Fig. 15.7 gives a measured output for the Bragg structure characterized. A very low level of background radiation was observed from images captured over the entire measurement wavelength range, including on resonance. This indicates a very low level of scattering. In fact the measured outputs of the grating are as good as those observed for straight 8 μm wide waveguides.

Various types of grating patterns can be implemented just as easily as a uniform one, including, apodised, sampled and interleaved patterns. Models for these types of gratings were constructed in the manner described above[37] and example responses produced for Au in SiO_2 near $\lambda_0 = 1550$ nm.[38] A peak reflectance greater than 90% is achievable using 1st-order designs.

Bragg gratings based on stepping the thickness of the metal film over a prescribed length were proposed and demonstrated using Au in polymer near $\lambda_0 = 1550$ nm,[39] then used to implement add/drop wavelength filters.[39,40]

*A slight adjustment of the theoretical center (Bragg) wavelength was necessary in order to align the responses spectrally.

15.6. CONCLUDING REMARKS

Presently, all of the fundamental passive elements required in an integrated optics technology based on thin metal films of finite width and operating in the long-range ss_b^0 mode have been demonstrated. Robust theoretical models have also been constructed and validated experimentally. Most of the implementation to date has been done using Au with low index claddings targeting operation near $\lambda_0 = 1550$ nm. Operation at different wavelengths is possible as is the use of other metals and higher index cladding materials including semiconductors.

Depending on the claddings to be used, fabrication can be relatively straightforward, as is the case with deposited claddings, or more complex if non-depositable claddings are needed. High quality materials are desirable and particular fabrication requirements include achieving high quality interfaces and matched claddings. Challenges and solutions associated with fabrication have recently been reviewed, and example structures fabricated in SiO_2 and $LiNbO_3$ using wafer bonding described.[41]

Three important conclusions can be drawn from the work conducted to date: (i) thin metal films of finite width embedded in a homogeneous dielectric can be used as the basis of a practical integrated optics technology operating in the ss_b^0 mode, (ii) the fabrication of high quality devices can be achieved, and (iii) simple yet rigorous models based on a modal description can be constructed for the necessary passive elements and used with confidence in the design and optimization of circuits.

This integrated optics technology is well suited to applications where the unique features of the plasmon-polariton wave are important or where significant advantage can be derived from a metal being co-located with the center of the optical mode. Exploiting non-linear effects in materials is an example of the former. Examples of the latter include using the metal as: a heating element in thermo-optic devices, an electrode in electro-optic devices, or a contact in charge carrier devices. In all cases, the principal advantage would stem from achieving a good overlap between the optical mode and the effect being exploited.

Materials effects of interest for device applications are generally weak. Significant phase changes for example must be accumulated over propagation lengths of about 1 to 10 mm using mainstream thermo-optic polymers and electro-optic crystals. Exploiting materials effects thus generally requires long-range waves, which is achievable using a judicious choice of metal geometry. Also, the propagation of a long-range (low loss) mode is essential if gain is to be eventually achieved, say, via optically pumped rare-earth doped claddings.

Within the context of subwavelength plasmonics,[42] one can envision using the ss_b^0 mode to convert far-field radiation to the near-field via appropriately designed integrated structures. Such structures would convert the ss_b^0 mode from its long-range form where it couples well with large incident field distributions, to its collapsed version, thus potentially providing means for efficiently and conveniently interfacing with nanoparticles and nanoparticle arrays[43,44] or SPP photonic crystals,[45] for example. In its collapsed form, the ss_b^0 mode exhibits the same qualitative characteristics as a single interface plasmon-polariton, namely it is short ranging and highly confined to the metal dielectric interfaces, just like the modes typically studied using near-field techniques.[46]

ACKNOWLEDGMENTS

The author is grateful to past and present co-workers that have contributed to the subject matter presented herein: Robert Charbonneau, Stéphanie Jetté-Charbonneau, Ian Breukelaar, Christine Scales, Junjie Lu, Guy Gagnon, Nancy Lahoud, and Greg Mattiussi.

REFERENCES

1. P. Berini: Plasmon–polariton modes guided by a metal film of finite width, Opt. Lett. **24**, 1011–1013 (1999).
2. P. Berini: Plasmon-polariton waves guided by thin lossy metal films of finite width: bound modes of symmetric structures, Phys. Rev. B **61**, 10484–10503 (2000).
3. M. Fukui, V.C.Y. So, R. Normandin: Lifetimes of surface plasmons in thin silver films, Phys. Status Solidi B **91**, K61–64 (1979).
4. D. Sarid: Long-range surface-plasma waves on very thin metal films, Phys. Rev. Lett. **47**, 1927–1930 (1981); Phys. Rev. Lett.**48**, 446 (1982).
5. J.J. Burke, G.I. Stegeman, T. Tamir: Surface-polariton-like waves guided by thin, lossy metal films, Phys. Rev. B **33**, 5186–5201 (1986).
6. F. Yang, J.R. Sambles, G.W. Bradberry: Long-range surface modes supported by thin films, Phys. Rev. B **44**, 5855–5872 (1991).
7. P. Berini: Plasmon-polariton modes guided by a metal film of finite width bounded by different dielectrics, Opt. Express **7**, 329–335 (2000).
8. P. Berini: Plasmon-polariton waves guided by thin lossy metal films of finite width: bound modes of asymmetric structures, Phys. Rev. B **63**, 125417 (2001).
9. E.D. Palik ed: *Electronic Handbook of Optical Constants of Solids (HOC), version 1.0*, (SciVision—Academic Press 1999)
10. R. Pregla, W. Pascher: The Method of Lines. In: *Numerical techniques for microwave and millimeter-wave passive structures*, ed by T. Itoh (Wiley, New York, 1989).
11. I. Breukelaar: Surface plasmon-polariton waveguiding and mode cutoff. MASc Thesis, University of Ottawa, Ottawa (2004).
12. M.P. Nerzhad, K. Tetz, Y. Fainman: Gain assisted propagation of surface plasmon polaritons on planar metallic waveguides, Opt. Express **12**, 4072–4079 (2004).
13. S.J. Al-Bader: Optical transmission on metallic wires—fundamental modes, IEEE J. Quant. Electr. **40**, 325–329 (2004).
14. R. Zia, A. Chandran, M.L. Brongersma: Dielectric waveguide model for guided surface polaritons, Opt. Lett. **30**, 1473–1475 (2005).
15. R. Charbonneau, P. Berini, E. Berolo, E. Lisicka-Shrzek: Experimental observation of plasmon-polariton waves supported by a thin metal film of finite width, Opt. Lett. **25**, 844–846 (2000).
16. R. Charbonneau: Demonstration of a passive integrated optics technology based on plasmons. MASc Thesis, University of Ottawa, Ottawa (2001).
17. R. Nikolajsen, K. Leosson, I. Salakhutdinov, S.I. Bozhevolnyi: Polymer-based surface plasmon-polariton stripe waveguides at telecommunication wavelengths, Appl. Phys. Lett. **82**, 668–670 (2003).
18. P. Berini, R. Charbonneau, N. Lahoud, G. Mattiussi: Characterization of long-range surface plasmon-polariton waveguides, J. Appl. Phys. **98**, 043109 (2005).

19. I. Breukelaar, P. Berini: Long-range surface plasmon-polariton mode cutoff and radiation in slab waveguides, J. Opt. Soc. Am. A **23** (8), 1971–1977 (2006).
20. I. Breukelaar, R. Charbonneau, P. Berini: Long range surface plasmon-polariton mode cutoff and radiation, Appl. Phys. Lett. **88**, 051119 (2006).
21. R. Zia, M.D. Selker, M.L. Brongersma: Leaky and bound modes of surface plasmon waveguides, Phys. Rev. B **71**, 165431 (2005).
22. P. Berini: Optical waveguide structures, US Patent 6741782.
23. K. Leosson, T. Nikolajsen, A. Boltasseva, S.I. Bozhevolnyi: Long-range surface plasmon polariton nanowire waveguides for device applications, Opt. Express **14**, 314–319 (2006).
24. J. Lu: Modelling optical waveguide bends and application to plasmon-polariton waveguides. MASc Thesis, University of Ottawa, Ottawa (2003).
25. R. Pregla: The method of lines for the analysis of dielectric waveguide bends, J. Light Technol. **14**, 634–638 (1996).
26. R. Charbonneau, P. Berini, E. Berolo, E. Lisicka-Shrzek: Long-range plasmon-polariton wave propagation in thin metal films of finite-width excited using an end-fire technique, Proc. SPIE **4087**, 534–540 (2000).
27. R. Charbonneau, N. Lahoud, G. Mattiussi, P. Berini: Demonstration of integrated optics elements based on long-ranging surface plasmon polaritons, Opt. Express **13**, 977–984 (2005).
28. A. Boltasseva, T. Nikolajsen, K. Leosson, K. Kjaer, M.S. Larsen, S.I. Bozhevolnyi: Integrated Optical Components Utilizing Long-Range Surface Plasmon Polaritons, J. Light Technol. **23**, 413–422 (2005).
29. H.S. Won, K.C. Kim, S.H. Song, C.-H. Oh, P.S. Kim, S. Park, S.I. Kim: Vertical coupling of long-range surface plasmon polaritons, Appl. Phys. Lett. **88**, 011110 (2006).
30. G. Gagnon: Thermo-optic variable optical attenuators using plasmon-polariton waveguides. MASc Thesis, University of Ottawa, Ottawa (2004).
31. T. Nikolajsen, K. Leosson, S.I. Bozhevolnyi: Surface plasmon polariton based modulators and switches operating at telecom wavelengths, Appl. Phys. Lett. **85**, 5833–5836 (2004).
32. T. Nikolajsen, K. Leosson, S.I. Bozhevolnyi: In-line extinction modulator based on long-range surface plasmon polaritons, Opt. Commun. **244**, 455–459 (2004).
33. S.I. Bozhevolnyi, T. Nikolajsen, K. Leosson: Integrated power monitor for long-range surface plasmon polaritons, Opt. Commun. **255**, 51–56 (2005).
34. R. Charbonneau, C. Scales, I. Breukelaar, S. Fafard, N. Lahoud, G. Mattiussi, P. Berini: Passive integrated optics elements based on long-range surface plasmon-polaritons, J Light Technol. **24**(1), 477–494 (2006).
35. S. Jetté: A study of Bragg gratings based on plasmon-polariton waveguides. MASc Thesis, University of Ottawa, Ottawa (2003).
36. S. Jetté-Charbonneau, R. Charbonneau, N. Lahoud, G. Mattiusi, P. Berini: Demonstration of Bragg gratings based on long-ranging surface plasmon polariton waveguides, Opt. Express **13**, 4674–4682 (2005).
37. S. Jetté-Charbonneau, R. Charbonneau, N. Lahoud, G. Mattiusi, P. Berini: Bragg Gratings based on long-range surface plasmon-polariton waveguides: comparison of theory and experiment, IEEE J Quant. Electr.**41**, 1480–1491 (2005).
38. S. Jetté-Charbonneau, P. Berini: Theoretical performance of Bragg gratings based on long-range surface plasmon-polariton waveguides, J. Opt. Soc. Am. A **23**(7), 1757–1767 (2006).
39. S.I. Bozhevolnyi, A. Boltasseva, T. Søndergaard, T. Nikolajsen, K. Leosson: Photonic bandgap structures for long-range surface plasmon polaritons, Opt. Commun. **250**, 328–333 (2005).
40. A. Boltasseva, S.I. Bozhevolnyi, T. Søndergaard, T. Nikolajsen, K. Leosson: Compact Z-add-drop wavelength filters for long-range surface plasmon polaritons, Opt. Express **13**, 4237–4243 (2005).
41. G. Mattiussi, N. Lahoud, R. Charbonneau, P. Berini: Integrated optics devices for long-ranging surface plasmons: fabrication challenges and solutions, Proc. SPIE **5720**, 173–186 (2005).
42. W.L. Barnes, A. Dereux, T.W. Ebbesen: Surface plasmon subwavelength optics, Nature **424**, 824–830 (2003).

43. J.R. Krenn, A. Dereux, J.C. Weeber, E. Bourillot, Y. Lacroute, J.P Goudonnet, G. Schider, W. Gotschy, A. Leitner, F.R. Aussenegg, C. Girard: Squeezing the optical near-field zone by plasmon coupling of metallic nanoparticles, Phys. Rev. Lett. **82**, 2590–2593 (1999).
44. S.A. Maier, P.G. Kik, H.A. Atwater, S. Meltzer, E. Harel, B.E. Koel, A.A.G. Requicha: Local detection of electromagnetic energy transport below the diffraction limit in metal nanoparticle plasmon waveguides, Nat. Mater.**2**, 229–232 (2003).
45. A. Boltasseva, T. Søndergaard, T. Nikolajsen, K. Leosson, S.I. Bozhevolnyi, J.M. Hvam: Propagation of long-range surface plasmon polaritons in photonic crystals, J. Opt. Soc. Am. B **22**, 2027–2038 (2005).
46. J.R. Krenn, J.-C. Weeber: Surface plasmon polaritons in metal stripes and wires, Phil. Trans. Roy. Soc. Lond. A **362**, 739–756 (2004).

CHAPTER SIXTEEN

LOCALIZED SURFACE PLASMONS FOR OPTICAL DATA STORAGE BEYOND THE DIFFRACTION LIMIT

JUNJI TOMINAGA
Center for Applied Near-Field Optics Research (CAN-FOR) National Institute for Advanced Industrial Science and Technology (AIST) Tsukuba Central 4, 1-1-1 Higashi, Tsukuba 305-8562, Japan j-tominaga@aist.go.jp

16.1. INTRODUCTION: HIGH-DENSITY OPTICAL DATA STORAGE

Optical data storage technology and its storage capacity have gradually been improved over the last two decades, especially thanks to the development of shorter-wavelength semiconductor laser units and high-precision optical lithography (mastering). Currently the available storage capacity is beyond 5 GB in a 12-cm disc for DVD, and a 25 GB disc drive system with a 405-nm blue laser unit is also available as blu-ray disc or HD-DVD. However, the storage capacity has almost reached the optical limit because of far-field diffraction. At the moment, there is no alternative way to improve the laser spot size to less than 300 nm using far-field optics, even with the most advanced optical disc technology employing a 405-nm wavelength and a lens numerical aperture (NA) of 0.85.

In order to overcome this issue, near-field optics has attracted much attention over the last decade.[1–4] In the early years, several near-field fiber-probes or flying heads with a small aperture were designed and simulated computationally.[4] Such a nano-aperture allows for the confinement of light in a nanoscale area, enabling sub-diffraction limit data readout. However, the optical throughput of the fiber and the aperture were found to be too low (less than 0.01%) to be applied to optical data storage.

As an alternative, a near-field system using a *solid immersion lens (SIL)* was invented and recently nearly 100-GB storage capacity in a 12-cm disc was experimentally confirmed in a system with an effective NA > 1.0.[5–6] However, the spacing between the *SIL* and the storage medium was less than ~50 nm, and the system required a specially designed flying-height adjustment system under high-speed disc rotation.

M.L. Brongersma and P.G. Kik (eds.), Surface Plasmon Nanophotonics, 235–246.

The most recent candidate for eliminating the flying height issue in near-field recording is the super-resolution near-field structure (*super-RENS*) that was proposed in 1998.[7] This approach uses a nonlinear optical masking layer which is placed in close proximity to a recording layer. The optical properties of the masking layer are affected *locally* by the high intensity in the laser focus. This results in a modified field distribution in a nanoscale area, which in turn affects the field distribution in the nearby recording layer. In the early *super-RENS* design, an Sb masking layer was used. However, the near-field interference generated around a small aperture adversely affects the signal from specific mark sizes in the recording layer. This phenomenon makes it impossible to implement this approach in current optical data storage technology which uses *run limit-length (RLL) codes*. In order to maintain high signal intensity from short and long recording marks included in *RLL code* (for example, nine different mark lengths in a CD), a novel super-RENS with a single light-scattering center (*LSC-super-RENS*) was subsequently developed. The recent study of super-high density optical storage using near-field optics has rapidly shifted towards the LSC-super-RENS discs worldwide.[8] Using the advanced *super-RENS*, TDK Corp., Samsung Electronics, and our group realized 50-nm resolution optical data storage with an optical system using a 405-nm wavelength and a NA of 0.65 or 0.85: one-third of the far-field resolution (one-sixth of the spot size).[9] Astonishingly, a signal carrier-to-noise ratio (*CNR*) beyond 40 dB was stably obtained in repeated readout, at the same intensity level as that of commercialized DVDs. In *LSC-super-RENS* discs a nano-explosive platinum oxide (PtOx) thin layer (~4 nm) replaces the silver oxide layer (~20 nm) in order to generate a single scattering center. By the end of 2004, the *super-RENS* disc showed the potential for more than 200-GB storage capacity in a 12-cm disc.

16.2. SUPER-RESOLUTION NEAR-FIELD STRUCTURE, SUPER-RENS

The *Super-resolution near-field structure, super-RENS,* was initially designed as an optical near-field recording system, but without any needle-like probes or flying lenses.[7] The concept goes back to the idea behind obtaining super resolution in magneto-optical discs in the early 1990s. However, nobody knew at the time that this technology was related to future near-field recording with an optical nonlinear film that rapidly responds to a temperature profile. In 1997, we noticed that an Sb thin film deposited on a transparent substrate reversibly shows a curious optical nonlinearity with temperature when it is quickly moved or scanned at a velocity above 1.0 m/s. It was found that the film makes a transition from opaque to transparent, or vice versa at temperatures around 400°C. Following this discovery, we confirmed that a laser beam can create an optical window and that its size can be manipulated and controlled. In 1998, we designed a multilayer system consisting of $SiN(1)/Sb/SiN(2)/Ge_2Sb_2Te_5/SiN(3)$, which can generate the super resolution effect with high spatial frequencies: the optical near-field. Figure 16.1 shows the signal intensity obtained from small marks with

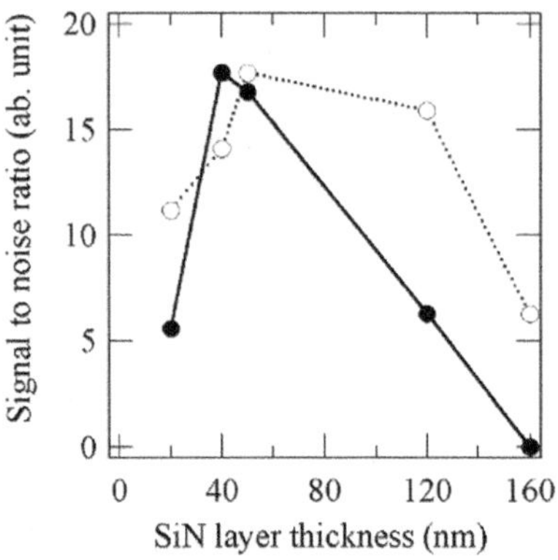

Figure 16.1. Super-resolution signal intensity from a super-RENS disc using a thin Sb film. Solid circles represent 100-nm size marks and open circles represent 300-nm marks. Signal recording and readout were done using a normal DVD drive unit with a 635-nm wavelength and a lens numerical aperture of NA = 0.60, operating at a standard linear velocity of 6.0 m/s. The diffraction limit was 540 nm: i.e., the resolution limit is 270 nm.

a size beyond the diffraction limit when the thickness of the *SiN(2)* spacer layer was varied.

As the thickness decreased, a high signal intensity was obtained beyond the resolution limit of far-field optics. At a thickness of 50 nm or less, the signal intensity obtained from large and small marks recorded in the phase-change film ($Ge_2Sb_2Te_5$) could not be improved any further. In addition, the signal intensity observed for a fixed mark size (for example, 100-nm phase-change marks) became larger and then smaller upon increasing the laser beam power. These observations could be explained with a thermal aperture model as shown in Fig. 16.2. The high intensity in the laser focus generates a small transparent window in the Sb layer. Upon increasing the aperture size, the signal gradually rises. The maximum response is obtained when the aperture size becomes similar to that of the mark size. As the aperture size is increased further by raising the incident laser power, the aperture gradually begins to include adjacent marks, resulting in the observed signal drop.

Initially the use of a thermal aperture appeared to enable sub-diffraction limit data storage. However, through experimental research on the *super-RENS* followed by computer simulations using the *finite difference time domain method (FDTD)*, we gradually understood a fatal issue with the optical nanoaperture. Optical near-fields are strongest at sharp edges because high spatial frequencies beyond the diffraction limit are required to re-construct the edge according to Fourier optics. This behavior is not specific to the *super-RENS*. In a two dimensional geometry, two edges are generated in the aperture. Due to the presence of these two edges, when the laser power is held constant for *super-RENS* near-field readout the signal intensity from one specific mark length used in *RLL* code can be less than that of other mark lengths. Optical recording systems inherited from CD and DVD technology utilize marks with

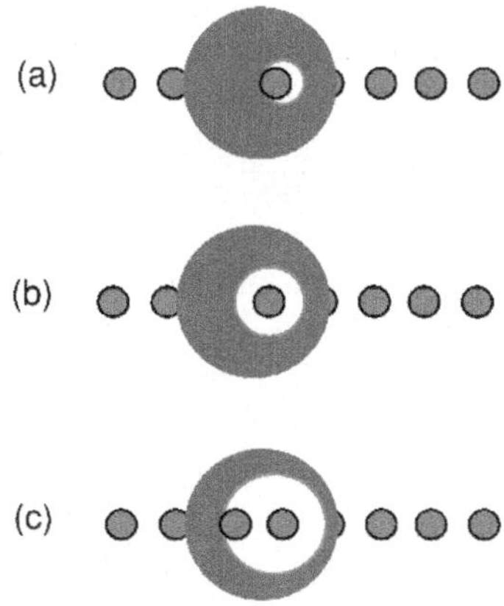

Figure 16.2. In an aperture type super-RENS disc, a transparent optical sub-window is generated in a focused laser spot. As the laser power is increased, the temperature of the Sb mask layer rises and generates a molten Sb region. At low power (a) the window is small and the signal of the small mark is also small. At increasing laser power, the window's size includes one mark: the signal reaches a maximum. As the power increases further, the window gradually begins to include the neighboring mark: the signal gradually drops.

different lengths: for example, for a clock time (T), nine different marks with lengths 3T, 4T, 5T, 6T, 7T, 8T, 9T, 10T, and 11T are utilized in CD systems.[10] In these systems all the signal intensities should be of the same level. Therefore, a single aperture will never be applied to near-field optical data storage systems that use the same CD or DVD data sequence. Almost all near-field recording approaches using an aperture will experience the same problem in the near future.

After we understood this problem, the *super-RENS* study rapidly shifted towards the use of an isolated light-scattering center. Fortunately, we found an alternative material, silver oxide, for this purpose. Silver oxide (AgOx) is not a new material; it has been produced in colloidal form using wet chemistry. However, a thin AgOx film can also be produced by vacuum deposition such as sputtering. Actually, AgOx films with different composition ratios are easily obtained by using a pure Ag target and a gaseous mixture of argon and oxygen. A new super-RENS disc using a AgOx layer instead of an Sb layer was designed and examined in 2000.[8] In 2003, we further improved the super-resolution characteristics by replacing the AgOx film with platinum oxide (PtOx). Since then, the newly designed super-RENS disc with a 4-nm thick PtOx film has reached a resolution of 50 nm, generating more than 40 dB of *CNR*.[9] We call this the 3rd generation super-RENS disc because this generation is different from the 1st and 2nd generations on several technical points: first, in the 3rd generation disc a data bit is stored as a nanometer-scale gas bubble, which is generated by the thermal decomposition of the PtOx film. Second, the phase-change film in turn plays a role in the super-resolution readout! Thanks to the amazing role of the phase-change film, *the 3rd generation super-RENS* disc can enhance and amplify the small near-field signal from a tiny pit more than 100 times.

16.3. THE ROLE OF THE OPTICAL PHASE-CHANGE FILM IN SUPER RESOLUTION

Readers may wonder why optical phase-change films consisting of *chalcogenide* can show such a dramatic super resolution effect. Actually, for the past two years, the mechanism had not been explained. By now we have gradually come to understand the important role of the phase-change film not only in super resolution systems, but also in future plasmonics applications.

In order to explain the observed phenomena, the refractive index in the active region must reach extremely large values compared to the surrounding masking area, since otherwise the scattered signal photons would be almost entirely obscured by photons reflected back from the masked area (giving rise to noise). One physical phenomenon that can induce index changes is known as the *Kerr effect*. It is caused by an optical nonlinearity of the 3rd order, in which the refractive index changes linearly with the laser intensity. However, this effect is generally not as large as that observed in the *super-RENS* disc, and the low threshold powers typically observed in the super resolution (SR) effect cannot be explained by this process either. Alternatively a 2nd order nonlinearity may cause second harmonic generation (*SHG*). However, the experimental results by Kim *et al.* ruled out this mechanism as a possible explanation with their observation of 80-nm pit patterns with a 40-dB CNR in a disc drive system using a 635-nm wavelength and a NA of 0.60: the theoretical resolution limit that can be obtained by *SHG* in this system is 132 nm.[9] Therefore, we conclude that 2nd and 3rd order optical nonlinearities are not essential to the operation of the *super-RENS* disc. To understand how large changes in the refractive index could occur in the phase change material, we will start with a classical description of the refractive index. In classical physics, the refractive index n is related to the *electronic polarizability* α through the *Clausius–Mosotti* equation (in Gaussian units):

$$\alpha_\infty = \frac{3}{4\pi N_A} \frac{n_\infty^2 - 1}{n_\infty^2 + 2} V, \tag{16.1}$$

where α_∞ and n_∞ are the *electronic polarizability* and refractive index at the wavelength $\lambda \to \infty$. N_A and V are *Avogadro's number* and the molar volume of the material. In quantum physics, this equation is further modified by the summation over the oscillator strengths of multiple transitions:

$$\alpha_m = \sum_k \frac{2e^2 \omega_{mk} \langle m | \hat{r} | k \rangle \langle k | \hat{r} | m \rangle E_0 \cos(\omega t)}{(\omega_{mk}^2 - \omega^2)\hbar}. \tag{16.2}$$

Equation (16.2) differs from Eq. (16.1) by the inclusion of a summation over the contributions from each band-transition $m \leftrightarrow k$. Here, ω, ω_{mk}, e, $\hat{r}$, E_0 are the applied laser frequency, the resonance frequency between bands m and k, the electron charge, the electron displacement (i.e., $e\hat{r}$ is the *dipole operator*), and the applied electric field strength, respectively. In a solid or liquid, the term $3V/4\pi N_A$ in Eq. (16.1) may be replaced by a constant, ρ, further simplifying Eq. (16.1) to:

$$n^2 = \frac{2\alpha + \rho}{\rho - \alpha}. \tag{16.3}$$

The parameter ρ can be thought of as the available space for each atom in the unit cell. When $\rho \sim \alpha$, the value of n^2 may diverge and reach large values close to the singularity. *Ferroelectric materials* are well known to show such behavior at a temperature called the Curie temperature, T_c. It is also known that GeTe exhibits *ferroelectric characteristics* with a 2nd order phase-transition at $T_c \sim 352°C$.[11,12] This ferroelectric effect is accompanied by Raman soft-mode phonons that were observed experimentally at ~3.5 THz (110 cm^{-1}). These phonons are attributed to the displacement of tellurium atoms in the unit cell. Yamada *et al.* discussed the crystalline lattice deformation of $Ge_2Sb_2Te_5$ (hereafter *GST*) and $Ag_{3.4}In_{3.7}Sb_{76.4}Te_{16.5}$ (hereafter *AIST*) systems as a function of temperature. They discovered relatively large lattice deviations at the Sb and the Ge sites $4(b)$, compared to the Te site $4(a)$ in the *NaCl-type fcc* unit lattice of GeSbTe.[13] It was found that the lattice transforms into a *hexagonal* lattice at ~260°C. On the other hand, AIST retains a hexagonal lattice (A-7, belonging to the $R\bar{3}m$ space group), which is similar to the original Sb lattice with the c-axis expanding from 11.2 to 11.6 Å at temperatures up to 350°C. At higher temperatures, the lattice transforms into *R3m rhombohedral*. These results support that AIST may also show a 2nd order phase-transition, albeit more anisotropically than that of the GST system. So far, many studies have revealed transition temperatures of optical phase-change alloys. However, most studies have focused on 1st order phase transitions only: transitions between the as-deposited amorphous phase and the crystalline phase, and the melting points. None of them has taken into account 2nd order phase-transitions because these only produce a very small discontinuity in the heat flow in differential scanning calorimetry (DSC) measurements, and a change in the optical reflectivity or transmittance that is too small to observe at the macro-scale.

Early in 2004 we proposed a readout model of the *PtOx super-RENS disc* that involved the ferroelectric properties of AIST and GST thin films, (published in reference 14). Hence, the SR effect is only active in a very narrow temperature range around the *ferroelectric catastrophe*. In the paper, we experimentally determine the relationship between the readout laser power and disc temperature, and clearly revealed that the threshold laser power required for the super resolution in the *super-RENS* discs agrees well with the temperatures required for the 2nd order phase-transition.

Here, let's discuss the relationship between ferroelectrics and the SR effect in more detail by using the *Landau theory.*[15] It is assumed that the free energy F_p of the material can be expanded into a power series of the dipole P. Because F_p must exhibit energy minima in P, it only contains even terms of P:

$$F_P = \tfrac{1}{2}\alpha P^2 + \tfrac{1}{4}\beta P^4 + \tfrac{1}{6}\gamma P^6 + \cdots. \tag{16.4}$$

The coefficient α is temperature dependent according to $\alpha = \alpha_0(T - T_0)$ with $\alpha_0 > 0$. Also, we assume $\beta > 0$. To find the energy minima of F_p, we require that the first derivative $\frac{\mathrm{d}F_p}{\mathrm{d}P} = 0$. Thus, we obtain:

$$\begin{aligned} \frac{\partial F}{\partial P} &= \alpha P + \beta P^3 + \gamma P^5 = E = 0 \\ \frac{\partial^2 F}{\partial P^2} &= \frac{\partial E}{\partial P} = \chi^{-1} \\ 4\pi\varepsilon^{-1} &= \frac{\partial E}{\partial P} = \alpha_0(T - T_0) + 3\beta P_s^2 = 2\alpha_0(T_0 - T) \end{aligned} \tag{16.5}$$

where χ in the second equation of (16.5) is the dielectric *susceptibility* and $P = \chi E$. Here, we neglected higher orders beyond the second derivative. As a result, we obtain the famous relationship of the Curie temperature:

$$\varepsilon \propto (T_0 - T)^{-1}. \tag{16.6}$$

In optical discs, an as-deposited amorphous film must be crystallized once before recording. In the process, the film volume is reduced by more than 5%. As a result the protective layers surrounding the phase-change film induce a high strain force. The strain force is generally anisotropic because the crystallization procedure is usually carried out along the tracks and the groove structure may modify or block the strain force across the tracks. From our previous experiments with a $ZnS\text{-}SiO_2/Sb/ZnS\text{-}SiO_2$ system, the strain force is estimated to be roughly 20–40 MPa.[16] As the temperature is increased, the crystallization is further accelerated due to grain growth. At a certain temperature the resulting volume change is compensated by thermal expansion. As the temperature is increased further, the thermal expansion changes the stress from tensile to compressive. Equation (16.4) is not applicable to situations involving stress and strain, and must be modified to include a uni-axial strain force (in general, we must consider bi-axial forces as well) and a coupling term with the dipole:[15]

$$F_P = \tfrac{1}{2}\alpha P^2 + \tfrac{1}{4}\beta' P^4 + \tfrac{1}{2}c(x - x_0)^2 + qxP^2. \tag{16.7}$$

The third term represents the free energy contribution due to strain and the fourth term is due to coupling with the dipole. The terms c, x_0 and q are *Young's modulus*, the original position of an atom, and a coupling constant, respectively. It should be noted that $\Delta x = (x - x_0)$ is equivalent to $\sim\rho^{1/3}$ in Eq. (16.3). In addition to the local minimum of F_p to P, we now also obtain another minimum with respect to displacement x:

$$\frac{\partial F}{\partial x} = 0. \tag{16.8}$$

As a result, we can obtain an elegant relationship between the displacement $\Delta x_s = (x - x_0)$ and P_s:

$$\Delta x_s = -qP_s^2/c. \tag{16.9}$$

This *self-distortion* now may induce a dipole. Alternatively, a large electrical dipole may induce a very large displacement in the unit cell beyond the *yield point*, resulting in plastic deformation, material flow, or a transition to a more energetically stable crystalline state. At the transition point, the refractive index theoretically has no meaning or value: this means that a large index change is expected at the phase boundary as a result of the summation of many higher spatial frequencies. This ferroelectric catastrophe probably is the readout mechanism of the observed super-resolution in super-RENS discs. We recently confirmed that the $Ge_2Sb_2Te_5$ alloy exhibits ferroelectric behavior due to a small local displacement of Ge atoms in a *fcc* Te unit lattice, observed using X-ray analysis of fine structure, *XAFS*.[17] It should be noted that chalcogenides without Ge atoms also show a strong SR effect. For example, Sb-based phase-change alloys, typically AgInSbTe, do not include Ge at all, but show a very high super resolution effect similar to $Ge_2Sb_2Te_5$. So, why? Is the ferroelectric model not complete? To supplement the model, we are currently considering a *Peierls distortion* induced by internal stresses, which are generated at a boundary of two materials. Group V-elements As, Sb and Bi are known to generate *Peierls transitions.*[18] typically, As and Sb have a crystalline structure known as *A*-7 at atmospheric pressure. At increasing pressure or under compressive stress, the *A*-7 structure transforms to the *simple cubic* (*sc*) structure in a phase transition. The transition is reversible. Seifert *et al.* reported the electronic structure and the band gap of this state.[19] The *A*-7 structure plays a role in metallic properties, while the *sc* plays a role in dielectric properties. As a result, a Sb-rich alloy may generate the same type of phase transition that occurs in the ferroelectric catastrophe. These models involving the *ferroelectric catastrophe* and the *Peierls transition* are now believed to be a suitable *super-RENS* readout method providing a signal intensity 100 times as large as the noise level.

16.4. SURFACE AND LOCAL PLASMONS FOR OPTICAL STORAGE

So far, we have discussed the influence of the phase change layer on the *super-RENS* effect with a model involving a phase transition that is accompanied by a large change of the refractive index. But, how effective are *nanoparticles* for the signal intensity of a *super-RENS* discs? The *LSC-super-RENS* disc research has recently revealed many attractive and interesting characteristics that can help understand the behavior of surface and local *plasmon polaritons* in data storage as well. In order to elucidate the plasmon effect separately, one needs to observe local plasmons due to the metallic nanoparticles. For this purpose, a novel method to generate silver nanoparticles may provide a solution.[10] We are able to fabricate a nanostructured silver film on the surface of almost any material without heating or spin-coating particles in a wet chemical process. By varying the fabrication conditions, the nanostructures can be transformed from particles into wires. Figure 16.3 shows an image of a nanostructured layer fabricated on an optical fiber. Figures 16.4 and 16.5 show other nanostructures on polycarbonate optical disc substrates.

An optical spectrum from the nanostructure mostly shows a sharp absorption at wavelengths around 380–420 nm due to the *localized plasmon resonance* of the condensed nanoparticles. Another sharp absorption is clearly observed at a wavelength

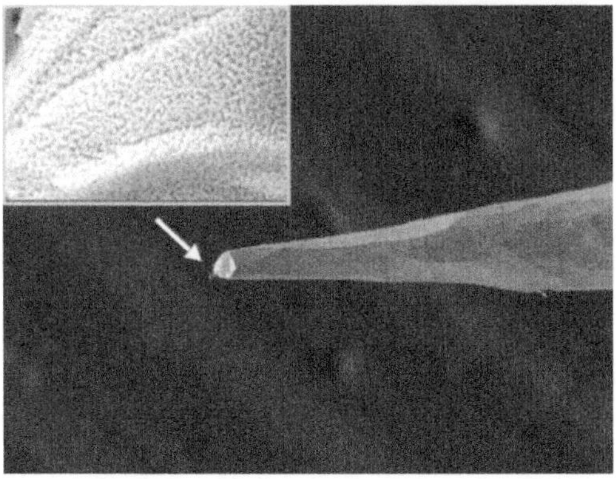

Figure 16.3. Ag nanostructure fabricated on the surface of an optical fiber by the deoxidation of silver oxide film. The silver oxide film was first deposited on the surface and then it was reduced by hydrogen and oxygen with small amount of seed gas CF_4. The diameter is less than 50 nm.

of 325 nm, which is known as the surface plasmon frequency of a planar Ag film. The localized plasmon resonance from the nanostructure fabricated using our method is very useful in the industry because the wavelength of light absorption coincides with that of a solid-state blue laser unit. As is well known, the localized plasmon resonance is very sensitive to the local geometry. Therefore, once the structure is fabricated uniformly on a disc surface, it can be disturbed or deformed by a pulsed laser beam to generate recording patterns. Since this modifies the local structure one can expect a large spectral shift and a large optical nonlinearity due to the plasmon enhanced absorption. Figure 16.6 shows one example of recorded pit patterns on the optical

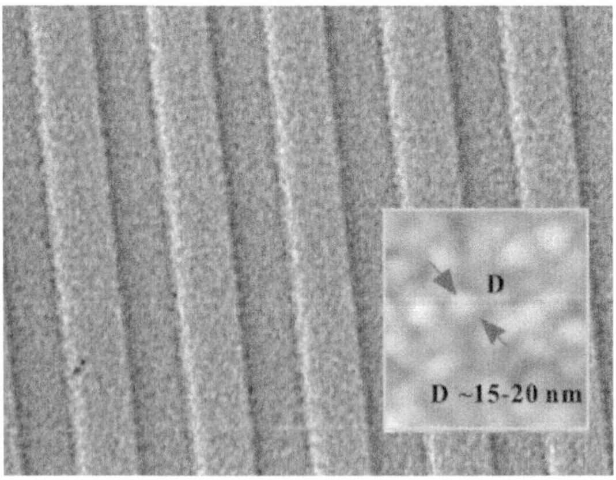

Figure 16.4. Ag nanoparticles fabricated on a recordable DVD substrate. The five vertical lines are called 'grooves', which guide a laser beam for recording. The groove width is 600 nm. The size of each particle is about 15–20 nm (see *Inset*).

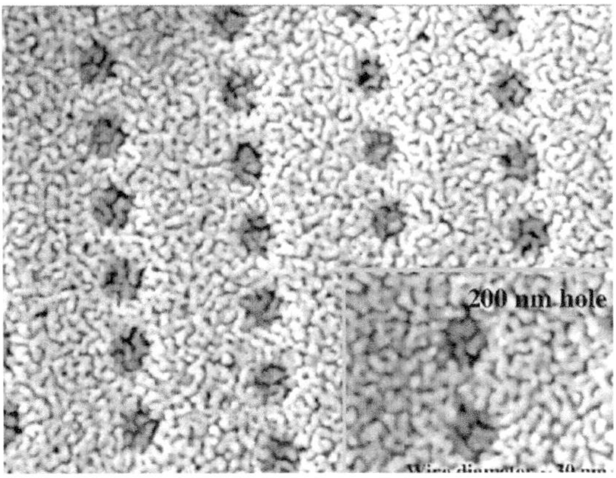

Figure 16.5. Ag nanowires fabricated on a pre-patterened disc surface. The pit diameter is 200 nm. The nanostructure is rigidly generated in the pits (see *Inset*).

disc. Here, a 200-nm AgOx film was deposited on a polycarbonate disc and then transformed into a nanostructured film. The sample disc was placed on a typical digital versatile disc (DVD) test bench and rotated at a 6.0-m/s linear velocity. The laser beam and optics used in the work employed a 635-nm wavelength and a numerical aperture of 0.60, respectively. The recorded areas are seen to be thermally damaged or condensed, losing the fine structures observed in Fig. 16.5. The super-resolution effect from the disc is shown in Fig. 16.7. It should be noted that no phase-change layer or *super-RENS* active films were used: only the Ag nanostructure film was fabricated on the disc surface. Therefore, the super-resolution effect observed is really the result of a nonlinearity due to the nanostructure itself.

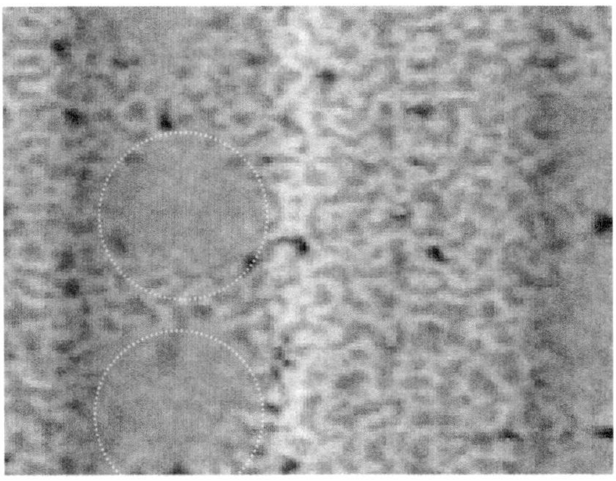

Figure 16.6. A groove recorded on the Ag nanostructured optical disc. In the recorded area, the image of the nanostructure becomes unclear because of thermal condensation (see white two circles).

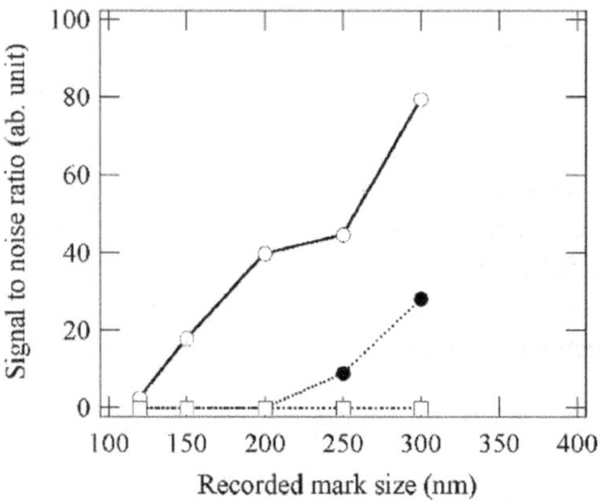

Figure 16.7. Signal intensity of super-resolution marks recorded in a Ag nanostructure optical disc. The resolution limit was 270 nm. Open circles represent the signal from a 100-nm thick silver oxide film, solid circles represent the 50-nm thick film, and open squares represent a 50-nm Ag film. Note that the thickness of the real Ag nanostructures was not estimated here because the thickness is always reduced by the deoxidation.

The nonlinearity seems to depend not only on the film thickness of the nanostructure but also on the oxygen content of the initial AgOx layer. However, as is clear from Fig. 16.7, the super-resolution effect is not as large as that of *super-RENS* discs. This may be because the nanoparticles or nanowires are randomly oriented and partly connected. For high resolution, especially, the size of the particles and the nearest neighbor distance would be key parameters. In *super-RENS*, as discussed above, the largest refractive index change is expected at the boundary between two phases, where the width must be less than 30 nm. Therefore, we can say that the application of metallic *nanoparticles* to optical data storage will be technically challenging in the near future, even using state-of-the-art nanofabrication technology.

16.5. SUMMARY

Through the super-RENS disc and its spin-off technology, the silver nanoparticle structure, we have learned that a giant refractive index variation plays an important role in enhancing the *CNR* in optical data storage. In our experimental work we find that, rather than using surface or localized plasmons, it is preferable to make use of the nonlinear characteristics of materials for the manipulation of localized photons. Although I want to believe that in the future someone might invent a technology that enables the manipulation of individual atoms or molecules at high speed and over a wide area, data storage technologies will have to make use of atoms as a group in the coming one or two decades. Until that day, lots of data will be accumulated experimentally and computationally.

REFERENCES

1. E. Betzig, J.K. Trautman, R. Wolfe, E.M. Gyorgy, P.L. Finn, M.H. Kryder, C.H. Chang: Near-field magneto-optics and high density data storage, Appl. Phys. Lett. **61,** 142 (1992).
2. E. Betzig, S.G. Grubb, R.J. Chichester, D.J. DiGiovanni, J.S. Weiner: Fiber laser probe for near-field scanning optical microscopy, Appl. Phys. Lett. **63,** 3550 (1993).
3. B.D. Terris, H.J. Mamin, D. Rugar, W.R. Studenmund, G.S. Kino: Near-field optical data storage using a solid immersion lens, Appl. Phys. Lett. **65,** (1994) 388.
4. H. Ukita, Y. Katagiri, H. Nakada: Flying head read/write characteristics using a monolithically integrated laser diode/photodiode at a wavelength of 1.3 μm, SPIE **1499,** 248 (1991).
5. S.M. Mansfield, G.S. Kino: Solid immersion microscope, Appl. Phys. Lett. **57,** 2615 (1990).
6. I. Ichimura, S. Hayashi, G.S. Kino: High-density optical recording using a solid immersion lens, Appl. Opt. **36,** 4339 (1997).
7. J. Tominaga, T. Nakano, N. Atoda: An approach for recording and readout beyond the diffraction limit with an Sb thin film, Appl. Phys. Lett. **73,** 2078 (1998).
8. J. Tominaga, D.P. Tsai eds: *Optical Nanotechnologies—The Nanipulation of Surface and Local Plasmons* (Springer, Berlin, Heidelberg, 2003).
9. J.H. Kim, I. Hwang, D. Yoon, I. Park, D. Shin, T. Kikukawa, T. Shima, J. Tominaga: *Technical Digest of Optical Data Storage 2003* (Vancouver, Canada, May 11–14, 2003), p. 24.
10. J. Tominaga, T. Nakano: *Optical Near-Field Recording—Science and Technology* (Springer, Berlin, Heidelberg, 2005).
11. T. Chattopadhyay *et al*: Neutron diffraction study on the structural phase transition in GeTe, J. Phys. C: Solid State Phys. **20,** 1431 (1987).
12. M.E. Lines, A.M. Glass: *Principles and Applications of Ferroelectrics and Related Materials* (Oxford Univ. Press, Oxford, 1977).
13. T. Matsunaga, Y. Umetani, N. Yamada: Structural study of a $Ag_{3.4}In_{3.7}Sb_{76.4}Te_{16.5}$ quadruple compound utilized for phase-change optical disks, Phy. Rev. B **64,** 1184116 (2001).
14. J. Tominaga *et al*: Ferroelectric catastrophe: beyond nanometre-scale optical resolution, Nanotechnology **15,** 411 (2004)
15. V.M. Fridkin: *Photoferroelectrics* (Springer, Berlin, Heidelberg, 1979).
16. J. Tominaga *et al.*: The characteristics and the potential of super resolution near-field structure, Jpn. J. Appl. Phys. **39,** 957 (2000).
17. A. Kolobov, P. Fons, A.I. Frenkel, A.L. Ankudinov, J. Tominaga, T. Uruga: Understanding the phase-change mechanism of rewritable optical media, Nat. Mater. **3,** 703 (2004).
18. R.E. Peierls: *Quantum Theory of Solids* (Calarendon Press, Oxford, 1955).
19. K. Seifert, J. Hafner, J. Furthmuller, G. Kresse: The influence of generalized gradient corrections to the LDA on predictions of structural phase stability: the Peierls distortion in As and Sb, J. Phys. Condens. Matter **7,** 3683 (1995).

CHAPTER SEVENTEEN

SURFACE PLASMON COUPLED EMISSION

ZYGMUNT GRYCZYNSKI[1], EVGENIA G. MATVEEVA[1], NILS CALANDER[2], JIAN ZHANG[3], JOSEPH R. LAKOWICZ[3] AND IGNACY GRYCZYNSKI[1]

[1]Department of Molecular Biology and Immunology, Department of Cell Biology and Genetics, University of North Texas, HSC, 3500 Camp Bowie Blvd. Fort Worth, TX 76107
[2]Department of Physics, Chalmers University of Technology, Goteborg, Sweden
[3]Center for Fluorescence Spectroscopy, University of Maryland at Baltimore, MD 21201

17.1. INTRODUCTION

Metallic particles and surfaces display diverse and complex opto-electrical properties. These properties, such as intense colors of noble metal colloids, strongly depend on metal and colloid size, and have been a subject of studies for centuries. Thin metal surfaces display strong absorption of light impinging under a very well defined angle that strongly depends on physicochemical properties of dielectrics on both sides of the metal film. For the last 20 years surface plasmon resonance (SPR) technology has been widely utilized in biochemical and biophysical analyses and is now extensively used for studying bioaffinity reactions on surfaces.[1–5] A typical experimental configuration for SPR analysis is shown in Fig. 17.1. A thin metal film (typically gold or silver ~50 nm thick) is illuminated through the glass prism at an angle θ. The electromagnetic light wave induces a periodic oscillating electric field that forces collective planar oscillations of free charges in the metal film (surface plasmons). At a very precisely defined angle, when the lateral component of the impinging light wavevector, k, matches the wavevector of the surface plasmons k_{sp}, these surface plasmon oscillations are in resonance with the frequency of the incident light. Under these conditions an electromagnetic field efficiently couples to the surface plasmons, which results in highly attenuated light reflection. This phenomenon (SPR) is extremely sensitive to small changes of the dielectric constant above the metal film and has been used to measure biomolecule binding to surfaces, as in the Biacore apparatus (http//www.biacore.com).

Excited surface plasmons in the metal film create highly enhanced evanescent fields penetrating the dielectric above the metal surface up to several hundred nanometers

M.L. Brongersma and P.G. Kik (eds.), Surface Plasmon Nanophotonics, 247–265.

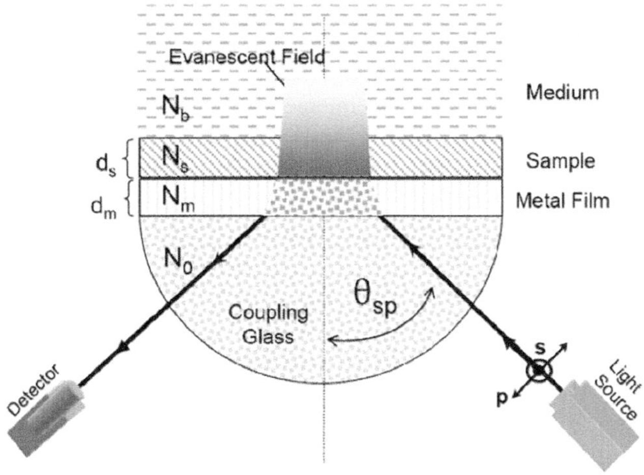

Figure 17.1. SPR configuration. At the θ_{sp} angle the reflectivity is strongly attenuated.

into the sample, see Fig. 17.1. Conversely, excited fluorophores present within this distance from the surface create an electromagnetic field that may strongly interact with free charges in the metal film inducing surface plasmons. The frequency of these plasmons corresponds to the emission frequency of the fluorophores. As a result, we observe a strong directional emission, which we call Surface Plasmon Coupled Emission (SPCE) (Fig. 17.2). The resulting SPCE exhibits the same spectral shape as the fluorophore emission, but is highly polarized with a sharply defined emission angle back into the glass substrate. A technology based on SPCE may provide 50% light collection efficiency and high intrinsic wavelength resolution with the use of a very simple optics. Such desirable properties can result in wide range of simple, inexpensive and robust devices of general use to biology and medicine. We stress that

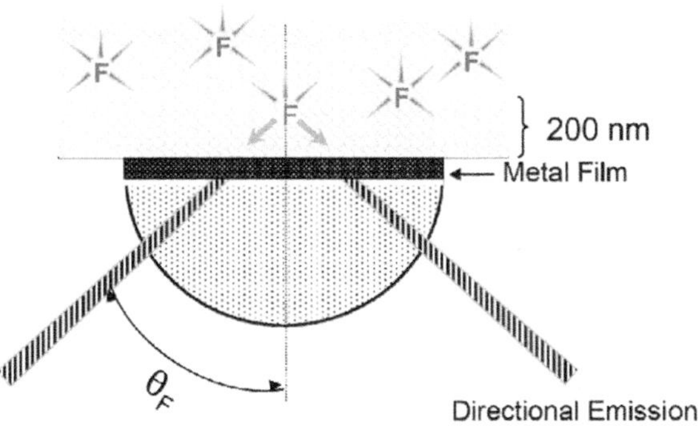

Figure 17.2. Concept of surface plasmon-coupled emission (SPCE). F represents a fluorophore. The excitation energy of the fluorophore couples to surface plasmons and radiates into the glass prism.

the directional SPCE is **not** due to reflections, but due to coupling of the oscillating dipoles of the excited fluorophores with surface plasmons on the metallic surfaces, which in turn radiate into the glass substrate.

In this chapter we describe the interaction of fluorophores with metallic surfaces that results in surface plasmon coupled emission (SPCE). SPCE significantly increases detection sensitivity and allows for many novel sensing applications. This new approach allows directing of a large fraction of the emission towards the detector by coupling the emission to surface plasmons on the metallic surfaces. The coupling angle and efficiency of SPCE strongly depend on the interface conditions. Because of such spectacular sensitivity to the refractive index above the metallic surface, similar to SPR technology, this new approach will be highly useful for detecting biomolecular binding on the surfaces, interaction and conformational changes of biomolecular systems, and provides a novel platform for biomedical assay development.

17.2. THEORY OF SURFACE PLASMON COUPLED EMISSION (SPCE)

The SPCE effect is closely related to surface plasmon resonance (SPR). Both the phenomenon of SPR as well as SPCE are complex and the underlying principles are new to many individuals with experience in biochemistry and/or fluorescence spectroscopy. In general, plasmon resonances on planar structures can be described by Maxwell's equations, solved using Fresnel theory. The electromagnetic fields from oscillating dipoles are expressed as integrals over plane waves in order to fit into the Fresnel scheme. This is a convenient way to describe the interaction of an oscillating electromagnetic field with planar structures supporting surface plasmons. In particular, the interaction of the excited fluorophore with surface plasmons results in the directional emission and can be theoretically investigated.

Surface plasmons in a thin metal layer, excited by fluorophores, can radiate into a glass prism at sharply defined angles determined by the emission wavelength and the optical properties of the respective layers of sample and the glass. In spite of the high theoretical complexity of the SPR and SPCE processes, basic Maxwell theory and theoretical modeling is a very useful tool to design experimental setups and structures, to fine-tune measurements, predict results, and explain experimental findings. Calculated quantities are SPCE angles, power levels, radiative decay enhancements and decay times (fluorescence lifetimes).

In this chapter we present an abbreviated version of the theoretical approach. Interested readers are referred for a detailed description of the theoretical method to references 6–9. The radiation from a dipole can be decomposed into an integral over plane waves (see Fig. 17.3), that is a two-dimensional Fourier transform. The decomposition is also called the Weyl identity.[10] The electromagnetic field is also divided into a p- and an s-polarized part. Fresnel theory, which is the theory of the refraction of plane waves in a dielectric planar structure, can then be applied[11,12] (see Fig. 17.4). Proper matching of the plane waves has to be done at all boundaries and also at the dipole location. A detailed schematic of the layer configuration and the angular notation is shown in Fig. 17.5. First a pre-integration (pre-summation) in

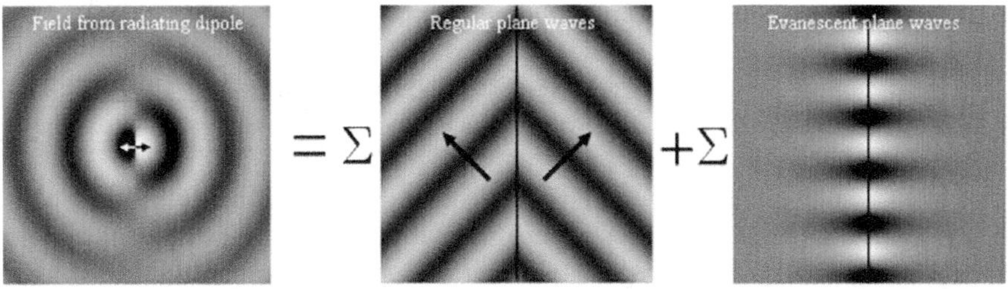

Figure 17.3. The electromagnetic field from an oscillating dipole is fit into the Fresnel scheme by decomposing it into an integral (or sum) of plane waves in the two main directions of the planar structure. The evanescent waves, i.e. non-propagating plane waves or plane waves with imaginary normal components of the wave-vector have to be included. When incorporated into the layered structure (Fig. 17.5), appropriate matching of the electromagnetic fields at the surface indicated by the straight vertical line above, has to be done, with the assumption of including an electromagnetic source.

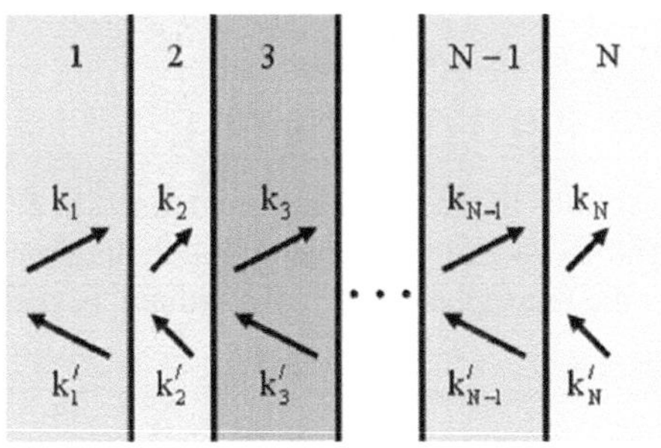

Figure 17.4. Maxwell's equations for the electromagnetic fields are solved by using Fresnel's theory of plane wave propagation in a planar dielectric structure. Two plane waves with mirrored wave-vectors are assumed to propagate in each homogenous layer and boundary-matched to the plane waves in the neighboring layers. The in-plane components of all wave-vectors are the same in all layers (Snell's refraction law). Appropriate matching of the electromagnetic fields has to be done at all boundaries. When including an oscillating dipole in one of the layers as a radiating source, appropriate matching of the electromagnetic fields also has to be done at the dipole location.

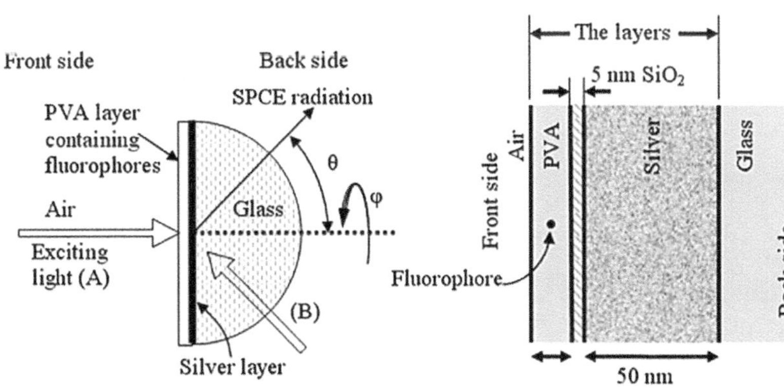

Figure 17.5. The basic configuration of an SPCE experiment. The excitation light is either coming from the left (front side, A) at normal incidence, exciting the fluorophores in the PVA layer directly, or from the right (back side, B) at the surface plasmon angle, exciting the fluorophores in the PVA layer via the SPR evanescent wave.

azimuth (φ) is done by using Bessel integrals, see Ref. 13, which means that only a one-dimensional numerical integration (in θ) has to be done in order to calculate the electromagnetic fields at any point.

The method can also be explained in terms of the Sommerfeld identity.[14] No integration is needed for the calculation of the far-fields. Parseval's equation is used for the power flow. The integrals for the electromagnetic fields at an arbitrary position in the layered structure, from a dipole normal to the plane are:

$$\underline{H} = -\frac{k_0^3 c_0 \mathbf{p}}{4\pi}\hat{\varphi}_r \int_0^\infty dn_\rho \frac{n_\rho^2}{\sqrt{n_d^2 - n_\rho^2}} J_1\left(k_0 n_\rho \rho\right)\left(a e^{i k_0 n_z z} + b e^{-i k_0 n_z z}\right) \tag{17.1}$$

$$\underline{E} = \frac{i\mu_0 c_0^2 k_0^3 \mathbf{p}}{4\pi n n_d} \int_0^\infty dn_\rho \frac{n_\rho^2}{\sqrt{n_d^2 - n_\rho^2}} \left[i\hat{\rho}_r n_z J_1\left(k_0 n_\rho \rho\right)\left(a e^{i k_0 n_z z} - b e^{-i k_0 n_z z}\right)\right.$$
$$\left. -\hat{z} n_\rho J_0\left(k_0 n_\rho \rho\right)\left(a e^{i k_0 n_z z} + b e^{-i k_0 n_z z}\right)\right] \tag{17.2}$$

The physical quantities in the equations are defined in Table 17.1. The power per solid angle for the far-fields from a dipole normal to the plane, normalized to the total output power from the fluorophore in a bulk medium, is given by:

$$p = \frac{3n^5 \sin(\theta)^2 \cos(\theta)^2}{8\pi n_d^3 \left|n_d^2 - n^2 \sin(\theta)^2\right|} |a|^2 \tag{17.3}$$

For a dipole parallel to the plane the p-polarized part and the s-polarized part of the normalized power per solid angle are given by:

$$p_p = \frac{3n^3 \cos(\theta)^2}{8\pi n_1^3} |a|^2 \cos(\varphi)^2 \tag{17.4}$$

$$p_s = \frac{3n^3 \cos(\theta)^2}{8\pi n_1 \left|n_1^2 - n^2 \sin(\theta)^2\right|} |a|^2 \sin(\varphi)^2 \tag{17.5}$$

The power flow in the layered structure for a dipole normal to the plane is given by:

$$\frac{P_z}{P_1} = \frac{3}{4}\Re\left[\frac{n^*}{n_1^3 n}\int_0^\infty dn_\rho \frac{n_\rho^3\sqrt{n^2 - n_\rho^2}}{\left|n_1^2 - n_\rho^2\right|}(a-b)(a^* + b^*)\right] \tag{17.6}$$

For a dipole parallel to the plane, it is given by:

$$\frac{P_{xy}}{P_1} = \frac{3}{8}\Re\left[\int_0^\infty dn_\rho\, n_\rho \left(\frac{n^*}{n_1^3 n} + \frac{1}{n_1\left|n_1^2 - n_\rho^2\right|}\right)\sqrt{n^2 - n_a^2}\,(a-b)(a^* + b^*)\right] \tag{17.7}$$

The power flow is normalized to the total radiated power from a fluorophore in a continuous bulk medium with a refractive index identical to that of the layer in which the fluorophore radiates. Free vacuum is not chosen for the normalization because

Table 17.1. Descriptions or definitions of the quantities occurring in the equations

Notation	Description or definition
k_0	Wave-vector in vacuum
n	Refractive index
n_d	Refractive index of layer containing dipole
k	nk_0
k_d	$n_d k_0$
n_ρ	$k_\rho = k_0 n_\rho$, where k_ρ is in-plane component of wave-vector
n_z	$k_z = k_0 n_z$, $n^2 = n_\rho^2 + n_z^2$, where k_z is z-component of wave-vector
c_0	Speed of light in vacuum
μ_0	Vacuum permeability
a	Electric field coefficient for the forward electromagnetic component in the layered structure
b	Electric field coefficient for the backward electromagnetic component
$\boldsymbol{p}$	Radiating dipole moment
$\hat{\varphi}$	Unit vector in azimuth
$\hat{\theta}$	Unit vector in direction of incidence
J_m	Bessel function of order m
$\underline{H}$	Magnetic field
$\underline{E}$	Electric field
x, y, z	Cartesian coordinates
ρ	$(x^2 + y^2)^{1/2}$
p	Normalized power per steradian
P	Power flow in z-direction
P_d	Total power from dipole in bulk medium

there seems to be an uncertainty about radiative decay enhancement in a dielectric medium compared to vacuum. By classical electromagnetic theory the radiated power of a dipole in a homogenous dielectric medium should be proportional to the refractive index, but it is very sensitive to the local environment. In the empty-cavity model[15,16] the dipole is considered to be inside an empty spherical cavity in the dielectric of refractive index n, giving an enhancement relative to vacuum of

$$n\left(\frac{3n^2}{2n^2+1}\right)^2 \tag{17.8}$$

This model is comparable to other models in Refs. 15 and 16. References 15 and 16 refers to a model that predicts a square dependence on the refractive index, which for refractive indices between values of one and two compares well with the empty cavity model.

17.3. COMPARISON OF EXPERIMENTAL STUDIES AND THEORETICAL PREDICTION

Figure 17.6 presents the layout of a typical experimental setup for SPCE experiments. The PVA layer contains fluorophores, in this case sulforhodamine 101 (R101). Excitation light with a wavelength of 514 nm from an argon ion laser is either coming from

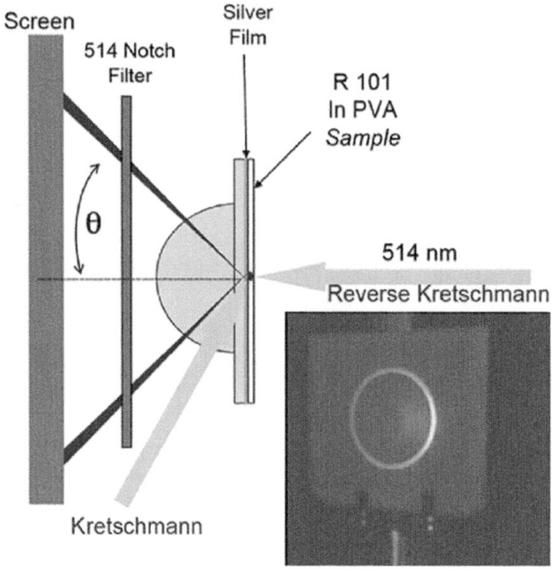

Figure 17.6. Schematics for the measurements. Kretschmann and Reverse Kretschmann configurations. The inset shows the SPCE ring.

the right (front side, Reverse Kretschmann configuration, RK) at normal incidence, directly exciting the fluorophores in the PVA layer, or from the bottom-left (prism hemi-sphere side, Kretschmann configuration, KR) at the surface plasmon angle, exciting the fluorophores in the poly(vinyl alcohol) (PVA) layer via the SPR evanescent wave. The fluorescence emerges in the shape of a cone through the hemi-sphere. The picture insert in the Fig. 17.6 shows the actual projection of the ring emitted by the R101 on a white screen.

A typical configuration for angular measurements is shown in Fig. 17.7. Instead of a hemi-sphere it is convenient to use hemi-cylinder. A picture of an actual apparatus used for measurements of the angular intensity distribution is shown below the experimental schematic. The PVA thickness used in this model experiment was either 15 or 30 nm. Fluorescence light with a wavelength near 600 nm is either detected at a specific angle using movable fiber optics or projected onto a screen, as shown in the inset in Fig. 17.6. The intense fluorescence light observed at a sharply defined angle corresponding to the ring at the prism side is the SPCE originating from surface plasmon resonances in the metal layer that were excited by the near field interaction with the excited fluorophores.[6–9] The right side of Fig. 17.7 presents the fluorescence intensity detected by fiber optics as a function of azimuthal angle θ. A sharp intensity increase is observed at an angle of 47° for a PVA thickness of 15 nm and at a 50° angle for a PVA thickness of 30 nm, respectively.

The approximately 3° change for this 15 nm film thickness difference is very large and easily measurable. It is important to stress that in routine SPR experiments

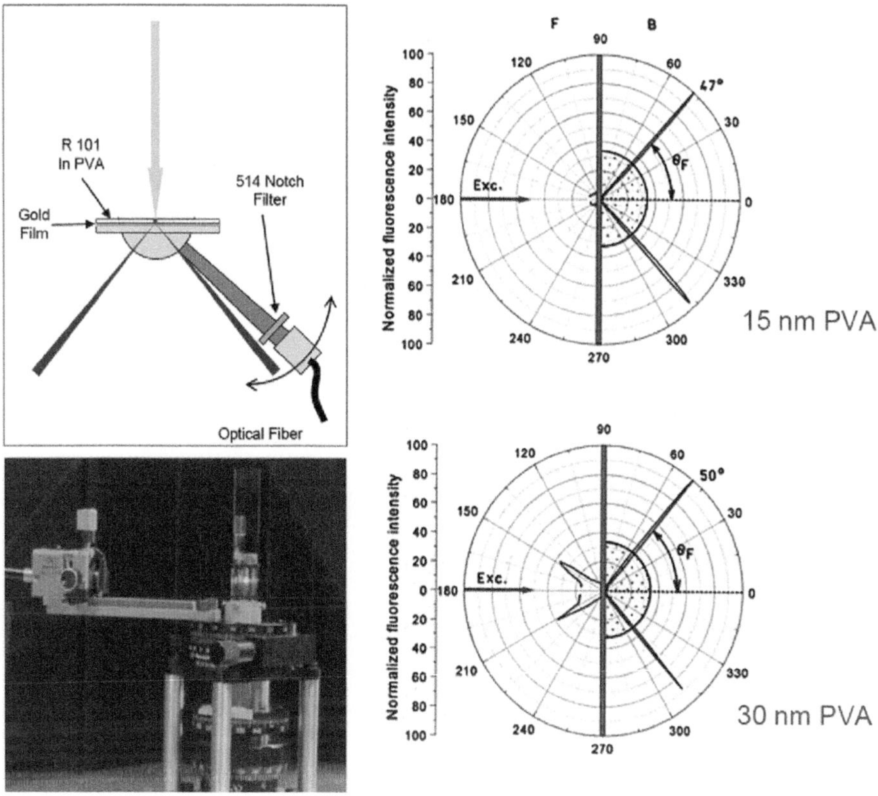

Figure 17.7. Experimental configuration for angular measurement. Right—angular emission intensity distribution.

the usual changes detected are in the millidegree range. Typical SPR and SPCE experiments are capable of detecting subnanometer changes in the layer.

To theoretically explain these experimental results, let's consider the detailed layer configuration shown in Fig. 17.5. The layered structure is built on a hemispherical prism. At the planar surface of the prism a glass substrate is attached with an index-matching fluid. A metal layer (silver of 50 nm thickness) is deposited on the glass slide. The metal is protected by a 5 nm thick layer of SiO_2, which also serves as a spacer that protects the fluorophores from quenching by avoiding direct contact with the metal.[6] A layer of polyvinyl-alcohol (PVA) doped with R101 is spin-coated on top of the SiO_2. The calculated far field radiation pattern is illustrated in Fig. 17.8 (top), and the azimuthal dependence of the fluorescence intensity in Fig. 17.8 (bottom). The light field from the fluorophore in the layered structure is calculated and illustrated in Fig. 17.9. The refractive indexes are assumed to be 1.5 for the glass prism, the SiO_2 and the PVA layers, and are taken from Ref. 17. The complex refractive index for the silver layer is assumed to be $0.1243 + 3.7316i$, and is taken from Ref. 18.

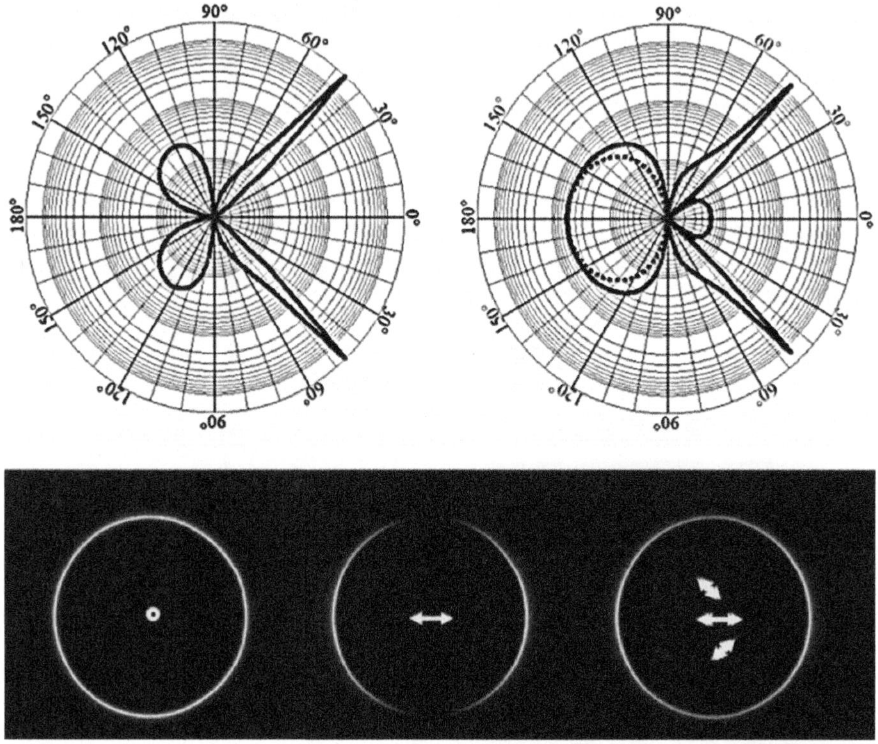

Figure 17.8. Top—Calculated far field emission radiation. Bottom—Azimuthal dependence of emission intensity for different orientations of the dipole transition.

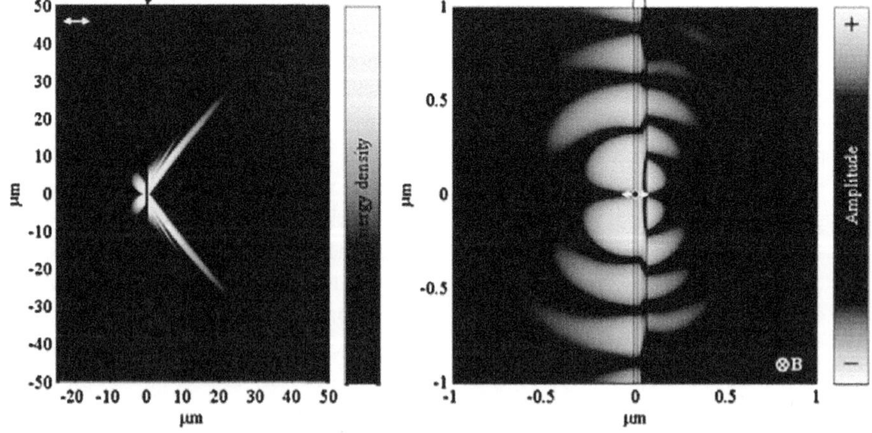

Figure 17.9. The calculated emission field from the fluorophore in the layered structure.

The far-field power density calculated from Eqs (3), (4) and (5) is illustrated in Fig. 17.8 (top) for a fluorophore in the middle of the 15 or 30 nm thick PVA layers, with the dipole oriented either normal or parallel to the layered structure. For the 15 nm thick PVA layer the SPCE is seen to be highly directed at the back side, at angles of 46.8°, an almost perfect match to the 47° reported in Ref. 17. For the 30 nm thick PVA layer, the SPCE direction is 50.4°, also an almost perfect match to the 50° reported in Ref. 17. The SPCE angle is nearly independent of the fluorophore position and orientation within the PVA layer. The full widths at half maximum (FWHM) are 1.1° for the 15 nm thickness and 1.8° for the 30 nm thickness of the PVA layer, for all orientations and positions of the fluorophores. We were pleasantly surprised by the perfect agreement of the experimental results and the independently done theoretical prediction.

17.4. BIOMEDICAL APPLICATIONS OF SPCE

Biological sciences are changing rapidly through the development of better analytical techniques and quantitative approaches stimulated by exact sciences such as physics. The repertoire of physical and optical phenomena that can be exploited in this context has been growing rapidly for many years. SPCE is an example of a new emerging technology that provides an excellent general platform for developing a surface based assay. As a practical application, in the following we describe novel approaches to immunoassays using surface plasmon coupled emission (SPCE). Many existing immunoassays are designed on transparent substrates with a layer of antibodies bound to the surface. These designs are intrinsically compatible with SPCE technology. In normal free-space conditions, fluorescence is mostly isotropic in space, and the detection sensitivity is limited in part by the light collection efficiency. When using the SPCE we can convert a significant part of the emission into a cone-like directional beam into a glass substrate and highly increase the signal collection efficiency. Also, we can maintain angular information that will reveal information regarding the biomolecular mass bound to the surface. For our experiments we used a model affinity assay using labeled anti-rabbit IgG antibodies against rabbit IgG bound to a 50 nm thick silver film. A schematic of the immunoassay principle is shown in Fig. 17.10.

17.4.1. Background Suppression

Surface plasmon coupling allows to selectively collect emission from molecules near the metal, and is particularly useful for any assay chemistry that can be localized on the surface. The fluorescence background from various biological components is a significant limitation for many immunoassays. Since SPCE exclusively emerges from the thin layer closely localized near the metallic surface we anticipated that a valuable attribute of SPCE will be background suppression.

We tested this desirable characteristic of the SPCE assay configuration for a highly absorbing and highly fluorescent background. Such a highly absorbing and scattering background is usually encountered with assays in whole blood or serum. The highly fluorescent background becomes a very difficult problem when working with shorter

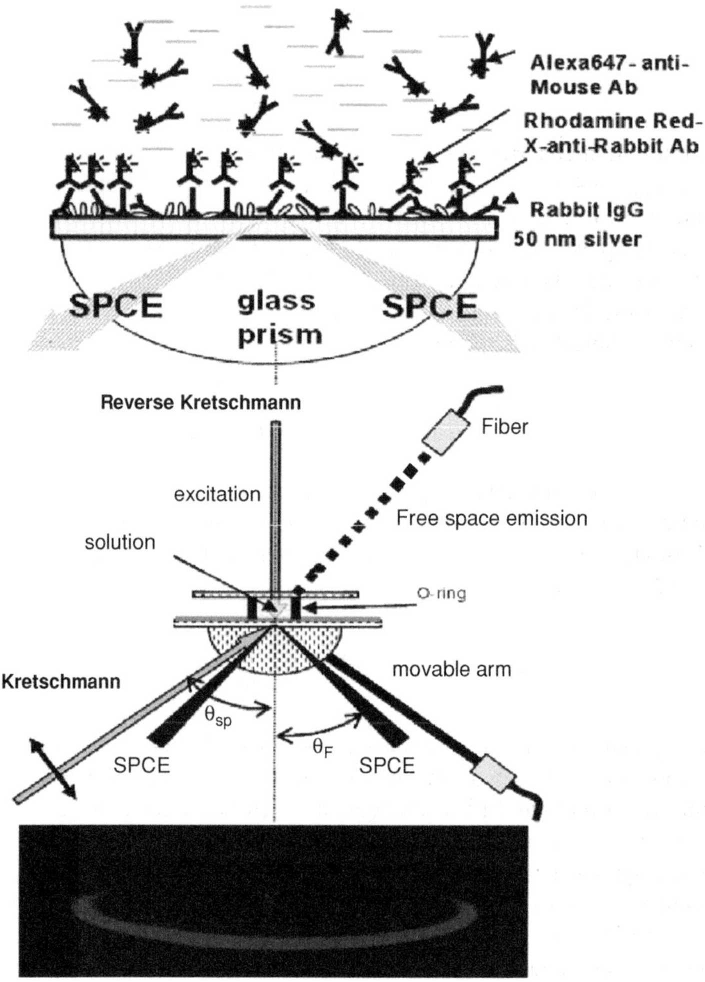

Figure 17.10. Schematic of the immunoassay (Top). Experimental configuration for biological samples. The sample is located in the space created by the O-ring spacer. (Middle) Photograph of the SPCE ring (Bottom).

excitation wavelengths where intrinsic fluorescence of biological specimens cannot be ignored.

Highly Absorbing Background

The optical properties of physiological samples are an important limitation for biomedical assays (high absorption, scattering, etc.). In practice, for medical testing it is often desirable to perform homogeneous assays without separation steps, sometimes in whole blood. We reasoned that SPCE should be detectable in optically dense media because the signal arises from material that is located within 200 nm of

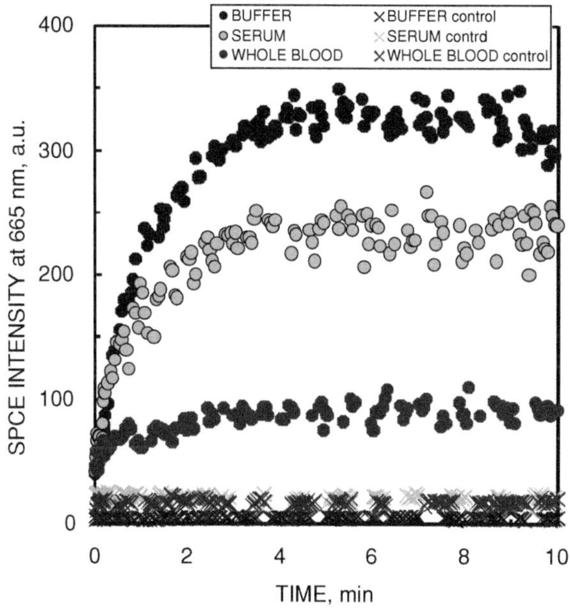

Figure 17.11. The kinetic traces for an immunoassay in buffer, serum, and whole blood.

the surface. We tested the kinetics of an SPCE-based immunoassay in whole blood and serum. It is important to realize that a 0.2 mm-thick blood sample has an optical density that would attenuate the fluorescence signal more than thousand-fold in the 550 nm wavelength range. Using SPCE, the signal was attenuated about 3-fold only (Fig. 17.11).[19] These results show the potential for using SPCE in optically dense and scattering samples.

Highly Fluorescent Background

We tested three optical configurations to determine the relative intensities and the extent of the background rejection using SPCE. These configurations are shown in the top panels of Fig. 17.12. The sample consisted of a surface saturated with rhodamine-labeled antibody. We then added Alexa 647-labeled antibody (not binding to the surface) to mimic autofluorescence from the sample. The 0.03° μM concentration of this antibody (0.13° μM of Alexa dye) resulted in a clearly visible free-space fluorescence signal from the sample. First, the sample was excited using the RK configuration, and the free space emission observed from the same water/buffer side of the sample. Compared to subsequent measurements the intensity of the desired rhodamine antibody (below) was weak. The free-space emission was dominated by the emission from the Alexa-647 dye at 670 nm, with only weak rhodamine emission contribution at 595 nm. We then measured the emission spectrum of the SPCE signal (middle panel), while still using RK illumination. The emission spectrum was dramatically

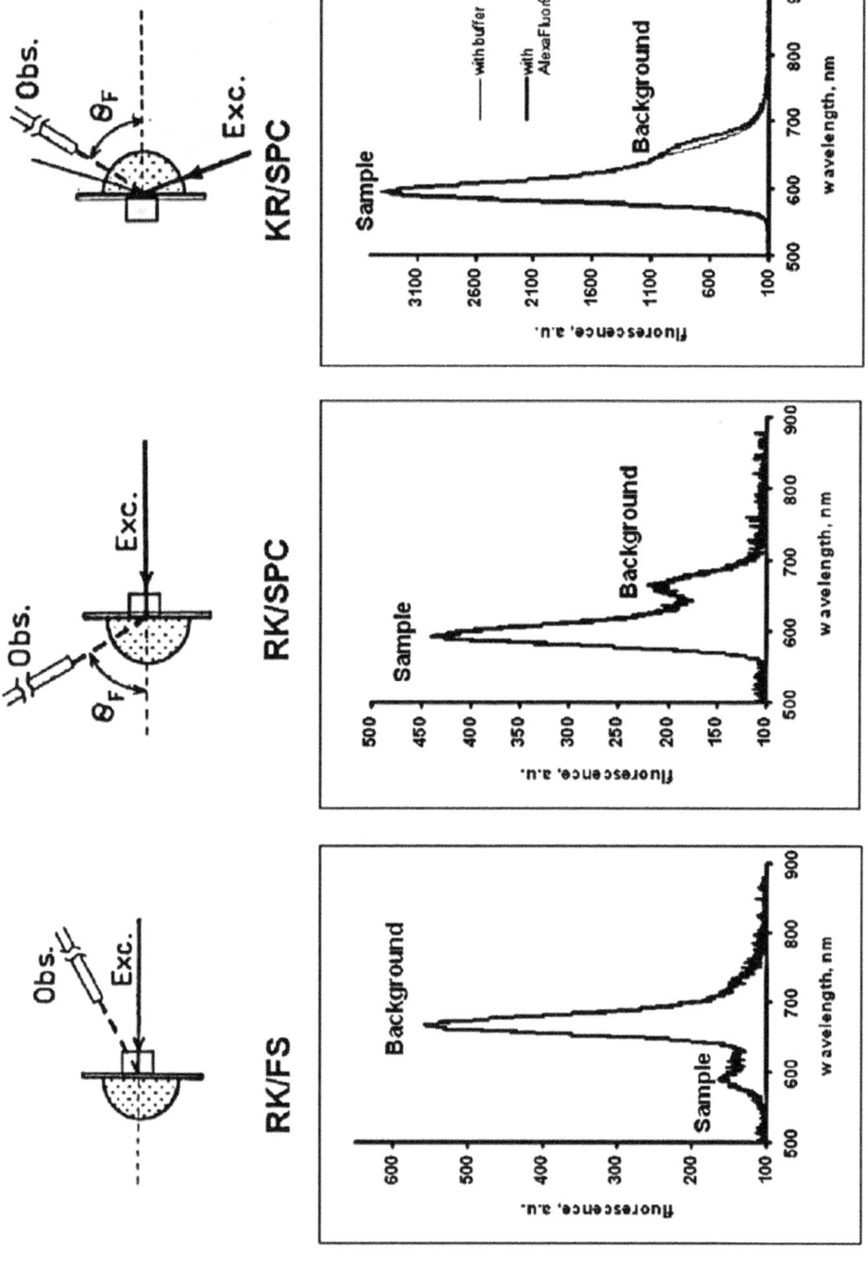

Figure 17.12. Observed fluorescence with a) free space emission in reverse Kretschmann excitation mode b) SPCE in reverse Kretschmann excitation mode and c) SPCE in Kretschmann excitation mode.

changed from a 10-to-1 excess of the unwanted background to a 5-to-1 excess of the desired signal.[20] Hence, the use of SPCE resulted in selective detection of the rhodamine-labeled antibody near the silver film.

We then changed the mode of excitation to the KR configuration (Fig. 17.12, right panel). In this case the sample was illuminated at θ_{sp} creating an evanescent field in the sample. The overall intensity was increased ~10-fold while further suppressing the unwanted emission from the Alexa-647 dye. The increased intensity and decreased background is the result of localized excitation by the resonance-enhanced field near the metal. In this case the emission was due almost entirely to the rhodamine, with just a minor contribution from the Alexa-labeled protein.

17.4.2. Intrinsic Wavelength Resolution

A very attractive property of SPCE is the intrinsic wavelength resolution. The coupling angle for SPCE strongly depends on the refractive index of the medium adjacent to the metal. In turn, the refractive index of the medium depends on the light wavelength. We tested the wavelength resolution of SPCE by measuring the emission spectra of a mixture of different fluorophores at different observation angles. For that purpose we spin coated a mixture of rhodamine 123 (R123), S101 and pyridine 2 (Py2) in PVA on a silver film. The concentrations were adjusted to yield approximately similar intensities for each fluorophore using an excitation wavelength of 514 nm, which excites all three dyes. The free-space emission showed contributions from all three fluorophores (Fig. 17.13 top panel). For the surface-coupled emission the spectra changed dramatically as the observation angle was changed from 52 to 42° (Fig. 17.13), with the peak shifting from 540 nm (R123) to 590 nm (S101), and to 660 nm (Py2). This effect is so dramatic that one can see the change in color by moving one's eyes around the hemicylinder axis. References 17 and 19 show the image of the fluorescence ring for the system shown in Fig. 17.13 obtained with a hemispherical prism. An excellent color resolution can be achieved with a simple digital camera and the intrinsic wavelength dispersion of SPCE.

17.4.3. Multi-Wavelength Immunoassay

As a final example of a novel SPCE application in immunoassays we describe multi-wavelength immunoassays. The intrinsic wavelength resolution of the SPCE opens up new interesting possibilities for a multiplex assay that is based on different color labels (Fig. 17.14). The experimental configuration used for two-wavelength SPCE was similar to that shown in Fig. 17.10. The protein-coated silver surface was illuminated at the surface plasmon angle through the glass prism (the Kretschmann (KR) configuration), or from the sample side (reverse Kretschmann (RK) configuration). The angle at which the emitted radiation propagates through the prism depends on the surface plasmon angle for the relevant wavelength. These angles depend on the emission wavelength, allowing measurement of multiple analytes using multiple emission wavelengths. We demonstrated this possibility using antibodies labeled with either

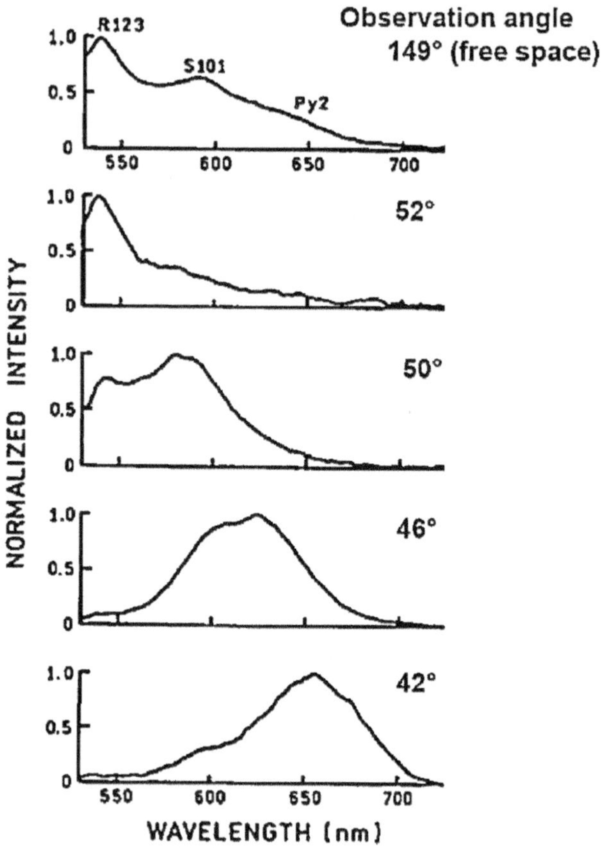

Figure 17.13. Emission spectra of a mixture of rhodamine 123 (R123), S (101), and pyridine 2 (Py2) in the PVA on the silver film. The emission spectra were measured at different observation angles.

Rhodamine Red-X or Alexa-Fluor-647.[21,22] These antibodies were directed toward an antigen protein bound to the silver surface. The emission from each labeled antibody occurred at a different angle on the glass prism, allowing independent measurement of the surface binding of each antibody. This method of SPCE immunoassays can be readily extended to 4 or more wavelengths.

To demonstrate the wavelength resolution for the emission we excited the sample through the aqueous phase (reverse Kretschmann). The observed angle dependent emission signal is shown on Fig. 17.15. The emission from each labeled antibody was strongly directional in the prism at the surface plasmon angle for the corresponding emission wavelength. Similar results were obtained using the Kretschmann configuration for excitation. This result demonstrates that SPCE is due to an interaction of the excited fluorophores with the metal surface and does not depend on creation of surface plasmon by the incident light.

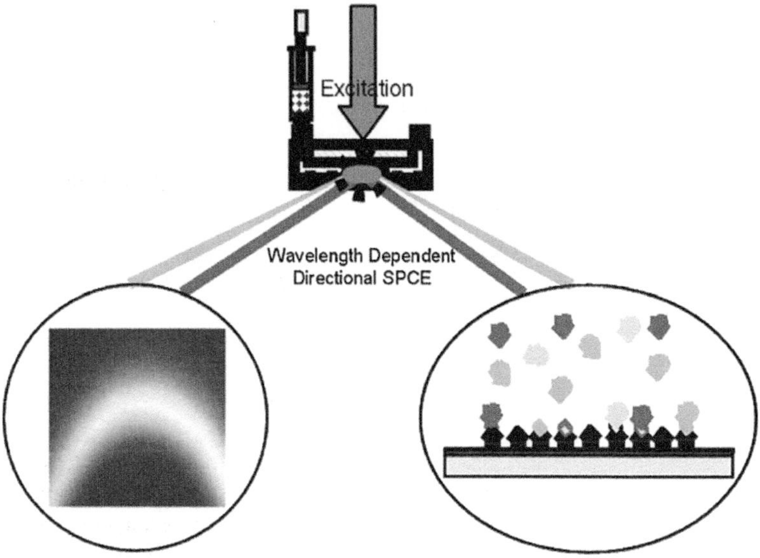

Figure 17.14. Concept of multi-wavelength (multiplex) assay.

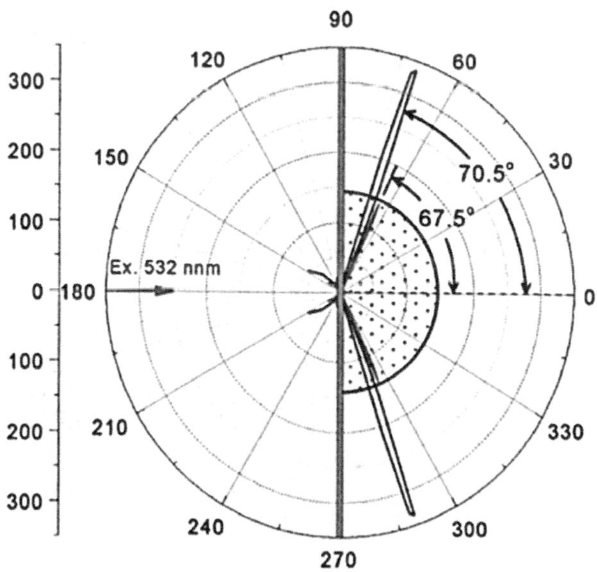

Figure 17.15. The Angle-dependent emission from surface-bound RhX-Ab and Alexa-Ab measured at 595 and 665 nm. The sample was excited at 532 nm using the RK configuration.

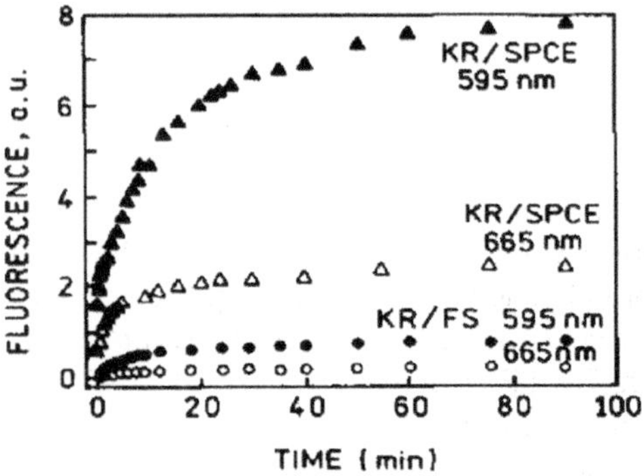

Figure 17.16. Binding kinetics observed for SPCE (▲,△) and free space fluorescence (●,○).

The surface binding kinetics are shown in Fig. 17.16 for KR excitation and two modes of observations - SPCE and free space (FS). Essentially, after about 30 minutes the binding is completed.

The angle-dependent intensity in Fig. 17.15 was collected through an emission filter to isolate the emission from each labeled antibody. However, these measurements did not resolve the emission spectra of each antibody. Figure 17.17 shows emission spectra collected using observation angles of 71, 69.5 and 68°. At 71° the emission is almost completely due to the Rhodamine Red-X-Antibody with an emission maximum of 595 nm. At 68° the emission is due mostly to the Alexa-Antibody at 665 nm, with a residual component from the Rhodamine Red-X-Antibody at 595 nm. At an intermediate angle of 69.5°, emission from both labeled antibodies is seen. These emission spectra show that the desired emission wavelength can be selected by adjustment of the observation angle.

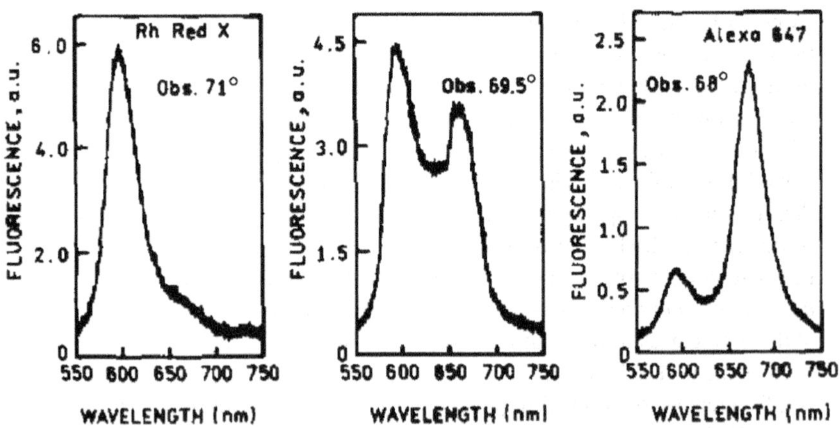

Figure 17.17. Emission spectra from Rh RedX-Ab and Alexa 647-Ab detected for different observation angles, using the KR configuration.

17.5. CONCLUSIONS

The Surface Plasmon Coupled Emission (SPCE) phenomenon offers new opportunities for high-sensitivity fluorescence detection. Its intrinsic simplicity and very efficient fluorescence signal collection enables novel approaches for simpler design of fluorescence based detection devices. In particular, SPCE may find application in detecting biomolecular binding on the surface (analogous to conventional SPR), medical assay development, and DNA hybridization. SPCE can be easily implemented in high-throughput screening (HTS) and fluorescence microscopy to facilitate single molecule detection.

ACKNOWLEDGMENTS

This work was supported by the NIH: NCRR, RR-08119, NCI, CA114460-1, BITC, and Philip Morris USA Inc.

REFERENCES

1. B.L. Frey, C.E. Jordan, S. Kornguth, R.M. Corn: Control of the specific adsorption of proteins onto gold surfaces with poly(l-lysine) monolayers, Anal. Chem. **67**, 4452–4457 (1995).
2. A.G. Frutos, R.M. Corn: SPR of ultrathin organic films, Anal. Chem. **70**, 449A–455A (1998).
3. Z. Salamon, H.A. Macleod, G. Tollin: Surface plasmon resonance spectroscopy as a tool for investigating the biochemical and biophysical properties of membrane protein systems. I: Theoretical principles, Biochim. Biophys. Acta **1331**, 117–129 (1997).
4. B.P. Nelson, A.G. Frutos, J.M. Brockman, R.M. Corn: Near-infrared surface plasmon resonance measurements of ultrathin films. 1. Angle shift and SPR imaging experiments, Anal. Chem. **71**, 3928–3934 (1999).
5. B. Liedberg, I. Lundstrom: Principles of biosensing with an extended coupling matrix and surface plasmon resonance, Sensors Actuators B **11**, 63–72 (1993).
6. N. Calander: Theory and simulation of surface plasmon-coupled directional emission from fluorophores at planar structures, Anal. Chem. **76**, 2168–2173 (2004).
7. N. Calander: Surface plasmon-coupled emission and Fabry-Perot resonance in the sample layer: a theoretical approach. J. Phys. Chem. B **109** (29), 13957–13963 (2005).
8. S. Ekgasit, C Thammacharoen, F. Yu, W. Knoll: Evanescent Field in Surface Plasmon Resonance and Surface Plasmon Field-Enhanced Fluorescence Spectroscopies, Anal. Chem. **76**, 2210–2219 (2004).
9. K. Vasiliev, W. Knoll, M. Kreiter: Fluorescence intensities of chromophores in front of a thin metal film, J. Chem. Phys. **120** (7), 3439–3445 (2004).
10. M. Born, E. Wolf: *Principles of Optics* (Pergamon, Oxford, 1980).
11. C.W. Chew: *Waves and Fields in Inhomogeneous Media* (Van Nostrand Reinhold, New York, 1995).
12. R.E. Benner, R. Dornhaus, R.K. Chang: Angular Emission Profiles of Dye Molecules Excited by Surface-Plasmon Waves at a Metal-Surface, Opt. Commun. **30** (2), 145–149 (1979).
13. M. Abramowitz, I.A. Stegun eds: *Handbook of Mathematical Functions*, 1st edn (Dover Publications Inc, New York, 1965).
14. A. Sommerfeld: *Partial Differential Equations in Physics* (Academic Press, New York, 1949).
15. F.J.P. Schuurmans, A. Lagendijk: Luminescence of Eu(fod)(3) in a homologic series of simple alcohols. J. Chem. Phys. **113** (8), 3310–3314 (2000).
16. R.M Amos, W.L Barnes: Modification of the spontaneous emission rate of Eu^{3+} ions close to a thin metal mirror, Phys. Rev. B **55** (11), 7249–7254 (1997).

17. I. Gryczynski, J. Malicka, Z. Gryczynski, J.R Lakowicz: Radiative decay engineering 4. Experimental studies of surface plasmon coupled directional emission, Anal. Biochem. **324**, 170–182 (2003a).
18. E.D. Palik: *Handbook of Optical Constants of Solids* (Academic, New York 1985).
19. E.G. Matveeva, Z. Gryczynski, J. Malicka, J. Lukomska, S. Makowiec, K.W. Berndt, J.R. Lakowicz, I. Gryczynski: Directional surface plasmon-coupled emission—application for an immunoassay in whole blood, Anal. Biochem. **344**(2), 161–167 (2005).
20. Z. Gryczynski, I. Gryczynski, E. Matveeva, J. Malicka, K. Nowaczyk, J.R. Lakowicz J: Surface-plasmon-coupled emission: New technology for studying molecular processes. In: *Cytometry: New Developments, Methods in Cell Biology*, vol. 75, 4th edn, ed by Z. Darzynkiewicz, M. Roederer, H.J. Tanke (Academic Press, New York, 2004), pp. 73–104.
21. E. Matveeva, J. Malicka, I. Gryczynski, Z. Gryczynski, J.R. Lakowicz: Multi-wavelength immunoassays using surface plasmon coupled emission, Biochem. Biophys. Res. Commun. **313**, 721–726 (2004).
22. J.R. Lackowicz, J. Malicka, I. Gryczynski, Z. Gryczynski: Directional surface plasmon-coupled emission: a new method for high sensitivity detection, Biochem. Biophys. Res. Commun. **307**, 435–439 (2003).

INDEX

Springer Series in

OPTICAL SCIENCES

Volume 1

1 **Solid State Laser Engineering**
By W.Koechner, 5th revised and updated ed. 1999, 472 figs., 55 tabs., XII, 746 pages

Published titles since volume 90

100 **Quantum Interference and Coherence**
Theory and Experiments
By Z. Ficek and S. Swain, 2005, 178 figs., XV, 418 pages

101 **Polarization Optics in Telecommunications**
By J. Damask, 2005, 110 figs., XVI, 528 pages

102 **Lidar**
Range-Resolved Optical Remote Sensing of the Atmosphere
By C.Weitkamp (Ed.), 161 figs., XX, 416 pages

103 **Optical Fiber Fusion Splicing**
By A.D. Yablon, 2005, 137 figs., XIII, 306 pages

104 **Optoelectronics of Molecules and Polymers**
By A. Moliton, 2005, 229 figs., 592 pages

105 **Solid-State Random Lasers**
By M. Noginov, 2005, 131 figs., XII, 238 pages

106 **Coherent Sources of XUV Radiation**
Soft X-Ray Lasers and High-Order Harmonic Generation
By P. Jaeglé, 2006, 332 figs., XIII, 416 pages

107 **Optical Frequency-Modulated Continuous-Wave (FMCW) Interferometry**
By J. Zheng, 2005, 137 figs., XVIII, 254 pages

108 **Laser Resonators and Beam Propagation**
Fundamentals, Advanced Concepts and Applications
By N. Hodgson and H.Weber, 2005, 587 figs., XXVI, 794 pages

109 **Progress in Nano-Electro Optics IV**
Characterization of Nano-Optical Materials and Optical Near-Field Interactions
By M. Ohtsu (Ed.), 2005, 123 figs., XIV, 206 pages

110 **Kramers–Kronig Relations in Optical Materials Research**
By V. Lucarini, J.J. Saarinen, K.-E. Peiponen, E.M. Vartiainen, 2005, 37 figs., X, 162 pages

111 **Semiconductor Lasers**
Stability, Instability and Chaos
By J. Ohtsubo, 2005, 169 figs., XII, 438 pages

112 **Photovoltaic Solar Energy Generation**
By A. Goetzberger and V.U. Hoffmann, 2005, 139 figs., XII, 234 pages

113 **Photorefractive Materials and Their Applications 1**
Basic Effects
By P. Günter and J.P. Huignard, 2006, 169 figs., XIV, 421 pages

114 **Photorefractive Materials and Their Applications 2**
Materials
By P. Günter and J.P. Huignard, 2006, 370 figs., XVII, 640 pages

115 **Photorefractive Materials and Their Applications 3**
Applications
By P. Günter and J.P. Huignard, 2007, 316 figs., X, 366 pages

Springer Series in
OPTICAL SCIENCES

116 **Spatial Filtering Velocimetry**
Fundamentals and Applications
By Y. Aizu and T. Asakura, 2006, 112 figs., XII, 212 pages

117 **Progress in Nano-Electro-Optics V**
Nanophotonic Fabrications, Devices, Systems, and Their Theoretical Bases
By M. Ohtsu (Ed.), 2006, 122 figs., XIV, 188 pages

118 **Mid-infrared Semiconductor Optoelectronics**
By A. Krier (Ed.), 2006, 443 figs., XVIII, 751 pages

119 **Optical Interconnects**
The Silicon Approach
By L. Pavesi and G. Guillot (Eds.), 2006, 265 figs., XXII, 389 pages

120 **Relativistic Nonlinear Electrodynamics**
Interaction of Charged Particles with Strong and Super Strong Laser Fields
By H.K. Avetissian, 2006, 23 figs., XIII, 333 pages

121 **Thermal Processes Using Attosecond Laser Pulses**
When Time Matters
By M. Kozlowski and J. Marciak-Kozlowska, 2006, 46 figs., XII, 217 pages

122 **Modeling and Analysis of Transient Processes in Open Resonant Structures**
New Methods and Techniques
By Y.K. Sirenko, N.P. Yashina, and S. Strom, 2007, 110 figs., XIV, 353 pages

123 **Wavelength Filters in Fibre Optics**
By H. Venghaus (Ed.), 2006, 210 figs., XXIV, 454 pages

124 **Light Scattering by Systems of Particles**
Null-Field Method with Discrete Sources: Theory and Programs
By A. Doicu, T. Wriedt, and Y.A. Eremin, 2006, 123 figs., XIII, 324 pages

125 **Electromagnetic and Optical Pulse Propagation 1**
Spectral Representations in Temporally Dispersive Media
By K.E. Oughstun, 2007, 74 figs., XX, 456 pages

126 **Quantum Well Infrared Photodetectors**
Physics and Applications
By H. Schneider and H.C. Liu, 2007, 153 figs., XVI, 250 pages

127 **Integrated Ring Resonators**
The Compendium
By D.G. Rabus, 2007, 243 figs., XVI, 258 pages

128 **High Power Diode Lasers**
Technology and Applications
By F. Bachmann, P. Loosen, and R. Poprawe (Eds.) 2007, approx. 535 figs., VI, 546 pages

129 **Laser Ablation and its Applications**
By C.R. Phipps (Ed.) 2007, approx. 300 figs., XX, 586 pages

130 **Concentrator Photovoltaics**
By A. Luque and V. Andreev (Eds.) 2007, approx. 242 figs., XV, 298 pages

131 **Surface Plasmon Nanophotonics**
By M.L. Brongersma and P.G. Kik (Eds.) 2007, approx. X, 298 pages

132 **Ultrafast Optics V**
By S.Watanabe and K. Midorikawa (Eds.) 2007, approx. 303 figs., XII, 520 pages

133 **Frontiers in Surface Nanophotonics**
Principles and Applications
By D.L. Andrews and Z. Gaburro (Eds.) 2007, approx. 89 figs., X, 250 pages

Springer Series in
OPTICAL SCIENCES

134 **Strong Field Laser Physics**
By T. Brabec, 2007, approx. 150 figs., XV, 500 pages

135 **Optical Nonlinearities in Chalcogenide Glasses and their Applications**
By A. Zakery and S.R. Elliott, 2007, approx. 60 figs., IX, 180 pages

GPSR Compliance

The European Union's (EU) General Product Safety Regulation (GPSR) is a set of rules that requires consumer products to be safe and our obligations to ensure this.

If you have any concerns about our products, you can contact us on ProductSafety@springernature.com

In case Publisher is established outside the EU, the EU authorized representative is:

Springer Nature Customer Service Center GmbH
Europaplatz 3
69115 Heidelberg, Germany

Batch number: 10614940

Printed by Printforce, the Netherlands